"The multiple impacts of climate change – from r.
quent hurricanes to droughts and changing migration patterns – are all
too obvious to our world. This book, focused on South and Southeast
Asia, examines the risks from our warming planet along with the various
factors that shape political, technical, and social responses to these
threats. Using a variety of methods and drawing from many disciplines,
this book illuminates the disproportionate effect of climate change on
vulnerable populations and how socio-political frameworks influence
decision making on government response to the threat. An important
addition to the growing body of literature on this critical topic."

–**Daniel Aldrich**, Author of *Building Resilience and Black Wave*
(Director, Security and Resilience Studies Program; Professor,
Political Science and Public Policy, Northeastern
University, Boston, USA)

"This volume seeks to invoke the critical socio-political perspectives
within a regional context that help shape the challenges imposed by climate change. Such uncharted consequences will dominate natural and
human landscapes and will have key geopolitical implications moving
forward. Indeed, the scalar political responses across the regions. It will
become crucial to help negotiate responses to such an inexorable challenge and change imposed by changing climatic regimes across South
and Southeast Asia. This evidence-based volume will help navigate policies, practices, and responses across nation-states. The balance between
developmental imperatives versus carbon neutrality will avoid the much-
vaunted tipping point in this discourse."

–**Dr. M. Satish Kumar**, Queen's University Belfast,
Belfast, Northern Ireland

"This edited volume focuses exactly on what is needed to understand the
impacts of climate change in South and Southeast Asia – power and
politics. This provocative and well-grounded text explains how vulnerable communities and nations will respond in the coming decades to
forced relocation, ecosystem change, and political upheaval. It goes
beyond explaining the impacts to explore practical solutions, including
communication."

–**Patrick Christie**, Professor, School of Marine and Environmental
Affairs and Jackson School of International Studies, University of
Washington, Seattle, USA

"This book is a timely and welcome contribution to the existing and fast expanding body of literature on the human and non-technical dimension of climate change. The focus falls on the challenges of and responses to climate change in the most populated region of the world, casting the spotlight on burning issues such as vulnerability, livelihoods, gender, risk, and politics in the Asian region. Contributions by a multiplicity of Asian scholars and other authors familiar with the dynamics of the region ensure that the experiences and observations of insiders are voiced. The 14 chapters make for insightful reading for anyone interested in and concerned about the inseparable interaction between climate change and human dynamics, as well as the growing complexities unleashed by this relationship."

–**André Pelser**, Ph.D., Professor & Research Fellow:
Sociology; Faculty: The Humanities, University of the Free State,
Bloemfontein, South Africa

"A critical read for those interested in hearing about the experiences of climate change in South and Southeast Asia from scholars with deep knowledge of the place."

–**Divya Chandrasekhar**, Ph.D., Associate Professor & PhD Program
Coordinator, City & Metropolitan Planning,
University of Utah, Salt Lake City, USA

Climate Change and Risk in South and Southeast Asia

This book, focuses on South and Southeast Asia, upgrades our understanding of the influence of multiple sociopolitical and governance factors on climate change and risks. Moving beyond science and technology-oriented discussions on climate change, it argues that the real solutions to climate change problems lies in societies, governance systems, non-state actors, and the power and politics underpinning these systems.

It presents a range of detailed conceptual, empirical, and policy-oriented insights from different nations of South and Southeast Asia, including India, Bangladesh, Sri Lanka, Indonesia, Vietnam, Maldives, and Bhutan. The chapters bring forth critical discussions of climate change, covering a diverse range of topics including livelihoods, gender, community perspectives, relocation, resilience, local politics, climate change communication, governance, and policy responses. By investigating climate change vulnerabilities as well as offering feasible solutions to the states and other non-state actors in responding to climate change and risks, this book deepens our existing knowledge of the social and political dimensions of climate change.

With interdisciplinary perspectives, this book will appeal to all students, researchers, and scholars of environmental studies, geography, disaster studies, sociology, policy studies, development studies, and political science. It provides valuable reading to practitioners, policymakers, and professionals working in related fields.

Devendraraj Madhanagopal (Ph.D.) is an Assistant Professor (II) in the School of Sustainability at XIM University (Odisha, India). He holds a Ph.D. in Sociology from the Department of Humanities and Social Sciences, Indian Institute of Technology Bombay (Mumbai, India). He is the recipient of several international travel grants and fellowships. His works appear in Environment, Development and Sustainability and Metropolitics journals. He is the author of "Local Adaptation to Climate Change in South India: Challenges and the Future in the Tsunami-hit Coastal Regions" (Routledge, UK. 2023 (Forthcoming, in Press). Also, he is the corresponding editor of the following edited books: i. Environment, Climate, and Social Justice: Perspectives and Practices from the Global South (2022). ii. Social Work and Climate Justice: International Perspectives. (Routledge, UK. 2023).

Salim Momtaz (Ph.D.) is an Affiliate Associate Professor at the University of Newcastle, Callaghan, Australia. He teaches in the area of Sustainable Resource Management. He received his B.Sc. and M.Sc. degrees in Geography from the University of Dhaka, Dhaka, Bangladesh. He did a Ph.D. in Development Studies from the University of London under a Commonwealth Scholarship. His academic career spans over 35 years. He published 13 books, 8 book chapters, and many articles in international journals. He was a member of the Scientific Advisory Committee – a Netherlands government research organization – between 2007 and 2010; currently, he is serving the European Climate, Infrastructure and Environment Executive Agency (CINEA) as an Expert Evaluator.

Routledge Studies in Hazards, Disaster Risk and Climate Change

Series Editor: Ilan Kelman, Professor of Disasters and Health at the Institute for Risk and Disaster Reduction (IRDR) and the Institute for Global Health (IGH), University College London (UCL)

This series provides a forum for original and vibrant research. It offers contributions from each of these communities as well as innovative titles that examine the links between hazards, disasters and climate change, to bring these schools of thought closer together. This series promotes interdisciplinary scholarly work that is empirically and theoretically informed, with titles reflecting the wealth of research being undertaken in these diverse and exciting fields.

Health, Wellbeing and Community Recovery in Fukushima
Edited by Sudeepa Abeysinghe, Claire Leppold, Alison Lloyd Williams and Akihiko Ozaki

Slow Disaster
Political Ecology of Hazards and Everyday Life in the Brahmaputra Valley, Assam
Mitul Baruah

Empowerment and Social Justice in the Wake of Disasters
Occupy Sandy in Rockaway after Hurricane Sandy, USA
Sara Bondesson

Climate Change and Risk in South and Southeast Asia
Sociopolitical Perspectives
Edited by Devendraraj Madhanagopal and Salim Momtaz

Gender-Based Violence and Layered Disasters
Place, Culture and Survival
Nahid Rezwana and Rachel Pain

For more information about this series, please visit: www.routledge.com/ Routledge-Studies-in-Hazards-Disaster-Risk-and-Climate-Change/book-series/HDC

Climate Change and Risk in South and Southeast Asia

Sociopolitical Perspectives

Edited by Devendraraj Madhanagopal and Salim Momtaz

Routledge
Taylor & Francis Group

LONDON AND NEW YORK

First published 2023
by Routledge
4 Park Square, Milton Park, Abingdon, Oxon OX14 4RN

and by Routledge
605 Third Avenue, New York, NY 10158

Routledge is an imprint of the Taylor & Francis Group, an informa business

British Library Cataloguing-in-Publication Data
A catalogue record for this book is available from the British Library

ISBN: 978-1-032-10670-0 (hbk)
ISBN: 978-1-032-10673-1 (pbk)
ISBN: 978-1-003-21647-6 (ebk)

DOI: 10.4324/9781003216476

Typeset in Times New Roman
by SPi Technologies India Pvt Ltd (Straive)

Contents

Figures

Tables

Acknowledgments

Editing this book has been an incredible journey. When I approached Dr. Salim Momtaz (Affiliate Associate Professor, University of Newcastle, Callaghan, Australia) with the idea of working together to produce a book on climate change and risks, he shared great interest and appreciation. Discussions with him were helpful in shaping the themes and geographical scope of our book, and I thank him for his continued support. Early exchanges with Dr. Ilan Kelman (Professor of Disasters and Health, University College London) helped shape the thematic focus and the structure of this book. We were encouraged by his insightful comments and constructive suggestions to develop this book with a vast scope. We are grateful for his valuable comments and suggestions.

As the aims and scope of this book are vast and diverse, we expected to face certain challenges in bringing together potential contributors. However, thankfully, we received immense responses from senior and emerging scholars of climate change from different regions of the world to contribute chapters to this book. I thank all our contributing authors for their valuable contributions and commitments throughout the process. Their contributions have made this book a truly "inter-disciplinary enriching volume."

I thank Dr. Rebekah Humphreys (Lecturer in Philosophy, University of Wales Trinity Saint David for providing review support to one of the chapters of this book. My big thanks to Dr. Andreas Rechkemmer (Professor of Global and Public Policy, College of Public Policy, Hamad Bin Khalifa University, Doha, Qatar) for providing an excellent foreword to this book.

I thankfully acknowledge Faye Leerink, Commissioning Editor at Routledge, for her continuous editorial support to make this project successful. I thank all the members of the Routledge Editorial team for their support in bringing this book to the world.

I thank Suba Ramya Durairaj, the Senior Project Manager at Straive for copy-editing and production support.

Along with Dr. Salim Momtaz, I happily thank our families, friends, colleagues, and the academic institutions (XIM University, Odisha, India; University of Newcastle, Callaghan, Australia) that we belong to for their support in various ways to develop and produce this book.

Devendraraj Madhanagopal
Odisha, India
29 April 2022

Foreword

The Global Climate Crisis and South and Southeast Asia: A Matter of Justice

Andreas Rechkemmer

Climate change has been known as an anthropogenic global issue of great concern to scientists, interested individuals, and activists for decades. Yet, broad public awareness, stringent government action, and juridification have been virtually absent until quite recently. The past few years, however, have made the international community finally realize that complex global challenges and crises – including climate change and the ensuing risks – will not go away easily but are likely to be the norm rather than the exception in this already turbulent 21st century. Three examples: First, the SARS-CoV-2 pandemic is obviously far from over. Global vaccine distribution continued to be uneven and a matter of economic and political privilege rather than equality and fairness, but also, in many Western countries, an odd case of responsibility and reason vs. delusion and denial. At the same time, new variants of the virus continued to emerge, all but suggesting that the largest global health crisis in at least a century is here for years to stay. It is tragic that the often short-sighted, irresponsible attitude to just, equitable, and effective global vaccine distribution became a root cause for a seemingly infinite loop of viral mutations and spread. As such, the policies that were adopted – or not adopted – by numerous countries allowed new variants to incubate where vaccines are either scarce or rejected, only to soon boomerang back.

Second, the rapidly deteriorating situation, the stunning collapse of societal status quo, security, and public order, and the ongoing and worsening humanitarian crisis in Afghanistan reminded us of the inherent vulnerability and fragility of what is commonly called the international order and once more challenged hegemonic stability theory, signifying the transition toward a messy, disjointed version of multipolarity. Hence, Afghanistan is but one example of a fundamental shift in the tenets of global and regional geopolitics and balance of power – a shift that is now ubiquitous in the world, it seems. Moreover, one of the insights from the Afghanistan case is that human security and justice seem to have become even more negotiable and disposable than before.

Third, Russia's war on Ukraine marks an epochal shift in international law and politics and ends the paradigm of post-WWII international rules-based governance and multilateralism for good. This ongoing tragedy on the global legal, political, and economic order can only be guessed at the time of writing, but its repercussions on almost any field and subject of global policy and law – including climate change negotiations, deals, and treaties – will be massive.

So then, what does all of this have to do with climate change? It doesn't bode well for countries already significantly affected by the effects of climate change – as all of the countries in South and Southeast Asia are, with no exception. It also provides a difficult, complicated prospect for countries that are dependent on international support and assistance for both mitigating and adapting to climate change – be it through finance and investment means, especially aid and attractive loans and bonds, technological, humanitarian, legal, or policy support – as the foundations of the global institutions we know – UN, World Bank, IMF, WTO, and others – have been questioned, attacked, and disrupted (whether those have ever been just and equitable, is another question).

The 6th Assessment Report of the Intergovernmental Panel on Climate Change (IPCC) affirmed that the global climate crisis is not only real and already impactful but certain to increase, perhaps exponentially, and become even much more destructive, disruptive, and deadly than previously projected. At the same time, the UN Climate Change Conference in Glasgow (COP26) and its aftermath reinforced widespread fear that it is becoming increasingly unlikely that the 1.5-degree goal sealed in the Paris Agreement – and probably even the 2.0-degree fallback position – can still be reached, meaning that unimaginable threats like mega heat waves, wet bulbs, floods and storm surges, droughts, hurricanes and blizzards, food crises and famines, forced mass migration and violent conflicts are to be expected to rise continuously throughout this century. COP 26 was more of the same: cynical delegations of certain industrialized or industrializing countries, as well as ruthless fossil fuel lobbyists, coerced poor countries already hit hard by climate change into defensive mode and dictated a watered-down compromise document that is simply far from adequate. Despite some mitigation pundits praising COP 26 for "keeping the 1.5-degree goal alive" by pointing to unrealistically ambitious (and expensive) carbon sequestration and storage ideas, the point is not about what's technically and hypothetically feasible, but it is very much about accountability and justice.

What has been done and continues to be done to this world's poor, marginalized, underdeveloped, oppressed, disenfranchised, and remote people? Much of the UNFCCC COP process has carried the handwriting of neoliberalism and neo-colonial rule. If those people in South and Southeast Asia, but also in the South Pacific, Africa, the Middle East, Latin America, and elsewhere count: why has the 2009 promise of COP 15 in Copenhagen of making available $100 billion in support of adaptation needs still not been met even to 50 percent? Why do the world's worst and most notorious greenhouse gas emitters still refuse to pay a single penny for the loss and damage to developing nations that they are responsible for? Why do wealthy carbon-emitting countries still refuse to commit to immediate, drastic, and tangible emission reductions knowing that their selfishness will kill millions of people, wipe out entire species, and make much of this planet uninhabitable?

Most people in South and Southeast Asia are in dire need of support, and the international community has largely failed them. But many of their own governments have failed them, too. It is tempting but dishonest and cynical when elected governments of certain major carbon-emitting and fossil fuel exporting nations in the region point fingers at the "West" and excuse their own inaction (or even counteraction) by blaming "colonializers" for their (real or perceived) faults. Governments in South and Southeast Asia must be held accountable, too. The largest economies and greenhouse gas emitters in the region are yet to become known for doing their due diligence when it comes to decarbonization governance, energy transition, or even adaptation and resilience building.

What COVID-19, Afghanistan, Ukraine, and climate policy as a global crisis phenomenon have in common is the toxic mix of short-sightedness, selfishness, and ruthlessness with which international norms and rules, the principles of solidarity and collective action, and the noble cause of pursuing equality, dignity, justice, and sustainability in international relations are being sacrificed for short-term gain, dominance, victory, or privilege. Forty-plus years of largely unregulated capitalism, profit-driven economic globalization, neoliberal rule, and tribalism have not furthered the spirit and the goals of the Charter of the United Nations nor the principled notion of multilateralism. They have ruined our planet, its ecosystems, and habitats and contributed to humanity finding itself in a state of shock, turmoil, and disintegration – closer to what Hobbes' Leviathan described as the State of Nature. By the way, climate change adds to other global risks and threats: it is intersectional, cross-cutting, and compounding. Zoonoses, and therefore epidemics and pandemics, are on the rise also because of changing climates, temperatures, precipitation, humidity, biomes, and human habitat. Wars such as in Sudan, Yemen, or Syria are also due to climate change, desertification, water shortage, crop failure, and hunger – as is forced migration as a mass phenomenon. And the list goes on.

We simply can no longer afford a business-as-usual approach, not even a moderately progressive approach – let alone a backward approach. This century of complex crises requires a whole new type of global action and response unlike anything before it, as with climate change, even peace, security, prosperity, and statehood are globally at risk. New, innovative, and disruptive economic, legal, political, and societal tools are needed, paired with technological advances, ethical and sustainable investments, social movements, and large-scale behavioral change. Ultimately, the climate agenda – and with it, many other agendas of global concern – is a matter of global justice, accountability, and survival. Measures and instruments must be designed to yield the safety and well-being first and foremost of the poor, the marginal, the disenfranchised, and the underserved – many of whom reside in South and Southeast Asia. The resilience of the weak will determines the fate of the whole. If that is the case, humanity – and with it, other species, ecosystems, and the planet – will benefit as a whole.

This wonderful edited volume by Dr. Devendraraj Madhanagopal and Dr. Salim Momtaz and all the excellent chapter authors is an important and timely contribution that addresses and advises the multiple challenges, change needs, and opportunities that are manifest and tangible in South and Southeast Asia today. The volume assembles a unique collection of diverse perspectives – both methodologically and subject matter-wise – on people, communities, countries, ecosystems, and the risks and threats they face from climate change, along with pathways to positive change and action.

Dr. Andreas Rechkemmer
Professor of Global and Public Policy, College of Public Policy,
Hamad Bin Khalifa University, Doha, Qatar

Abbreviations

ADB	Asian Development Bank
ADRC	Asian Disaster Reduction Center
AIR	All India Radio
ANOVA	Analysis of Variation
BBS	Bangladesh Bureau of Statistics
BCCSAP	Bangladesh Climate Strategy and Action Plan
BMI	Body Mass Index
CARE	Cooperative for Assistance and Relief Everywhere
CBA	Community-Based Adaptation
CBO	Community-Based Organizations
CCA	Climate Change Adaptation
CEDAW	Convention on the Elimination of All Forms of Discrimination Against Women
CI	Critical Infrastructure
CIDCO	City and Industrial Development Corporation
CMC	Colombo Municipal Council
CMFRI	Central Marine Fisheries Research Institute (India)
COVID-19	Coronavirus disease 2019
CRZ	Coastal Regulation Zone
CZMA	Coastal Zone Management Authority
CZMP	Coastal Zone Management Plan
DFID	Department for International Development
DRR	Disaster Risk Reduction
DRR	Disaster Risk Reduction
DS	Divisional Secretariat
EAC	Expert Appraisal Committee
EC	Environmental Clearance
ECCC	Executive Committee on Climate Change (India)
EIA	Environmental Impact Analysis
EMF	Environmental Management Framework
EPPA	Environmental Protection and Preservation Act
EQ-5D	EuroQol-5 Dimension instrument for measuring the quality of life
FA	Fishery Association

FAO	Food and Agricultural Organization
FCI	Financial capital Index
FGD	Focus Group Discussion
FR	First Response
FYP	Five Year Plan
GAM	Free Aceh Movement
GDh	Gaafu Dhaal Atoll
GDP	Gross Domestic Product
GEO	Global Environment Outlook
GHG	Green House Emissions
GNH	Gross National Happiness
GNI	Gross National Income
GOB	Government of Bhutan
GoB	Government of Bangladesh
GoM	Government of Maharashtra
GOs	Governmental organizations
GoSL	Government of Sri Lanka
HCI	Human Capital Index
H-C-R	Health Communication Resources
ICZMP	Integrated Coastal Zone Management Project
ICZMPWB	Integrated Coastal Zone Management Project (Government of West Bengal)
IDMC	Internal Displacement Monitoring Center
IIT	Indian Institute of Technology
ILO	International Labor Organization
IMD	India Meteorological Department
INR	Indian Rupee
IPCC	The Intergovernmental Panel on Climate Change
JICA	Japanese International Cooperation Agency
JNPT	Jawaharlal Nehru Port Trust
JVP	Janatha Vimukthi Peramua
KI	Key Informants
LDC	Less Developed Country
LKR	Sri Lankan Rupee
LTTE	Liberation Tigers of Tamil Eelam
MaRHE	Marine Research and High Education Center (Maldives)
MDGs	United Nations Millennium Development Goals
MEE	Ministry of Environment and Energy (Maldives)
MGI	Membership in Groups Index
MICI	Mass Communication and Individual Communication Index
MLVI	Macro Level Vulnerability Index
MMR	Mumbai Metropolitan Region
MNSHE	National Mission for Sustaining the Himalayan Eco-system (India)
MoDM	Ministry of Disaster Management
MoEF	Ministry of Environment and Forests (Government of India)

MoEFCC	Ministry of Environment, Forest and Climate Change
MoMWD	Ministry of Megapolis & Western Development
MoNPEA	Ministry of National Policies and Economic Affairs
MoT	Ministry of Tourism (Maldives)
MOTAC	Ministry of Tourism Arts and Culture (Maldives)
MPA	Marine Protection Area
MPCB	Maharashtra Pollution Control Board
MSAAPC	Maharashtra State Adaptation Action Plan on Climate Change
NAPA	National Adaptation Plan of Action (Government of Bangladesh)
NAPCC	National Action Plan on Climate Change (Union Government of India)
NBRO	National Building and Research Organization
NBSAP (Maldives)	National Biodiversity Strategy and Action Plan 2016–2025
NCI	Natural Capital Index
NCZMA	National Coastal Zone Management Authority
NECS	National Environment Commission Secretariat Bhutan
NFP	National Policy Framework
NGO	Non-Governmental Organisation
NIC	National Intelligence Council (United States)
NLB	National Legislative Bodies
NMSKCC	National Mission on Strategic Knowledge for Climate Change (India)
OA	Overall Assessment
ODI	Overseas Development Institute
OECD	Organization for Economic Cooperation and Development
ONGC	Oil and Natural Gas Corporation
PACI	Participation and connection index
PAP	Project Affected Person
PCI	Physical Capital Index
PDO-ICZMP	Program Development Office for Integrated Coastal Zone Management Plan
Ph.D.	Doctor of Philosophy
PKK	*Pemberdayaan Kesejahteraan Keluarga* (Family Welfare and Empowerment Program)
PMCCC	Prime Minister's Council on Climate Change (India)
PPP	Public Private Partnership
PRA	Participatory Rural Appraisal
REDD	Reducing emissions from deforestation and forest degradation
RMTWG	Resilience Measurement Working Group
RPCETI	Regulation on the Protection and Conservation of Environment in the Tourism Industry (Maldives)
SAARC	South Asian Association for Regional Cooperation

SANDRP	South Asian Network on Dams, Rivers, and People
SAPCC	State Action Plans on Climate Change (India)
SCI	Social capital Index
SDGs	United Nations' Sustainable Development Goals
SEZ	Special Economic Zone
SMF	Social Management Framework
SoE Report	State of Environment Report
SPSS	Statistical Package for Social Sciences
SPSS	Statistical Packages for Social Sciences
SSF	Small-Scale Fishers
T. Aman	Transplanted Aman
T. Aus	Transplanted Aus
TERI	The Energy and Resources Institute
TG-CH	Tam Giang – Cau Hai
TMP	Tourism Master Plan
TNSDMA	Tamil Nadu State Disaster Management Agency
TOI	Times of India
UDA	Urban Development Authority
UDMC	Union Disaster Management Committee (Bangladesh)
UN	United Nations
UN ESCAP	United Nations Economic and Social Commission for Asia and the Pacific
UNDP	United Nations Development Programme
UNDRR	United Nations Office for Disaster Risk Reduction
UNESCO	United Nations Educational, Scientific and Cultural Organization
UNFCCC	United Nations Framework Convention on Climate Change
UNHCR	United Nations High Commission for Refugees
UNICEF	United Nations International Children's Emergency Fund
UNIPRS	University of Newcastle International Postgraduate Research Scholarship
UNISDR	United Nations Office for Disaster Risk Reduction
UNO	Upazila Nirbahi Officer (Bangladesh)
UNOCHA	United Nations Office for the Coordination of Humanitarian Affairs
UNOPS	United Nations Office for Project Services
UNRSC	University of Newcastle Research Scholarship Central
URP	Urban Regeneration Project
USAID	United States Agency for International Development
WAI	Weighted Average Index
WB	World Bank
WBPCB	West Bengal Pollution Control Board.
WBSAPCC	West Bengal State Action Plan Climate Change
WBSCZMA	West Bengal State Coastal Zone Management Authority
WHO	World Health Organization
WMR	World Migration Report

Contributors

Randall S. Abate is the Assistant Dean for Environmental Law Studies and a Professorial Lecturer in Law at The George Washington University Law School. From 2018-2022, he served as the inaugural Rechnitz Family and Urban Coast Institute Endowed Chair in Marine and Environmental Law and Policy, a Professor in the Department of Political Science and Sociology, and the Director ofthe Institute for Global Understanding at Monmouth University in WestLong Branch, New Jersey, USA. He teaches courses in U.S. environmental law,climate change law and justice, human rights and environmental protection, and animal law. Prior to joining the Monmouth University faculty in 2018, he had 24 years of full-time law teaching experience at six US law schools. He has delivered lecturesand taught courses on environmental and animal law topics in more than20 countries. He has published six books including Climate Change andthe Voiceless: Protecting Future Generations, Wildlife, and NaturalResources and Climate Justice: Case Studies in Global and RegionalGovernance Challenges. He holds a B.A. from the University of Rochesterand a J.D. and M.S.L. (Environmental Law and Policy) from VermontLaw School.

Alankrita Anand is a Ph.D. candidate at the Department of Sociology at the University of York, United Kingdom. Her doctoral research examines the influence of household gender dynamics on young married women's access to reproductive healthcare in Eastern Bihar, India. It is a collaborative study being carried out in partnership with Project Potential, a Bihar-based NGO, and is being supported by the Phil Strong Memorial Prize awarded by the British Sociological Association. Alankrita's larger research interests include the sociology of health and illness, reproductive justice, feminist agency, and participatory methodologies. Before starting her Ph.D., she worked in the development sector in India, focusing on media for development, and subsequently as a program manager and facilitator at Ashoka University. Alankrita holds a Master's in Women's Studies from the Tata Institute of Social Sciences, Mumbai.

Cynthia M. Caron (Ph.D.) is an Associate Professor of International Development, Community, and Environment (IDCE) at Clark University (Worcester, USA), where she also directs the undergraduate program in

International Development and Social Change. She is a development and political sociologist (Ph.D., Cornell University) with interests in natural resource governance, gender, post-disaster reconstruction, and development practice. Before joining academia in 2012, she worked for international organizations in Sri Lanka, Bhutan, Rwanda, and Ethiopia on projects to improve peoples' access to land and natural resources. Her research employs qualitative methodologies and is in two domains: (1) Energy, Forests, and Climate Change and (2) Land Rights and Land Tenure. She is an associate editor of the journal *Society and Natural Resources* and is on the Reviewer Board for *Land*. In addition to numerous book chapters, her publications have appeared in *Development in Practice, Contemporary South Asia* and *The Journal of Extreme Events*.

Anita Chalmers (Ph.D.) is an academic and lecturer at the University of Newcastle, Callaghan, Australia, where she has taught a broad range of natural resource management courses. She was awarded a Ph.D. from the University of New England (Armidale) in ecosystem management and botany. Dr. Chalmer's research interests focus on the ecology and management of vegetation to reduce the risks associated with environmental decision-making, thereby providing environmental and social benefits. She has also contributed to environmental policy development and implementation through participation in Advisory bodies at both a State and Federal level within Australia.

Marcella Schmidt di Friedberg (Ph.D.) is a Professor of Geography, at the "Riccardo Massa" Department of Human Sciences for Education (University of Milano-Bicocca, Milan, Italy). She is Chair since 2016 of the Commission on the History of Geography of the International Geographical Union (UGI) and Vice-Director of the Marine Research and High Education Center (MaRHE) in Faaf-Magoodhoo (Republic of Maldives). Her research interests concern Cultural Geography, Gender Geography, Hazard and Resilience, and the History of Geographical Thought. She has been working and publishing extensively on the relations between nature, culture, memory, and landscape in different contexts, from the Mediterranean to Japan. Her current research focuses on the culture and heritage of the Maldives.

Kasturi Gandhi is a research fellow with the Bosserman Center for Conflict Resolution (Maryland, USA), working on a project titled "Where we learn about Environmental Conflict: Analyzing Environmental Education and Conflict Resolution Curricula in India." She has a Masters in Public Policy and her M.A. dissertation focuses on coastal regulation in Mumbai, the city in which she was born, raised, and currently lives; it covers the socio-ecological impacts of flawed government data and marginalization in coastal governance processes. She uses qualitative and archival research methods in her work, having previously worked as an archival assistant in Kutch (Gujarat), where she created documentaries about craft and pastoralist communities. She's also worked with agrarian clusters in south and northeast India, having made photo stories, essays, and films about their community-driven livelihood interventions. Her documentation serves as a digital archive of intangible culture and oral community heritage from pockets of rural India.

Ha Dung Hoang (Ph.D.) is a senior lecturer and researcher at the University of Agriculture and Forestry, Hue University, Vietnam. He is also a member of the Centre for Climate Change Study in Central Vietnam. His main research interests are climate change mitigation and impacts, natural resource management, and sustainable development, especially in developing countries. He is the author of several publications, including peer-reviewed articles, book chapters, and international conference proceedings.

Md. Faruk Hossain (Ph.D.) is an assistant professor of Geography and Environment at the University of Dhaka, Dhaka, Bangladesh. He obtained B.S. Honours and M.S. degrees in Geography and Environment from the University of Dhaka in 2002 and 2003, respectively. He received his second M.Sc. degree in Disaster Preparedness Mitigation and Management (DPMM) program from the Asian Institute of Technology (AIT), Khlong Nueng, Thailand in 2014. He was awarded a Ph.D. degree in the Sustainable Resource Management Program from the University of Newcastle, Callaghan, Australia, in 2021. He started his career as a lecturer in the Geography and Environmental Studies department of the University of Chittagong in 2008. Later, he joined as a lecturer in the Department of Geography and Environment of the University of Dhaka in 2011 and is currently working as an assistant professor. He has a research interest in environment, climate change, disaster management, natural resource management, migration geography, GIS and remote sensing, etc.

Juhi Huda (Ph.D.) is an assistant professor (Environmental Studies) at FLAME University, Pune, India. She received her doctoral degree in Environmental Studies from the University of Colorado, Boulder, USA. She also holds two Master of Arts degrees in English from the University of Nevada, Reno, USA, and the University of Pune, India; and a B.A. in English from the University of Pune, India. Dr. Juhi has a total work experience of over nine years. She was a Postdoctoral Researcher at NASA Jet Propulsion Laboratory (JPL), California Institute of Technology, USA, and has taught at the University of Colorado, Boulder, and the University of Nevada, Reno, USA.

Janaka Jayawickrama (Ph.D) is a social anthropologist with a background in conducting research in multi-disciplinary and culturally diverse contexts, trained in India, the USA, and the UK. His work has a particular focus on understanding the well-being of communities within uncertainty and dangers of life presented by crises. Since 1994, Janaka has worked within and between academia, policy, and practice in disasters, conflicts, and uneven development. His pioneering work on concepts of care and wellbeing has influenced the policy and practice of the global humanitarian and development intervention discourse. Janaka continues to collaborate with scholars from Asia, Africa, Europe, the Americas, and the Middle East to examine mechanisms from ancient wisdoms to improve wellbeing within modern challenges. He is a distinguished professor at the Centre for Community Wellbeing of the Department of History at College of Liberal Arts, Shanghai University, Shanghai, China.

Derek Kauneckis (Ph.D.) is an associate research professor at the Division of Earth and Ecosystem Sciences in Desert Research Institute, USA. He conducts research on community and regional adaptation to climate change, interjurisdictional cooperation on environmental issues, and environmental governance. His work examines how localized policy systems interact with climate information, formal and informal mechanisms of cooperation, and the institutional design of regional-scale climate collaborations. Recent research has looked at the use of climate modeling in river systems management, local government's climate adaptation networks, work with native communities on developing climate adaptation and resilience plans, organic waste systems, and scenario development around extreme weather events. He works on integrated analysis of complex systems where human decision-making interacts with natural and built systems. His work draws on institutional analysis, network theory, and socio-technical ecological systems analysis. He is the recipient of multiple teaching and mentoring awards and has received funding from the National Science Foundation (NSF), National Oceanic and Atmospheric Administration (NOAA), and the National Renewable Energy Laboratory (NREL). He teaches courses on resilience theory and practice, science, technology and innovation policy, environmental policy, and policy analysis.

Siti Kusujiarti (Ph.D., University of Kentucky) is a professor of sociology and Chair of the Department of Sociology and Anthropology at Warren Wilson College, Asheville, North Carolina, USA. Dr. Kusujiarti has conducted extensive research and teaching on gender relations, development, disaster, and climate change in Indonesia and Southeast Asia. Her publications include a co-authored book titled *Power, Change, and Gender Relations in Rural Java: A Tale of Two Villages* (2012), an article titled "Riskscapes of Gender, Disaster, and Climate Change in Indonesia" (with Ann R. Tickamyer) in *Journal of Regions, Economy, and Society* (July 2020), and the chapter "Indonesian Family" in *World of Equality: How 25 Couples Around the Globe Share Housework and Childcare (the "Work")*, Edited by Francine M Deutsch; Ruth Gaunt published by Cambridge University Press (2020).

Devendraraj Madhanagopal (Ph.D.) is an Assistant Professor (I) in the School of Sustainability at XIM University (Odisha, India). He holds a Ph.D. in Sociology from the Department of Humanities and Social Sciences,

Indian Institute of Technology Bombay (Mumbai, India). He is the recipient of several international travel grants/fellowships. His works appear in *Environment, Development and Sustainability* & *Metropolitics* journals. He is the author of "Local Adaptation to Climate Change in South India: Challenges and the Future in the Tsunami-hit Coastal Regions" (Routledge, UK. 2023 (Forthcoming, in Press). Also, he is the corresponding editor of the following edited books: i. Environment, Climate, and Social Justice: Perspectives and Practices from the Global South (2022). ii. Social Work and Climate Justice: International Perspectives. (Routledge, UK. 2023).

Stefano Malatesta (Ph.D.) is an Associate Professor in Geography at the "Riccardo Massa" Department of Human Sciences for Education (University of Milano-Bicocca, Milan, Italy). In 2008, he completed his Ph.D. program in Models, Languages, and Tradition of the Western Culture at the University of Ferrara. He is a member of the Board of the Marine Research and High Education Center (MaRHE) in Faaf-Magoodhoo (Republic of Maldives) and of the Executive Committee of the International Small Islands Studies Association (ISISA). He teaches Human Geography of Small Island Systems, Geography of Tourism, and Coastal and Maritime Tourism. He served (2018–2021) as the Chair of the Working Group on Small Island and Archipelagos (AGeI). In 2017, he was awarded by the Italian Geographical Society (Società Geografica Italiana) with a special mention for under-40 geographers in Italy. He serves as an advisor for UNESCO. His main research topics are human geography of small islands, the geopolitics of the Indian Ocean, Ocean Literacy, children geographies, and education for sustainable development.

Salim Momtaz (Ph.D.) is an affiliate associate professor at the University of Newcastle, Callaghan, Australia. He teaches in the area of Sustainable Resource Management. He received his B.Sc. and M.Sc. degrees in Geography from the University of Dhaka, Dhaka, Bangladesh. He did a Ph.D. in Development Studies from the University of London under a Commonwealth Scholarship, working under the supervision of Professor Richard Munton. His academic career started at the University of Dhaka in 1986. He moved to Australia in 1994. From 1995 to 1998, he taught Geography and Environmental Studies at Central Queensland University. He joined the University of Newcastle in 1999 and became an Affiliate in 2021. He had a stint in the USA, teaching Environmental and Social Impact Assessment at Georgetown University, Washington, DC, as a visiting professor. He received Rotary International Ambassadorial Fellowship to teach and conduct research in Bangladesh. His current research interests include Environment and Development, Climate Change, environmental governance, and Environmental and Social Impact Assessment. He led the team that conducted one of the first social impact assessment studies in Australia titled *Independent Social Impact Assessment: Proposed Castle Hope Dam – Calliope River, and Awoonga Dam upgrade, Queensland*. He has published 13 books, 8 book chapters, and many articles in international journals. He was a member of the Scientific Advisory Committee – the Netherlands government research organization – between 2007 and 2010. Currently, he is serving the European Climate, Infrastructure and Environment Executive Agency (CINEA) as an Expert Evaluator.

Mohammed Moniruzzaman Khan (Ph.D.) is a professor in the Department of Sociology at Jagannath University as well as adjunct faculty of the Institute of Disaster Management and Vulnerability Studies (IDMVS), the University of Dhaka, and North South University, Dhaka, Bangladesh. His two books *Disaster and Gender in Coastal Bangladesh: Women's changing roles, risk, and vulnerability* and *Coastal Disaster Risk Management in Bangladesh: Vulnerability and Resilience* are soon to be published. Taking teaching and research as a career with expertise in "disaster, gender, and environment," Dr. Khan's one and only mission and vision are to "empower and develop agency of women" in DRR institutionally through sustainable development in the most disaster-prone coastal area and for most vulnerable people which will ultimately assist the government to achieve SDG. Dr. Khan's research interest/expertise covers disaster, gender, environment, climate change, women empowerment, development of agency of women, social class, indigenous community, etc. Mr. Khan received his Ph.D. in "Disaster and Women: The Wave of Change in the Position and Status of Women in Coastal Bangladesh" from the Department of Sociology and Anthropology, School of Language, Social and Political Sciences, University of Canterbury, New Zealand; Master of Science in Sociology from London School of Economics and Political Science (LSE), London, United Kingdom, and Bachelor of Social Sciences in Sociology from the University of Dhaka.

Tien Dung Nguyen has been working as a lecturer regarding extension training at the University of Agriculture and Forestry, Hue University. He is pursuing second-year Ph.D. in Rural Development at Hue University. His research focuses on organic agriculture, agricultural commercialization, multidimensional poverty, and climate change.

Van Chung Nguyen (Ph.D.) is a researcher and lecturer at the Faculty of Rural Development of the University of Agriculture and Forestry at Hue University in Vietnam. His main research interest includes the development planning and livelihood strategies on topics such as agri-food value chains, capacity building for farmers, and community development in Vietnam.

Shitangsu Kumar Paul (Ph.D.) is a professor in the Department of Geography and Environmental Studies, University of Rajshahi, Bangladesh. He joined the University of Rajshahi as a lecturer in October 1999 and regularized as an Assistant Professor in 2002, Associate Professor in 2009, and Professor in 2014 after external reviews. He completed his Bachelor of Science in Geography and Environmental Studies from the University of

Rajshahi in 1993, followed by a Masters of Science from the same university in 1994. He obtained another Master's in Regional and Rural Development Planning from the Asian Institute of Technology, Khlong Nueng, Thailand, in 2005, followed by a Ph.D. in 2010 from the same institute. He has more than 45 publications in national and international peer-reviewed journals and books.

Eila Romo-Murphy (Ph.D.) is a research associate with H-C-R, with a Ph.D. at the University of Jyväskylä, Finland, in 2014. Her interest includes disaster preparedness of citizens and organizing media campaigns to increase the disaster preparedness skills of vulnerable communities. She aspires to help communities at risk to mitigate climate change, including natural disasters. She believes in the survival capabilities of citizens helping each other as a community, in conjunction with effective communication and collaboration of government, non-government, and media stakeholders. Her doctoral dissertation dealt with the evaluation of Aceh community preparedness during the Indian Ocean tsunami, followed by an assessment of media disaster preparedness campaigns in Padang, Indonesia. She has been involved in evaluating the outcomes of First Response Radio, whose purpose is to promote health and survival during natural disasters. One such evaluation was conducted in Bihar in the aftermath of the 2017 floods.

Jayant K. Routray (Ph.D.) is a Professor Emeritus at the Asian Institute of Technology (AIT), Khlong Nueng, Thailand. He joined AIT as an Assistant Professor in June 1988 and was promoted to the rank of Associate Professor in 1993, and Professor in 2005. He has served as an AIT faculty member for 29 years, out of which, he served as a professor for about 12 years. He completed his Bachelor of Science in Geography (Hons) from Utkal University, Bhubaneswar, India, in 1974, followed by a Master of Science in Geography from the same university in 1976. He obtained his Master's in Regional Planning from Indian Institute of Technology Kharagpur, India, in 1979, and a Ph.D. from Utkal University, Bhubaneswar, India, in 1987. He has 62 papers in international refereed journals, 20 research papers in national and regional journals, 22 book chapters, 25 research monographs/books, manuals, and proceedings, 12 papers in conference proceedings, 5 projects/workshop reports, and 8 non-refereed publications.

Lisha Samuel is a native of Mumbai, India. She holds a B.A. in Psychology from the University of Connecticut and an M.A. in Communication with a Strategic Public Relations and Social Media Concentration from Monmouth University, West Long Branch, New Jersey. She coordinated communication activities as a Graduate Assistant for the Institute for Global Understanding at Monmouth University. She also received a research grant from the Urban Coast Institute at Monmouth University to conduct research and prepare a paper on climate change communication. She has held positions as a Marketing Copywriter and a Public

Relations Associate for Red Rooster PR. She plans to pursue a career in climate change communication where her communication and media skills can help with new strategies to raise awareness of climate change.

Johannes Dragsbaek Schmidt (Ph.D.) is a senior expert at the Nordic Institute of Asian Studies, Copenhagen University, and an Adjunct Associate Professor in Development and International Relations at Aalborg University, Denmark; Visiting Fellowships Asia Research Centre, Murdoch University, Australia; Institute for Political Economy, Carleton University, Canada; and Asia-Europe Institute, University of Malaya, Kuala Lumpur, Malaysia. Furthermore, he was previously a fellow at ISEAS, Singapore; Centre for Strategic and International Studies, Jakarta, Indonesia; Social Research Institute, Chulalongkorn University, Bangkok, Thailand; and Third World Studies Center, University of the Philippines, and has extensive research experience from Southeast Asia, India, China, and Bhutan. He offered consultancy services for UNESCO, World Bank, ASEM, Nordic Consulting Group, and served as a lead researcher on research projects funded by the Danish Development Agency, Danida, Ministry of Foreign Affairs, and funding by the Danish Agency for Science, Technology. He has published about 250 publications with a primary focus on Asian issues.

Maria Schreider (Ph.D.) is a marine ecologist with broad interests in the ecology of estuarine habitats. She studied the effects of anthropogenic impacts in estuaries, in particular, on epifaunal assemblages in seagrasses. Her more recent research interests focus on the ecological effects of invasive species in estuaries and on the dynamics of algal blooms in small coastal lagoons. She is also studying the population ecology of ghost shrimps with a view of their potential use for aquaculture.

Ann R. Tickamyer (Ph.D., University of North Carolina, Chapel Hill, USA) is Professor Emerita of Rural Sociology and Demography at Penn State University. Her research and scholarship center on rural poverty, livelihoods and inequalities, gender and development, disaster and climate change, and research methods. She has published six books and more than 80 journal articles and book chapters. She and Siti Kusujiarti published *Power, Change, and Gender Relations in Rural Java: A Tale of Two Villages* (2012) and several recent articles on gender, disaster, and climate change in Indonesia.

Farhana Zaman (Ph.D.) is a professor in the Department of Sociology at Jagannath University, Dhaka, Bangladesh. She holds a Ph.D. degree from the Institute of Disaster Management and Vulnerability Studies and obtained her undergraduate and a graduate degree in Sociology from the University of Dhaka, Dhaka, Bangladesh. She has completed Advanced Graduate Studies in Social Entrepreneurship from Salve Regina University, Rhode Island, USA. She has been serving as an associate editor for the *Jagannath University Journal of Social Sciences* (JUJSS) since 2018. Her research interests span the areas of climate change, disaster, vulnerability, coping, adaptation, and building resilience across the risk-prone coastal regions. She has published several articles on climate change and disaster-related issues in both national and international journals. Her first book *Sociology of Disaster: Concepts, Perspectives, and Theories* provide an excellent introduction to the sociological study of disaster, and her second book *Social Dynamics of Salinity: Problems and Prospects in Coastal Bangladesh* is committed to serving as a useful tool for disaster scholars who are planning to pursue their studies further in the field of the disaster from the social science perspective.

1 The social and political dimensions of climate change

Focus on South and Southeast Asia

Devendraraj Madhanagopal and Salim Momtaz

The climate crisis: an ever-growing threat in South and Southeast Asia

The countries that are contributing less to global warming and that can least afford the impacts of climate change are at the forefront of climate change. The Intergovernmental Panel on Climate Change (IPCC) has long listed the small island developing states (SIDS) as well as South and Southeast Asia as the most vulnerable countries to climate change. In its report of 2022, IPCC has further reiterated its view about South Asian countries along with the SIDS and many developing countries of Africa, and Central and South America by identifying them as "hotspots of high human vulnerability" (IPCC, 2022). The populations of South and Southeast Asia regions stood at around 2.5 billion. This is also home to the majority of the world's poor and marginalized groups. In recent decades, adverse effects of climate change and associated extreme weather events are already starkly visible across South and Southeast Asia. Climate change aggravates the pressures on natural resources and the environment of Asia. It is more likely to bring unprecedented challenges to human lives, health, livelihoods, development, infrastructure, and geopolitics in the near future (Hijioka et al., 2014; ADB, 2017). A large proportion of the Asian population resides in urban and coastal regions. More than a billion people in South and Southeast Asia reside in vulnerable regions to climate change and risks, and their livelihoods are increasingly dependent on climate-sensitive sectors.

Neumann et al. (2015) found that in 2000, Asia had the largest low-elevation coastal zone population, that is, 461 million people – almost 73% of the world's total low-elevation coastal zone population. Their analysis further showed that, under all scenarios, the range of percentage of people in Asia who live in low-elevation coastal zones will be almost similar (or more) in 2030 and 2060. In 2000, India, the densely populated maritime nation in South Asia, had 64 million who lived in the low-elevation coastal zones. This number will be 216 million in 2060, which means around 10.3% of the world's total low-elevation coastal zone populations are from India. This range is also almost similar in the case of Bangladesh. However, Bangladesh has a very large non-urban population as compared to India. The largest share of the people in the low-elevation coastal zone populations worldwide are from

DOI: 10.4324/9781003216476-1

five Asian nations: China, India, Bangladesh, Indonesia, and Vietnam. It should be noted that four are from South and Southeast Asia.

In 2016, around 50% of Asian people were being urbanized, and it is expected that 64% of Asian populations will be urbanized by 2050 (ADB, 2017). Such growing dynamics of urbanization and the increasing concentration of populations and economic centers across coastal regions of Asia have already created more complexities in the face of climate change. It is likely to bring massive challenges to Asia in the future. In the face of changing climate, such growing dynamics of urbanization and the resulting consequences have already created several complexities, and it is likely to bring massive challenges to Asia in the future. Like this, the growing depletion of Himalayan glaciers and the subsequent river flow changes potentially increase the vulnerability of millions of Asians. In addition, the livelihoods of around a billion people belonging to South and Southeast Asian regions depend on climate-sensitive sectors. Hence, a large proportion of South and Southeast Asia are more prone to climate change and associated water-related extremes (SAARC 2008; Vinke et al., 2017; Alam & Sawhney, 2019; Mall et al., 2019; Aryal et al., 2020).

The growing complexities of climate change, society, and politics: where do the solutions lie?

Climate change challenges and solutions to climate change have been among the pressing topics that increasingly received attention worldwide for the past one to two decades across physical, natural, and social sciences. By realizing the extraordinary problems of climate change worldwide, in recent years, there has been a growing but insufficient attention given to embedding "social sciences" in climate policy and emphasizing the politics of climate change and adaptation (for example, Yearley, 2009; Giddens, 2011; Dunlap & Brulle, 2015; Victor, 2015; Zehr, 2015). In recent years, there has been a paradigm shift to consider the research knowledge closely from social, political, and anthropological specializations. Primarily, after recognizing the social vulnerability to climate change, there has been a growing recognition of climate change knowledge, focusing on Asia. The Global North has realized a need to unravel the multiple intricacies of the social and political dimensions of climate change in Asia and Africa to have a better understanding of climate change governance and politics. Climate change knowledge on South and Southeast Asia meets the vast demands of knowledge gaps not just in Asia but also in the entire globe. This is because the social, political, cultural, and governance systems of South and Southeast Asia are highly diverse and facing radical changes, particularly in recent decades.

Against the backdrop of South and Southeast Asian societies, there is also a growing body of literature that discusses the multifaceted nature of social and political dimensions of climate change and disasters (some selected recent works include Dubash, 2015; Kelman, 2015; Leal Filho, 2015; Momtaz & Shameem, 2015; Kelman et al., 2017, 2019; Kelman & Colbourn, 2017; Sternberg, 2017;

Momtaz & Asaduzzaman, 2018; Pal & Shaw, 2018; Barua et al., 2019; Chou et al., 2020). However, within the existing literature on South and Southeast Asian nations, there are numerous gaps in our knowledge of societies' responses and the different range of governance systems across scales, and the underpinning politics of policies, programs, and actions of the state and non-state actors in responding to climate change and disasters. Vulnerabilities and adaptive capacity of societies to climate change are highly localized and largely vary on different social and political factors. Effective leadership and proactive governance systems across all levels play crucial roles in shaping the adaptive capacity to climate change, readiness, and resilience of societies.

The social and political forces of the societies, communities, regions, and nations of Asia could intersect with the adverse effects of climate change and decrease the adaptive capacity of Asians. However, these trends are not similar across the South and Southeast nations. Climate change impacts may increasingly pose more severe challenges to societies that lack proactive policies, strategies, and support from non-state actors (including NGOs) to respond to climate change, further aggravating the vulnerabilities of South and Southeast Asian societies and marginal communities, in particular. In addition to climate change, Asian societies must adapt to multiple vulnerabilities, including coastal hazards and other sociopolitical complexities. Overall, given the growing scientific evidence, it is established that climate change threats and their sophistication will continue to loom at the forefront in the coming years and decades.

Through structured institutions and political leadership, social inclusion efforts will add value to the resilient nature of the marginal societies in planning and responding to climate change at the local and institutional levels. It is essential to be aware that inclusive sociopolitical systems, gender-sensitive governance and leadership, and community-centric adaptation plans are the starting point for many rural and coastal societies engaging with the state and other external actors in adapting to climate change. However, in South and Southeast Asia, little knowledge is available to understand and discuss the responses of social and political systems to climate change (Rahman et al., 2018; Wang et al., 2019). Smooth regional interactions, effective and coordinated communication, and cooperation among the governments across scales are crucial in promoting socio-economic development while simultaneously adapting to climate change. Like this, local leaders and active participants of civil society can provide climate-informed education practice and help communities develop their capacity to manage the effects of climate change by reducing the social, cultural, political, and institutional barriers that hamper adaptation interventions to climate change. However, the available body of knowledge is inadequate, particularly in the context of South and Southeast Asia, to deal with the magnitude of climate change and associated disasters. To support this, in the following section, we list a few interdisciplinary edited volumes discussing the intricacies of sociopolitical systems of South and Southeastern nations in responding to climate change and highlight the scope of those volumes and existing knowledge gaps.

Some recent knowledge additions: scope and existing knowledge gaps

The book titled *Climate Change Adaptation and Disaster Risk Reduction: An Asian Perspective*, edited by Rajib Shaw, Juan M. Pulhin, and Joy Jacqueline Pereira, is one of the essential works discussing the problems and solutions of climate change and disasters in Asia. A good proportion of the chapters are case studies, and the remaining ones are based on reviews. Compared to the other books listed in this section, this book provides substantial empirical evidence to discuss disaster risk reduction, local vulnerability, and adaptation. Employing academic research and field practice explores the linkages between climate change adaptation and disaster risk reduction in various sectors of Asia, including agriculture, urban, agroforestry, fisheries, coastal systems, water insecurities, health, etc. This book provides a blend of academic analysis, field-based evidence, and practices. However, many chapters of this book discuss the context of the nations, but are not exclusively focused on a particular region of the country. This book is recommended for students, scholars, and practitioners. It provides less focus on policy-oriented discussions as the objectives of this book are mainly focused on documenting the responses and experiences of the nations in responding to climate change and disaster risks (Shaw et al., 2010).

Focusing on India, the book titled *Handbook of Climate Change and India Development, Politics and Governance* edited by Navroz K. Dubash critically explores and discusses how development, politics, and governance systems shape and influence climate change discourses, policies, and negotiations, and it sets out the future agenda. The book has a wide range of scope, and it receives the attention of policymakers, academics, professionals, civil society activists, and practitioners. However, the book provides less focus on the social factors that shape the vulnerability of marginal populations at the local levels. In particular, no chapter provides evidence on how caste, gender, class, and social norms influence the politics of climate change governance in India. However, it is difficult to point out this aspect as one of the limitations of this edited volume as the book's scope lies in policy-oriented discourses. The other limitation lies in the exclusive focus on climate change, but not the challenges of disaster risks and extreme events (Dubash, 2015).

The compilation of 12 chapters in the book titled *Natural Disaster Management in the Asia-Pacific Policy and Governance* by Caroline Brassard, Arnold M. Howitt, and David W. Giles focuses on public policy, community engagement, and humanitarian systems of disaster management in Asia-Pacific regions. Some chapters of the book showcase the often-neglected issues in disaster recovery processes such as mental health and manpower support. Overall, the book argues for the need for the shift of governance systems in managing the disasters, focusing on the vibrant political environment and humanitarian systems that are highly needed and timely in the era of growing disasters and the resulting uncertainty (Brassard et al., 2015).

Indrajit Pal and Rajib Shaw's edited book titled *Disaster Risk Governance in India and Cross-Cutting Issues* provides a detailed account of disaster risk reduction and governance in India. It presents cases, insights, and discussions from national and regional programs, projects, and other field-based experiences. The book is particularly interested in the researchers looking at the issues of disasters through governance, development, legal, and administrative perspectives. Nevertheless, it provides limited attention to climate change adaptation and the efforts to integrate climate change adaptation with disaster risk reduction (Pal & Shaw, 2018).

The edited book by Anamika Barua, Vishal Narain, and Sumit Vij titled *Climate Change Governance and Adaptation Case Studies from South Asia* features a range of case studies focusing on climate change governance and adaptation interventions, adaptive capacity, water resources, and climate actions. The book is particularly interesting for researchers and practitioners interested in climate change governance and adaptation centers in the Himalayas and Nepal regions. A large portion of the book is devoted to discussion on this. This can be considered as a positive as well as a limitation of the book. Climate change governance and adaptation have always been a contested issue, and recent research has highlighted the conflicts surrounding marine governance in the face of climate change and extreme events. The book, however, has not discussed such climate change complications in coastal settings (Barua et al., 2019).

This recent volume, *Climate Change Governance in Asia*, edited by Kuei-Tien Chou, Koichi Hasegawa, Dowan Ku, and Shu-Fen Kao, exclusively focuses on climate change governance in Asia. A large proportion of the book provides comprehensive conceptual and policy-oriented discussions on climate change, risks, transitions, and the remaining ones deal with empirical evidence and discussions. Majority of policy discussions in the book deal with macro-level analysis. Research has repeatedly pointed out that Bangladesh and India are the most vulnerable nations to climate change and associated disaster risks. However, the book has not provided contributions to discuss the complications and opportunities of climate change governance in many South Asian countries such as Bangladesh, Nepal, India, and Sri Lanka (Chou et al., 2020).

Why this book and what is its scope?

Various sociopolitical forces and governance systems have shaped and influenced climate change policies and adaptation strategies, and they continue to influence climate change and risk in the future. This book aims to upgrade our understanding of the multiple contemporary and emerging sociopolitical factors, actions, and perspectives of South and Southeast Asia societies in shaping vulnerability and adaptation to climate change. It unravels the underlying social and political relations that shape and influence the governance actors, institutions, policies, programs, and actions in responding to climate change. Moving beyond science and technology-oriented discussions on climate change, this book, through cases, conceptual, theoretical, and empirical discussions, describes that the real solution to climate change problems lies in societies, governance systems, non-state actors, and the politics underpinning these systems. It looks that actions, programs, and policies across South and Southeast Asia in responding to climate change are not similar, and they vary across regions and space. In the era of climate change, it is essential to upgrade our understanding of how far the regional/local developmental plans have integrated climate policies and interventions to climate change of national/federal governments in responding to climate change. To understand this, it is not sufficient to focus on the relationship between the national/federal and regional/local governments in implementing climate policies and developmental plans. There is also a need to document the local climate actions and strategies followed in different countries to cope and adapt to climate change. The politics that shape and influence the key climate policies and lessons learned across different sectors and levels (national/federal, state, and local/regional) need to be explored to upgrade our knowledge to place and discuss the problems in a broader context. Through a sociopolitical lens, this book presents novel perspectives, discussions, and policy recommendations on the responses of the state, non-state actors, and communities to climate change and disasters in South and Southeast Asia.

The environmental and economic damages have been estimated, recorded, and researched from many directions in the recent past. References to many of these works can be found in the various chapters of this book. Recurring extreme weather events have also created multifaceted direct and indirect social problems through disproportionate effects on vulnerable people; involuntary displacement and migration of affected people; biased environmental planning and governance; and unequal distribution of benefits of climate change projects often influenced by sociopolitical dynamics at state and local contexts. While there have been numerous research works in South and South Asia on environmental and economic impacts of climate change; food security; livelihood and vulnerability assessments; and related topics, few have examined how sociopolitical perspectives influence climate change decision-making at various levels of administration and actions on the ground.

According to the Global Climate Risk Index developed by the think tank Germanwatch (Eckstein et al., 2019), India (5th) and Sri Lanka (6th) were among the ten most-affected countries in 2018. This organization has also developed a Long-Term Climate Risk Index that shows that seven out of the ten countries most affected by climate change effects from 1999 to 2018 belong to the South-Southeast Asian region, namely Myanmar, Philippines, Pakistan, Vietnam, Bangladesh, Thailand, and Nepal. Furthermore, half of the South Asian population has been affected by one or more climate-related disasters in the last decade (Fallesen et al., 2019). This book covers many of these climate-vulnerable countries of South and Southeast Asia.

The countries of South-Southeast Asia have experienced recurring extreme weather events in the last few decades including cyclones, unusual tidal surges, torrential rain and subsequent flood, and drought. Coastal areas of Bangladesh, India, Pakistan, and Sri Lanka are severely exposed to the damages of the sea-level rise where rapid and often unplanned development activities have compounded the effects of climate change. The countries, such as Nepal, Afghanistan, and Bhutan, that are away from the sea also suffered from climate change effects in the form of rising temperature, drought, and melting of glaciers. The island nation of the Maldives whose existence is dependent on tourism has been going through the dilemma of whether to protect the country from the damages of climate change or invest in tourism development activities that are often making the numerous islands more vulnerable to climate change. This edition attempts to focus on many of these countries and attempts to fill this knowledge gap.

By featuring a multiple and diverse range of contributions, including case studies and advanced research discussions, this edited book is a valuable resource for academics, practitioners, and policymakers interested in the debates on climate change issues in South and Southeast Asia. This book aims to bring together contributions from various scholars, including climate change, environmental science and policy, sociology, geography, fisheries, law, environmental philosophy, earth, and ecosystem sciences, international development, and development studies. All their contributions to this book are centered on the social and political dimensions of climate change. Focusing on different regions and conceptual themes, this book, against the backdrop of social and political dimensions, presents novel insights into the past and ongoing responses of the state actors, non-state actors, and communities in responding to climate change. Through their contributions, overall, they aim to address the following question: How have the socio-economic and political systems of the societies responded to climate change and associated disasters, and how such practices, institutions, and strategies continue to shape the climate policies, programs, and actions across multiple scales? Though the main focus of this book is on South and Southeast Asia, we hope that this book will grab the attention of the global climate change community beyond boundaries as the discussions, both conceptual and theoretical aspects of some chapters of this book, have much significance in international research, policy, and practice. It showcases case studies, field-based evidence, and policy discussions of climate change and provides valuable lessons.

We structure the following chapters of this book into three parts. The details of the chapter headings and summary of the chapters are the following.

Part I – the multiple challenges of climate change and risks

In the 1960s and 1970s, the nations of the world started to come to terms with the fact that the global population had increased exponentially in the previous century and natural resources had been severely depleted to humankind's detriment. Human activities became highly anthropocentric which failed to respect the fact that humans are an integral part of and therefore highly dependent on the health of the environment. Discussion on striking a balance between economic growth and environmental protection started to get momentum. This discussion led to the emergence of the sustainable development concept of the World Commission on Environment and Development (WCED), that is, "development that meets the needs of the present without compromising the ability of future generations to meet their own needs." The idea was promoted that humans were a part of the environment like other living and non-living things. Following in the same vein, Chapter 2 by Janaka Jayawickrama is a change of pace in this collection of contextually diverse chapters. It takes the reader to a deeper understanding of Hindu, Buddhist, Taoist, and Confucius philosophies that had been urging a balance in the environment where humans lived in harmony with nature. The chapter critiques human activities in the name of development and examines the knowledge and wisdom of ancient philosophies. While it acknowledges that the ancient religions may not have answers to all the anomalies and challenges of modern times; we can learn from them to live in peace with nature and create harmony.

In Chapter 3 by Shitangsu Kumar Paul and Jayant K. Routray, we see a second contribution to the central coast of Bangladesh. Bangladesh, as a deltaic region exposed to the Bay of Bengal, has been on the news in recent decades for the tremendous losses of lives and livelihoods it incurred due to heavy rainfall and flood during the Monsoon, drought during dry seasons, and the increasing number of cyclones. This study was conducted in three villages severely vulnerable to cyclones and storm surges. The purpose of the chapter is to identify major livelihood groups based on post-cyclone livelihood capital status; their comparative adversities and security status; and their livelihood strategies that are influenced by recurrent extreme weather events. The study employs scientific analysis of livelihood security indicators and indices to identify vulnerable locations and groups. It is hoped that the findings of this research would help decision-makers and proponents to develop and implement appropriate programs and projects that would lead to the enhanced socio-economic status of the targeted groups and reduced vulnerability and insecurity.

In the recent past, Bangladesh has made significant financial commitments to tackle climate change mostly through preparedness and adaptation. The country has mainstreamed climate governance into the national environmental agenda and implemented projects and programs. International partners and local non-governmental organizations (NGOs) are also playing an important role at the grassroots level to respond to climate change effects. In Chapter 4, Md. Faruk Hossain, Salim Momtaz, and Anita Chalmers explore various aspects of climate change governance practices in Bangladesh with an emphasis on how communities perceive government and NGOs' performance in implementing climate change programs and projects. This empirical study is based on case studies from a "climate change" affected region located in the south-central coastal zone. This study highlights the major adaptation and mitigation measures undertaken by the government and NGOs. The main aim of the chapter is to assess local communities' perception of climate change governance practices using five criteria – accountability, transparency, participation, governance effectiveness, and resources. This research follows a mixed-method approach. To evaluate climate change governance practices of the community people, an empirical survey was undertaken. The findings of this evaluation reveal that the national government has undertaken numerous measures to address extreme weather events related to climate change. However, the activities at the local level have not met community expectations. There are issues with funding and accountability. The NGOs seem to have performed better in the eyes of the recipients of benefits. The authors call for better incorporation of local knowledge in decision-making.

The Republic of Indonesia is the world's largest island country located in the Pacific Ring of Fire. The country has experienced rapid development in the recent past that brought people's lives and livings to a better place. Yet, the country faces challenges on multiple fronts. It has a complex population consisting of diverse backgrounds and cultures. From recent research, we come to know more and more that climate change effects are not equitably or ubiquitously distributed. And that effective response to climate change adaptation to address the goals of sustainable development can only be fulfilled by developing appropriate measures that would address this disparity. As in many developing countries, women and cultural and ethnic minorities are regarded most vulnerable to climate change. Chapter 5 by Ann R. Tickamyer and Siti Kusujiarti attempts to examine this issue. They identify the Indonesian gendered risk spaces where women are disproportionately vulnerable to death, destruction, and diseases during and in the aftermath of natural disasters. They also observed in their three case studies that women are less vulnerable in the Bantul district where women are empowered through various participatory programs of the government, NGOs, and volunteer organizations that promote social justice, gender equality, and participation, than in the two other case studies of Aceh and Merapi with less external support. Therefore, the authors emphasize gender analysis to develop and implement development policies and programs that would address women's vulnerability, address climate justice, and ensure sustainability.

Part II – responding to climate change: sociopolitical underpinnings of multiple actors and institutions

Chapter 6 by Ha Dung Hoang, Salim Momtaz, Maria Schreider, Van Chung Nguyen, and Tien Dung Nguyen is another case study based on empirical investigation. It is from Vietnam by the researcher Ha Dung Hoang and his team. Small-scale fishers (SSFs) play an important role in the Vietnamese economy by sustaining local communities in the coastal regions. This sector is at the forefront of climate change, experiencing the effects of sea-level rise, tidal surges, and recurrent cyclones that affect fishing efforts and the amount of catch. The chapter brings evidence from Tam Giang-Cau Hai lagoon, Central Vietnam, of how SSFs become aware and respond to climate shocks that influence their lives and livelihoods. It quantifies SSF's resilience using "absorptive capacity," "adaptive capacity," and "transformative capacity."

Based on SSFs' recollection of extreme climate events and their responses, the chapter demonstrates independent individual and sometimes collective community strategies employed to face climate shocks and stressors. The authors observed that the timely support from local authorities and other organizations helped the fishers in the post-disaster recovery. It is expected that a better understanding of fishers' perceptions and responses would lead to better intervention by the government to reduce the vulnerability of fishing communities in Vietnam.

Like many countries, Sri Lanka has been experiencing the effects of climate change – manifestations of which can be seen in recurring heavy rains, flash floods, and landslides. By examining three case studies of relocations in the aftermath of the 2004 Indian Ocean tsunami; heavy monsoonal rains of 2016 in central highlands; and the collapse of a landfill following heavy rains in 2017, respectively, Chapter 7 by Cynthia M. Caron explores how the relocation of affected people has been used as a measure to address climate-related disasters. These relocations have been seen as reactive actions of the government to address the immediate problem emanating from natural disasters rather than as a part of a comprehensive plan to address climate change effects in the long term. The author has placed their analyses in the country's historical ethnopolitical context of government interventions that are influenced by political agenda and characterized by disproportional distributions of benefits to the diverse ethnic communities. The author has asserted their observation of the Sri Lankan government's development agenda by quoting "ethnic and class politics and urban-rural dynamics shape the Sri Lankan government's development agenda, including relocating families to pursue them." The author calls upon the government for preventative or planned relocation programs according to the risk profiles created under Sendai Framework by UN agencies. In the end, the author emphasizes the role of the large activist and social justice-oriented community that exists in Sri Lanka.

India and Bangladesh are the densely populated maritime nations in South Asia. Extreme climate events have continuously hit the coastal regions of both countries for the past few decades, and it is highly expected that these climate change vulnerabilities will continue to rise in the future. To respond to this, both nations must strengthen their local climate change action programs – for this, it is essential to have a better understanding of the underpinned politics and power that influence these programs. Chapter 8 by Devendraraj Madhanagopal, Mohammed Moniruzzaman Khan, and Farhana Zaman – by taking the case of India and Bangladesh – provides comparative discussions and a better understanding of local politics and their influence on local climate action programs. The authors present a review of climate change policies and action programs in India and Bangladesh against their historical developments in the recent past. After this, through case studies from selected coastal regions of India and Bangladesh, the authors discuss the ways in which power-centric local politics influence climate change adaptation through field narratives and insights. The authors observe simultaneous development of climate change policies and action programs in both countries in the mid-2000s – almost similar periods – nevertheless, as they explain, adaptation programs in both nations are still evolving, and many nuances of power and local politics of climate action programs, and the resulting social and gender marginalization of certain social groups in responding to climate change, which leads them as the powerless victims of climate change and risks.

Digital accessibility is not equally distributed across Asia. Research has widely established that socioeconomic factors play a crucial role in determining the ability of the communities and regions in accessing to digital and electronic media. The state of Bihar is one of the backward regions of India. A large proportion of Bihar's population is trapped in poverty and social backwardness, and the move to the internet and social media has been slow across this state. Hence, they are largely dependent on "Radios" for accessing information, including emergency broadcasts. The 2017 Bihar floods were one of the worst disasters of the 2010s in India. Apart from human mortality, huge assets, and infrastructure loss, this flood had cascading effects in Bihar in multiple ways. By taking the case of the 2017 Bihar floods, through empirical research, Chapter 9 by Alankrita Anand and Eila Romo-Murphy discusses the roles of First Information broadcast content in helping the flood-affected citizens and the ways in which communication platforms (including Radios) helped citizens to strengthen their resilience during the floods. While discussing the roles of media in the 2017 flood disaster, the authors also provide an overview of the information flow during disasters and highlight how Radios acted as a powerful media platform in transferring flood and weather-related information, emergency broadcasts, and recovery activities. Through interviews and discussions with the flood-affected citizens and response network representatives, the authors bring the positives as well as flaws of the existing communication systems of the state and communities in mitigating and responding to the cascading effects of flood disasters effectively.

Though the international scientific community has established that climate change is one of the most pressing challenges humankind has ever faced, it has largely been unknown and less known to millions of the general public. These numbers are tremendously higher in underdeveloped and developing nations. Ironically, these populations are the worst climate change victims; given their social vulnerability, they continue to face drastic climate change effects in the future. Chapter 10 by Lisha Samuel and Randall S. Abate examines the importance of climate change communication in Kolkata of West Bengal, one of the densely populated cities in India. Through insights from law experts, journalists, and reputed academicians of India, the authors observe the complications of climate change communication in Kolkata and explore the way forward to respond to climate change through effective communication strategies. As identified by them, climate change issues are discussed (but limited) in English news outlets but not in the local language (Bengali) news outlets. The chapter also reports the insufficient coverage of climate change issues and their effects, including climate change migration, in almost all the news outlets focused on Kolkata. Having pointed out such insufficient attention and coverage, the authors explore the theories and strategies that bolster effective communication methods to publicize climate change migration to the masses.

Part III – power and politics of climate change: focusing on strategies and policies

Though South Asia is often considered one of the most vulnerable climate change hotspots on planet earth, Bhutan is often praised by the international scientific community for its initiatives in environmental protection and its progressive climate change actions. Whereas the reality is different – Bhutan continues to face significant climate change challenges, and Bhutan is more likely to be hit by multiple effects of climate change and risks due to the rise in temperature and its impact on glaciers, and its high dependence on climate-sensitive sectors for their livelihoods. Chapter 11 by Johannes Dragsbaek Schmidt provides a comprehensive overview of current and future climate change vulnerability in the context of Bhutan's geography, socioeconomic nature, and livelihoods and discusses the existing policy responses of the state to combat climate change. While discussing the implications of climate change on agriculture and other livelihoods of Bhutan, the author also highlights Bhutan's ambitious climate change goals and their ramifications on the agriculture sector. Being a mountainous country, water supply levels in Bhutan have always been unevenly distributed. The chapter discusses how climate change projections may seriously threaten Bhutan's existing water security crisis and how it intensifies electricity production in total. Overall, the author also observes the problems that policy initiatives and institutions of Bhutan are facing to achieve the balance between economic self-sufficiency, carbon neutrality, and other climate mitigation and adaptation activities, and thereby, he offers some remarks on those challenges.

For the past few decades, developmental complications across the coastal regions of India have been on the rise. The rising sea levels and the increasing frequency and intensity of extreme climatic events have already complicated the existing developmental complications all along the coasts of India, and it is more likely to be increased in the future. The coastal stakeholders, mainly the marine fishers, are the frontline victims of the combined effects of developmental complications and climate change. How has the Indian state responded to these challenges through policies, and whether those policies considered the demands of locals? Chapter 12 by Devendraraj Madhanagopal, Lisha Samuel, and Kasturi Gandhi critically examines these questions by taking three regions in three coastal states of India: Tamil Nadu, Maharashtra, and West Bengal. All the coastal areas that the authors focus on are increasingly facing the brunt effects of developmental complications and climate change; some of these areas had already witnessed the protests of the marine fishers against coastal laws. The chapter provides a holistic overview of the history, politics, and loopholes of a few important laws and policies that influence coastal India. By highlighting the multiple vulnerabilities of marine fishers in three different coastal regions of India, the authors draw attention to how the combined effects of climate change and developmental complications have pushed the coastal populations, particularly the marine fishers, on the verge of vulnerability, and questions the existing coastal regulation zone policies and state climate action plans in addressing their problems. Having set out these discussions, the authors go on to suggest the importance of having local-centric, human, and inclusive policy-oriented approaches and actions of the state to face the contemporary and future challenges in coastal India.

According to the IPCC, the nation of Maldives falls in the group of 38 SIDS. These member states are considered highly vulnerable to climate change due to their smallness and exposure to the oceans and severe climatic events. As in other SIDS and developing countries, Maldives is facing a dilemma of balance between economic return and environmental protection. This is particularly significant in the case of this country as its beautiful coral resources provide architectural materials as well as an important income source through tourism. Healthy coral reefs are also linked with the health of the country's environment. Is it possible to strike a balance between various uses of coral reefs through proper policy formulation and management? Chapter 13 by Marcella Schmidt di Friedberg and Stefano Malatesta discusses both the value and the function of corals in the context of Maldives' economy and society. It also reviews the importance of coral reefs' preservation both as heritage and ecosystems in environmental policies at national and local levels. The authors propose to initiate a discussion on the effectiveness of healthy coral reefs and tourist attractiveness.

Given the vast diversity of the Indian subcontinent, implementing the mandates laid out in the climate policy of the Union Government of India and the action plans of the state governments have always been complex and sometimes contentious as well. Governing climate change mitigation and adaptation activities, then, requires a better understanding not just of focusing on the ways in which the Union government of India respond but equally on the mechanisms and action plans at the sub-national levels. The top-down administrative structure of India's climate policymaking and its limitations have already been well documented. However, the lack of a theoretical approach or framework to understand the interactions amongst various institutions at the sub-national levels further delimits our understanding of India's policy-making. Chapter 14 by Derek Kauneckis and Juhi Huda addresses these knowledge gaps by introducing a polycentricity framework and provides insights into the emerging climate policies of India. The authors provide a comprehensive overview of the complexities of India's administrative structures and the diversity of climate change risks across the country and propose how and why the polycentricity framework will be highly beneficial to analyzing climate risks within the complex decentralized policy systems of India. Highlighting the greater efficiency of polycentricity framework, the authors advocate that it can act as a powerful diagnostic tool to unpack the complexities of policy systems and make the researchers understand how and why management systems fail to respond to climate change and the factors that hamper effective climate actions.

References

ADB Asian Development Bank (2017). A Region at Risk: The Human Dimensions of Climate Change in Asia and the Pacific. https://www.adb.org/sites/default/files/publication/325251/region-risk-climate-change.pdf

Alam, M., & Sawhney, P. (2019) Regional Overview. In: Alam, M., Lee, J., & Sawhney P. (eds) *Status of Climate Change Adaptation in Asia and the Pacific*. Cham: Springer Climate. https://doi.org/10.1007/978-3-319-99347-8_3

Aryal, J. P., Sapkota, T. B., Khurana, R., Khatri-Chhetri, A., Rahut, D. B., & Jat, M. L. (2020). Climate Change and Agriculture in South Asia: Adaptation Options in Smallholder Production Systems. *Environment, Development and Sustainability*, 22, 5045–5075. https://doi.org/10.1007/s10668-019-00414-4

Barua, A., Narain, V., & Vij, S. (2019). *Climate Change Governance and Adaptation Case Studies from South Asia*. Boca Raton: CRC Press.

Brassard, C., Howitt, A. M., & Giles, D. W. (eds) (2015). *Natural Disaster Management in the Asia-Pacific. Policy and Governance*. Springer Japan.

Chou, K-T., Hasegawa, K., Ku, D., & Kao, S-F (2020). *Climate Change Governance in Asia*. London: Routledge.

Dubash, N. K. (2015). *Handbook of Climate Change and India: Development, Politics and Governance*. London: Routledge.

Dunlap, R. E., & Brulle, R. J. (eds) (2015). *Climate Change and Society: Sociological Perspectives*. New York: Oxford University Press.

Eckstein, D, Künzel, V, Schäfer, L, & Winges, M, (2019) *Global Climate Risk Index 2020 – Who Suffers Most from Extreme Weather Events? Weather-Related Loss Events in 2018 and 1999 to 2018*. Bonn: Germanwatch.

Fallesen, D., Khan, H., Tehsin, A., & Abbhi, A., (2019). World Bank Blogs – South Asia needs to act as one to fight climate change, World Bank. https://blogs.worldbank.org/endpovertyinsouthasia/south-asia-needs-act-one-fight-climate-change

Giddens, A. (2011). *The Politics of Climate Change*, 2nd Edition. Cambridge, UK: Polity Press. pp. 272.

Hijioka, Y., Lin, E., Pereira, J. J., Corlett, R. T., Cui, X., Insarov, G. E., Lasco, R. D., Lindgren, E., & Surjan, A., (2014). Asia. In: Climate Change (2014). Impacts, Adaptation, and Vulnerability. Part B: Regional Aspects. Contribution of Working Group II to the Fifth Assessment Report of the Intergovernmental Panel on Climate Change [Barros, V. R., Field, C. B., Dokken, D. J., Mastrandrea, M. D., Mach, K. J., Bilir, T. E., Chatterjee, M., Ebi, K. L., Estrada, Y. O., Genova, R. C., Girma, B., Kissel, E. S., Levy, A. N., MacCracken, S., Mastrandrea, P. R., and White, L. L. (eds.)]. Cambridge, United Kingdom and New York, NY, USA: Cambridge University Press, pp. 1327–1370.

IPCC (2022). Summary for Policymakers [Pörtner, H.-O., Roberts, D. C., Poloczanska, E. S., Mintenbeck, K., Tignor, M., Alegría, A., Craig, M., Langsdorf, S., Löschke, S., Möller, V., Okem, A. (eds.)]. In: Climate Change 2022: Impacts, Adaptation, and Vulnerability. Contribution of Working Group II to the Sixth Assessment Report of the Intergovernmental Panel on Climate Change [Pörtner, H.-O., Roberts, D. C., Tignor, M., Poloczanska, E. S., Mintenbeck, K., Alegría, A., Craig, M., Langsdorf,

S., Löschke, S., Möller, V., Okem, A., Rama, B. (eds.)]. Cambridge, UK and New York, NY, USA: Cambridge University Press.

Kelman, I. (2015). Asia-Pacific Islander Responses to Climate Change. In Koh, K-L., Kelman, I., Kibugi, R., & Osorio, R. E. (eds) *Adaptation to Climate Change ASEAN and Comparative Experiences*. World Scientific. https://doi.org/10.1142/9789814689748_0001

Kelman, I., & Colbourn, T. (2017). Climate Hazards and Health in Asia. In: T. Sternberg (ed) *Climate Hazard Crises in Asian Societies and Environments*. London: Routledge. pp. 56–72.

Kelman, I., Orlowska, J., Upadhyay, H., Stojanov, R., Webersik, C., Simonelli, A. C., Prochazka, D., & Nemec. D. (2019). Does Climate Change Influence People's Migration Decisions in Maldives? *Climatic Change*, 153(1–2), 285–99. https://doi.org/10.1007/s10584-019-02376-y

Kelman, I., Upadhyay, H., Simonelli, A., Arnall, A., Mohan, D., Lingaraj, G., & Webersik, C. (2017). Here and Now: Perceptions of Indian Ocean Islanders on the Climate Change and Migration Nexus. *Geografiska Annaler: Series B, Human Geography*, 99(3), 1–20. https://doi.org/10.1080/04353684.2017.1353888

Leal Filho, W. (2015). *Climate Change in the Asia-Pacific Region*. Springer, Cham: Springer International Publishing. https://doi.org/10.1007/978-3-319-14938-7

Mall, R. K., Srivastava, R. K., Banerjee, T., Mishra, O. P., Bhatt, D., & Sonkar, G. (2019). Disaster Risk Reduction Including Climate Change Adaptation Over South Asia: Challenges and Ways Forward. *International Journal of Disaster Risk Science*, 10, 14–27. https://doi.org/10.1007/s13753-018-0210-9

Momtaz, S., & Asaduzzaman, M. (2018). *Climate Change Impacts and Women's Livelihood – Vulnerability in Developing Countries*. London: Routledge. https://doi.org/10.4324/9780429462474

Momtaz, S & Shameem, M. (2015). *Experiencing Climate Change in Bangladesh – Vulnerability and Adaptation in Coastal Regions*. 1st Edition. Academic Press.

Neumann, B., Vafeidis, A. T., Zimmermann, J., & Nicholls, R. J. (2015). Future Coastal Population Growth and Exposure to Sea-Level Rise and Coastal Flooding-A Global Assessment. *PloS One*, 10(3), e0118571. https://doi.org/10.1371/journal.pone.0131375

Pal, I., & Shaw, R. (Eds.). (2018). *Disaster Risk Governance in India and Cross Cutting Issues*. Singapore: Springer Nature.

Rahman, T. H. M., Hickey, G. M., Ford, J. D., & Egan, M. A. (2018). Climate Change Research in Bangladesh: Research Gaps and Implications for Adaptation-Related Decision-Making. *Regional Environmental Change*, 18, 1535–1553. https://doi.org/10.1007/s10113-017-1271-9

SAARC (South Asian Association for Regional Cooperation) (2008). Regional cooperation on climate change adaptation and disaster risk reduction in South Asia: Road map. *SAARC Workshop on Climate Change and Disasters: Emerging Trends and Future Strategies*. Kathmandu, Nepal. pp. 168–182.

Shaw, R., Pulhin, J., & Pereira, J. (2010). *Climate Change Adaptation and Disaster Risk Reduction: An Asian Perspective*. Bingley: Emerald Group Publishing.

Sternberg, T. (2017). *Climate Hazard Crises in Asian Societies and Environments*. London: Routledge.

Victor, D. (2015). Climate Change: Embed the Social Sciences in Climate Policy. *Nature*, 520, 27–29. https://doi.org/10.1038/520027a

Vinke, K., Martin, M. A., Adams, S., Baarsch, F., Bondeau, A., Coumou, D., et al. (2017). Climatic Risks and Impacts in South Asia: Extremes of Water Scarcity and

Excess. *Regional Environmental Change*, 17, 1569–1583. https://doi.org/10.1007/s10113-015-0924-9

Wang, G., Mang, S. L., Riehl, B., Huang, J., Wang, G., Xu, L., Huang, K., & Innes, J. (2019) Climate Change Impacts and Forest Adaptation in the Asia-Pacific Region: From Regional Experts' Perspectives. *Journal of Forestry Research*, 30, 277–293. https://doi.org/10.1007/s11676-018-0827-y

Yearley, S. (2009). Sociology and Climate Change After Kyoto: What Roles for Social Science in Understanding Climate Change? *Current Sociology*, 57 (3), 389–405. https://doi.org/10.1177/0011392108101589

Zehr, S. (2015). The Sociology of Global Climate Change. *WIREs Climate Change*, 6(2), 129–150. https://doi.org/10.1002/wcc.328

Part I

The multiple challenges of climate change and risks

2 "Those Who Make an Enemy of the Earth Make an Enemy of Themselves"

Climate Change and Human Activities from a South and Southeast Asian Perspective

Janaka Jayawickrama

Introduction

With its human-centered principle, scientific disposition, and materialistic worldview of life, contemporary civilization has uprooted the human relationship to the world with all the necessary comforts the human beings needed and beyond. At the same time, the human relationship with the natural world is becoming less than harmonious. South and Southeast Asian countries are not immune to this. According to UN ESCAP (2019) data, both South and Southeast Asia are home to over 2.6 billion people, which is more than a quarter of the world's population. Further, in terms of disasters (ADRC, 2019), Asia is leading the deaths, damage, and affected populations, with floods, extreme temperatures, storms, and epidemics as leading disasters, mostly related to climate change effects. The increase in population, rapid urbanization, unequal distribution of resources, and consumerism contribute to climate and environmental challenges.

This chapter aims to critically examine the ancient wisdom based on the quality of having experience, knowledge, and judgment, which emerge from Hindu, Buddhist, Taoist, and Confucius philosophies from Asia, to propose some solutions to contemporary challenges of climate change and environmental threats. This chapter has examined the Hindu, Buddhist, Taoist, and Confucius philosophies as they have a wider impact on societies in South and Southeast Asia today. These philosophies also have socio-political impacts on individuals, communities, and governance structures in the region. Within this chapter, there is an underlying acknowledgment of the adverse effects and impacts of human activities on all life and non-life forms on this planet. From the exploitation of animals to genetically modified plants to the overall destruction of this globe are important aspects, and much has been already critiqued and established. Therefore, this chapter focus on the importance of learning from ancient wisdom to find possible solutions to current challenges.

While critiquing the existing approaches to development and failures of human activities, this chapter directly dives into various ancient philosophies

DOI: 10.4324/9781003216476-3

to tease out their wisdom. While acknowledging the modern contribution to life, such as vaccination and mobile technology, this chapter questions the inability of contemporary societies to live in harmony with nature. In that, this examination of ancient wisdom provides a suggested framework to rebuild and create harmony. At the same time, this chapter does not claim that ancient philosophies provide all the answers to contemporary challenges.

Disconnection through Intelligence

Humans have evolved to be the most intelligent animal on this planet over the past 2 million years (Bingham, 2000). However, human beings have contributed to more destruction of nature than any other animal. Increasing conflicts, disasters, and uneven development activities, as well as individual human greed, have separated us from the rest of the animal kingdom (Smith and O'Keefe, 1980). Let's think this way. Suppose insects, earthworms, or birds on earth suddenly vanished. In that case, it is highly possible that ecosystems and human civilization would seriously struggle to exist. Humans would have less food, fewer resources, and more pollution to deal with. In return, however, if the humans on earth suddenly vanished, we can easily imagine that the rest of nature would flourish. The hard truth is that human beings are becoming less and less important on this planet. Over the past centuries, we have been disconnected from our planet by various human activities such as urbanization and technological advances.

This separation between humans and the environment means that we cannot remain a confused mass of needs, greed and impulses, and activities without any restraint or guided by awareness. Humans consider themselves separated from the rest of the animal kingdom and environment. We are further divided through skin color, geography, class, caste, gender, sexuality, and wealth; therefore, not just separate from nature but other humans. The world is divided today and facing many risks and challenges more than ever before. South and Southeast Asia face these more than any other region (ADRC, 2019). Increasing storms, sea-level rise, flooding, surges, and various human activities such as rapid urbanization are all reasons for worsening erosion in the region.

The origins of the contemporary risks and challenges in South and Southeast Asia and globally can be understood as the absence of adjustment between the process of life, which is one of the increasing interdependences, and the "system" of life, the integrated habits of mind, loyalties, and greed, embodied in laws and institutions.

We cannot address these challenges and issues alone through physical agreements or material development. We have seen enough evidence that they are insufficient to create a human community that can live in awareness and harmony with the environment (Horton and Horton, 2019). For this, we need a human community with awareness and relationships with themselves, each other, and nature because humans cannot live without a flourishing environment. Social, political, cultural, and economic activities must be re-evaluated and executed within this awareness and relationships. The hindrances to the human society to live in harmony with nature are in the mind of humans. We have not developed a sense of duty and responsibility we owe to each other and nature. Gore (1992) explains this tension between humans and nature as the obsession with consuming larger amounts of the earth's resources. He further argues that this consumption leads to recurring catastrophes, manifesting a clash between humans and nature. All these risks, such as species extinction, disappearing rainforests, climatic imbalances, and natural hazards, have a strong human contribution.

Within the current marketplace, humans are producing nature through urbanization, rapid infrastructure development, and replacing animal habitats with human settlements (Smith and O'Keefe, 1980). In this process of production of nature, people have changed their relationship to nature from an organic and harmonious one to a controlling one, which is never fully achieved because nature remains far above human capabilities. People also changed their relationships with themselves and each other as individuals through the creation of divisions and controlling measures. Contemporary science and technology, which is essentially part of mainstream development, is expected to solve the problems such as flooding, drought, earthquakes, and even the risks of climate change. Instead, science and technology contribute to increased risks. For example, land-use expansion increases flooding. Earthquake risks are increased by fracking and oil explorations. The global pandemic of coronavirus disease, Ebola hemorrhagic fever, and human immunodeficiency virus/acquired immunodeficiency syndrome are some of the examples that not only the physical world but even the biological world is also changing. Science and technology themselves create new risks and challenges. It is important to note that science is also a belief about how the world works.

At the same time, science and technology bring inadequate resolutions to these risks and challenges (Attfield, 2014). One reason for this is that mainstream positivist science is overloaded with a contradictory understanding of nature. On the one hand, humans are part of nature, and on the other hand, nature is seen to be external to humans. Within the discourse of post-normal science, attempting to respond to difficult challenges underpins the inconsistent understanding of nature through science (O'Brien and O'Keefe, 2014). As explained by scholars such as Carrozza (2015) and Funtowicz and Ravetz (1993), the discourse of post-normal science is a drive against the tendency toward giving sciences and scientists a crucial role in policymaking while disregarding laypeople. In the sense that the current marketplace produces risks and challenges that science cannot control. Essentially, a holistic understanding is needed of the human relationship with nature.

This means that the overall linear scientific analysis, which flows from European and North American knowledge, is insufficient with contemporary risks and challenges (Zelinsky, 1975). In this analysis, a few factors singled out from a complex and rich social reality become very abstract, so social aspects come after the science, not before. The separation of the researcher and the researched through independence, impartiality, and neutrality cannot produce a meaningful understanding of the experience of the research and phenomena that have been researched. There is very little justification for maintaining this problem of scientific objectivity, particularly when the objective is to bring about a change to the context. For example, the oppressed do not have access to the sources of the knowledge from which they draw strength and nourishment, which were also destroyed during the colonial period. To break out of this crisis of the modern scientific knowledge system, which penetrated value norms of the Global South, a major de-colonization of the minds of scholars and practitioners alike is required.

It is also important to note here that while examining mainstream science and technology failures, this chapter does not reject the positive advancements and contributions to modern civilization. At the same time, the attempt is not to romanticize the ancient traditions and move backward. This chapter emphasizes that we can learn from the ancient wisdom and adapt them to effectively deal with contemporary challenges of climate change and environmental threats. This is exactly what Mohandas Karamchand Gandhi (1869–1948) said:

> I do not want my house to be walled in on all sides and my windows to be stuffed. I want the culture of all lands to be blown about my house as freely as possible. But I refuse to be blown off my feet by any.
>
> (Quoted in Singh and Mani, 2012, p.68)

Learning from the Past to Live Forward

It is important to examine the past in contemporary societies to go forward. Building on historical understandings, we can learn what is useful for us today and the future and learn from our ancestors' mistakes. Our past is not a series of ad-hoc and secular happenings without any pattern or shape. It is essentially a meaningful process and significant development. Of course, human history is full of wars and battles, economic disorders, and political upheavals, so not harmonious. However, below in the depths of these to be found the truly impressive drama – the tension between the limited effort of human beings and the sovereign purpose of nature, which is to create, maintain, and destroy all life and non-life forms on this planet (Radhakrishnan, 1963). The human cannot rest in an unresolved dispute between themselves and nature. They must seek harmony and strive for adaptation. It is important to understand that a series of integrations marks human progress by developing more inclusive harmonies. When any particular integration is found inadequate to the new condition, the human breaks it down and advances to a larger whole. While human civilization is always on the move, certain periods stand out clearly as periods of intense change. The transition from destructive wars to the peaceful and prosperous Empire of Ashoka in the Indian Subcontinent in 260 BCE, and the establishment of trade routes from Eastern China to Southern Europe and North Africa in the 2nd century BCE, were such periods. However, none of these is comparable to the current tensions and anxieties that are worldwide in character and extend to every aspect of human life.

Understanding the past at the macro and micro levels must start with a clear perspective and goal for learning for positive change. This is fundamental for correct action. Before action is taken, the underlying values need to be critically understood. It is then that praxis, the action-reflection-action process, becomes an instrument for initiating and reinforcing the positive macro or micro changes that are desired.

The search for the underlying paradigm begins from the perspective that participatory democracy and development are two sides of the same holistic vision that has inspired human endeavor in different Asian socio-cultural settings over recorded history. The values that have emanated from this deeper interpretation of the vision, even though somewhat diffused in practice, have been implicitly incorporated, in one way or another, into these two fundamental cornerstones. In this connection, several fundamental values that existed in traditional Asian societies must be identified and re-examined. Some critical values relate to looking at life in its totality and all its richness; participation of the people in decisions that affected their lives; sharing and caring for the community, cooperative activities beyond individual self-interest; trust, innocence, simplicity, thrift; a work ethic with a fine-tuned balance between work and leisure; harmony with nature and rational use of both natural and financial resources; communal ownership of the commons; and complementarity between women and men, as well as, gender equity.

To elaborate further, historically, in Asian cultures, there were built-in methods of consensus-making. One example is the republican tradition of the Lichchavis in ancient South Asia, renowned for their republican government (Encyclopaedia Britannica, 2010). Another is the Krishna-Arjuna dialogue in the *Bhagavad Gita* in the Hindu tradition (Radhakrishnan, 1963), which also illustrates correct decision-making at a higher level of consciousness. The great Tamil poet, in describing Ram Rajya (a good society) in *Ramayanam* (Nandakumar, 1971), says that everyone had enough, and no one had more than enough. Confucius argued that people need a good environment with good representations and a good economy before thinking and acting appropriately (Rainey, 2010). The ancient book, the *Tao Te Ching*, explains Tao as a path people travel, the natural road, and the way of ultimate reality (Blakney, 1956).

The literature is full of instances of people's participation and self-governing practice at the base of the system alongside wise and compassionate leadership at the top. Democratic consensus-making was very much a part of the Asian tribal mode of governance (Wignaraja, 1991). Peer pressure was a major deterrent to corrupt practices, which established the checks and balances within a society. This also extended to everyday activities such as taking loans and credits, where peer pressure ensures repayment. There was an open information system at the community level, which further reinforced the principle of equity, checks, balances, and accountability. Democracy in labor relations, exchange labor, common ownership of land, and other productive assets further reinforced cooperative values, mutual aid, and equity. In all these, the most important aspect is that the realization of all-inclusivity paved the way for people to live in harmony with natural processes and the environment.

Based on this understanding that there is much to learn from ancient wisdom, this chapter presents the following value concepts with pointers to practical elements for the contemporary world.

- **Contentment versus Greed**

The second line of the 204th verse of the Buddhist text *Dhammapada* (Radhakrishnan, 1950, p.126) says,

Santushti paramaṃ dhanam

The simple translation of this line is that contentment is the greatest wealth. This is beyond merely economic wealth. Why is this so important to the contemporary world? In the dream of prosperity, growth, and progress, humans are invading all the corners of this planet – or at least attempting. Progress and growth are ambitions of the 18th century; development, an idea, which became a project of the 20th century that continues into the 21st century. Of course, humans have achieved so much technological and scientific advancement, which is expected to give them comfort. Compared to our grandparents, many people can be expected to live into our 90s. We can communicate with anyone across the world through our mobile technologies, and income is increasing dissatisfaction levels yet. Very easily, we could establish the argument that the current generation of humans is the most comfortable compared to previous generations. However, the World Health Organization (2017) estimated that 322 million people worldwide suffer from depression. South and Southeast Asia are not immune to depression. With all the comforts, technologies, and resources, we continue to suffer and be sad. This does not simply equate depression with sadness, hopelessness, and lack of motivation; however, the current generation seems unable to deal with the challenges and difficulties that life throws at them. Life is indeed uncertain and dangerous. Rising taxes and house bills, job losses, health problems and chronic illnesses, pandemics, technological dangers, natural hazards, evictions, and bankruptcy make life uncertain and dangerous. With rapid urbanization and increasing material development, South and Southeast Asia populations are facing these challenges.

A young man in a developed Southeast Asian country running a small business found that the business was collapsing due to the COVID-19 pandemic. His wife, who found the deteriorating living conditions to be hard, left him with their three children. He is now a bordering alcoholic and experiencing depression. A mother living in a conflict-affected South Asian country found out her son was killed in a bomb explosion at the marketplace. Her whole world has collapsed now, and there seems to be no hope. Two teenage girls from an island nation in Asia that lived reasonably comfortable and happy lives suddenly became orphans because a tsunami washed away their parents. Now they are physically unable to talk. These are a few examples of the uncertainty and danger of life. No matter where one lives and what living conditions they maintains, there is no guarantee of life. Explaining this uncertainty, the *Dhammapada* (Radhakrishnan, 1950) says all the elements of being are impermanent and unsatisfactory, which leads to sorrow. Further, it says that everything in existence is dependent on other things and causes at any time, which becomes impermanent.

According to *Bhagavad Gita* (Radhakrishnan, 1963), humans are unique from the rest of the animal kingdom because of the quality of making decisions on what is right. It further says that sleep, hunger, sex, and fear are common to all animals. What separates humans from animals is the knowledge of right and wrong. However, humans feel frustrated with the attractions of this world and yet attached to them. They become devoted to their personalities and are distressed by the conflicting dynamics of this world. In other words, Confucius explains this from a practical perspective. Confucius suggested that to live in this world with less sorrow, we must establish humanity, and humans should overcome greediness and self-interest (Rainey, 2010). Further, humanity has been defined as the opposite of gathering material possessions, selfishness, exploits, and profits. What is important to understand is that all these points toward mechanisms that enable humans to deal with uncertainty and danger of life.

However, contemporary societies continue to consume more and more materials due to material development, promoting selfishness, exploitation, and profits. This makes human beings incapable of dealing with life itself and harms the environment and planet. The general message that runs through societies is that people need to work hard to make a living and save enough to live happily in retirement. The challenge is sinking in personal loans to keep living and being unable to save much. However, our greed and consuming larger amounts of materials never stop.

In *Dhammapada* (Radhakrishnan, 1950), the Buddha explains that self-indulgence or greediness is the worst among all illnesses, and attachment creates the greatest sorrow (verse 203, p.126). It is explained in the Confucius philosophy that human nature is based on emotions and desires, which are stimulated by the five senses (eyes, nose, ears, tongue, and skin). These are natural to all humans; however, we will act badly when we act on these emotions (Rainey, 2010). To avoid all these, the Buddha has proposed the middle path. Similarly, the *Bhagavad Gita* (Radhakrishnan, 1963) explains that the path to freedom is not for the person indulging in too much food or eating too little, sleeping too much, or sleeping too less. In return, moderation in food and recreation, regulated sleep and waking, and control in action, which establishes total self-discipline, can overcome all sorrows. The great Indian saint Kabir Das explained that human happiness and joy are within themselves, and there is no need to search in the external world (Jhawar, 2004).

The concept of contentment is indeed important to understand within the individual human. Without individual contentment, there won't be social contentment, which can be understood as the core of sustainability. Whether Hindu or Buddhist or Confucius or Tao philosophies, they all point toward contentment as an important concept and practice for true freedom. In Buddhist thought, this freedom is the relaxed body, calm and focused mind, and living in the present with a conscious awareness of the past and future (Radhakrishnan, 1950). Ancient tools such as meditation, Yoga, Tai Chi, and many other contemplative and reflective practices are proven instruments that facilitate this freedom. The moment humans become content with their living, it may be possible to minimize the environmental risks and reverse the destruction of nature.

- **Humility versus Arrogance**

In Buddhist teachings, humility and reflections go together (Radhakrishnan, 1950). A humble and reflective person can easily recognize their own greed, hatred, and ignorance. *Dhammapada*, verse 201 claims that winning produces hatred. The defeated is suffering from sorrow and therefore gives up both victory and defeat to claim a happy life (Radhakrishnan, 1950, p.126). What this means is that the identity of this or that, where the individual is absorbed in status, qualifications, and presentation, can breed hatred and sorrow. This is the opposite of humility.

When we look around contemporary societies, winning or victory has been prominent. Boasting oneself as important, knowledgeable, intelligent, and accomplished has been recognized as an admired human quality. Modern societies are filled with celebrities – not just in the entertainment industry but also in academia, politics, sciences, and many other fields. This is, of course, due to the revolutionized communication by mobile technology today. People are boasting about their achievements everywhere and encouraging everyone to be like them. Attached to this comes the products for sale – promoting everlasting consumerism. In this, two processes are emerging: first, the imagination and creativity of people are being destroyed, and second, people believe without inquiry.

In many ways, scholars, such as Ivan Illich (1971), Bush and Saltarelli (2000), and Smith (2000), have commented on the repetitive nature and problems in knowledge transferring to students in the form of information without critical examination in mainstream education. There is increasing information flow, but Smith (2000) explains that contemporary universities have become corporate organizations everywhere. While traditional education aims to train the future labor force, contemporary education has become a profitable business (Smith, 2000). In this, any intellectual debate, scientific inquiry, or general responsibility toward society or nature is controlled by financial benefits. Illich (1973) predicted that instead of facilitating analytical thinking within students, this type of education produces frustration, hatred, self-indulgence, and greed. In return, this process takes out the joy of life and living.

The Buddha, in his teaching, challenged this illogical thinking or blind faith about 2,500 years ago.

> Do not go upon what has been acquired by repeated hearing; nor upon tradition; nor rumor; nor upon what is in a scripture; nor upon surmise; nor upon an axiom; nor specious reasoning; nor upon a bias towards a notion that has been pondered over; nor upon another's seeming ability; nor upon the consideration, "The monk is our teacher." Kalamas, when you yourselves know: "These things are good; these things are not blame-able; these things are praised by the wise; undertaken and observed, these things lead to harm and ill," abandon them.
>
> (Thera, 2008, p.10)

This is, in essence, the critical thinking that is needed to understand what is presented by society and the world at large. It is important to question things we do not understand, and we should not believe without understanding. In day-to-day life, we buy materials because they are advertised as important or improve the quality of life. However, do we examine and critically evaluate this information that is transferred into our heads? Suppose humans are to protect the environment or live harmoniously with nature. In that case, we need this critical thinking and questioning as the Buddha suggested. The risk of believing what is presented to us through various information channels, including educational institutions, is that we easily end up in arrogance, which leads to hatred, self-indulgence, and greed. Both these prevent the humility needed in engaging with our families, societies, and the natural world. *Isha Upanishad*, a principal Upanishad in the Vedantic tradition of Hindu philos-ophy, points toward the knowledge acquired without critical examination.

Andham tamah pravishanti ye avidyam upasate
tato bhuya iv ate tamo ya u vidyayam rataah.

They who worship ignorance enter into darkness.
They who worship knowledge enter into greater darkness.

(Sri, 2016, p.21)

The first line of this verse is logical and understandable. If one is worshiping ignorance, they will enter into darkness. However, the second line is confusing. This second line points out that while knowledge is useful in day-to-day sur-vival or for certain aspects of life when it becomes a belief as ultimate truth, it leads to darkness. This can be understood through the analogy of a boat. One can use a boat to cross a river. Once they have crossed the river, the journey can be heavy and difficult if they continue to carry that boat on their head. In a practical sense, one must leave the boat when one crosses the river. According to this Upanishad verse, contemporary society is filled with huge amounts of information in the form of knowledge, and people are carrying this by believ-ing them to be important, which does not facilitate peace and prosperity in societies and harmonious relationships between humans and nature.

The risk of worshiping knowledge is that the individual feels they know and cannot accept other perspectives. This is what happened in the colonial project in relation to the superiority of race, knowledge, and systems. The colonizers thought that their knowledge was superior to the wisdom in Asia, Africa, the Americas, Australia, and the Middle East. This arrogance is frequently expressed as I/we know more. Yours/theirs is a poorer knowledge. This is a barrier to the ability of the mind to evolve further, and people defend their limited information. This leads to an unreasonable sense of restlessness, followed by aggression, violence, destruction, and control of other people and nature.

In contrast to these contemporary education and knowledge systems. *Tao Te Ching*, the 6th century BCE classic text by Lao Tzu, suggests that everything that seems to be divided is interconnected and interdependent in this world.

> *Being and non-being produce each other;*
> *Difficult and easy complement each other;*
> *Long and short define each other;*
> *High and low oppose each other;*
> *Voice and sound harmonize each other;*
> *Front and back follow each other*
>
> (*Tao Te Ching*, chapter 2, Cited in Yang, 2019, p.120)

This critical and analytical approach provides an understanding that nothing is independent, and everything is part of a whole. That means humans are part of the natural world of this planet and, therefore, think and behave accordingly. According to *Dhammapada* (Radhakrishnan, 1950), *Bhagavad Gita* (Radhakrishnan, 1963), and *Tao Te Ching* (Yang, 2019), the highest level of wisdom is knowing oneself, and the most powerful success is conquering oneself. To achieve this, humility is a prerequisite.

- **Inclusivity versus Exclusivity**

The contemporary mainstream knowledge has created various compartmentalization. For example, this compartmentalized learning and treatments govern institutional education and health systems (Illich, 1978). Education is delivered in a classroom by a trained professional, and health has become specified and is delivered by a specialist (Illich, 1971). While there is a usefulness in the specialism, this also creates a disconnection. This notion of division or compartmentalization is promoted within societies as an achievement or self-made (Ngomane, 2019). This idea of achievement is so dominant within societies the individual becomes so self-centered and completely overlooks the connection to other human beings, animals, trees, and the entire planet. Grades, qualifications, and status as power and quality of knowledge become linked to the position.

In contrast, the *Karaniya Metta Sutta* – Hymn of Universal Love (Buddharakkhita, 1989), aspire that all beings be joyful and secure, happy within themselves, without any exception, movable or immovable, long or huge, medium, or small, subtle, or gross, visible, or invisible, residing far or near, born, or coming to birth. This aspiration or the wish comes from a realization that the individual human being is connected to all beings, including trees and plants on this planet. Any harm to them will affect the individual, so they not only aspire but work toward the protection of all living beings. This understanding provides a different view, an undivided perception that makes the individual human humble as well as content with what they have. The opportunity for hatred, greed, and destruction has little or no space within this awareness. Therefore, an inclusive worldview is facilitated.

Industrialization, technological advances, conquering, and colonial ideologies are sweeping through the contemporary world. Taming rivers, breaking plains, clearing forests, and claiming the earth for the benefit of humans have become a common theme in many development projects over the years. These self-destructive and earth-ravaging approaches to life and living will not suffice for humans to survive on this planet. In contrast, the purpose of the Buddhist approach is to eradicate suffering and the sources of that suffering. Freedom from suffering involves examining the source of suffering very closely, understanding its true nature, and behaving in such a way appropriate to that true nature (Buddhadasa, 1989). The Buddhist perspective is that humans are part of nature. If the environment is destroyed, humans cannot survive and enjoy life.

In line with the Buddhist perspective, the *Bhagavad Gita*, chapter 3, verse 27, claims that the grace of nature governs every human activity, and only arrogant people deny that (Radhakrishnan, 1963, p.143). Through this realization of humility, one leads an inclusive life. The understanding is that nature is the main ruler of every human activity. This means that destroying nature will affect its regular processes and activities, which in return will affect the human living. There is a warning to arrogant people who do not realize the power of nature. In chapter 7, verses 4–6, the *Bhagavad Gita* focuses on the relationship between humans and the environment (Radhakrishnan, 1963, pp. 2013–215). The responsibility of the human with mind, intelligence, and senses is to maintain the other elements of nature such as land, water, air, and space healthy. The harmonious relationship between humans and nature helps the planet to flourish.

The idea of inclusivity in both Buddhist and Hindu perspectives is not an intellectual exercise. It is an idea that must be realized by every human being and practiced. Even from the point of selfishness, human beings should protect and conserve the environment for their own survival. This is the ultimate inclusivity. What humans exhale are inhaled by plants, and what plants and trees exhale, humans inhale. In this realization of inclusivity, there is no question of division between humans and nature.

Taoism shares the same viewpoint of this idea of inclusivity. In chapter 42, the *Tao Te Ching* claims, "the Tao gives birth to One. One gives birth to Two. Two gives birth to Three. Three gives birth to everything." (Zhihua, 2012, p.279). This means humans have the same origins as everything. Further,

Zhihua (2012) claims that according to Taoism, the planet and human beings not only have the same origin and nature but also share the same structure and law. Based on this perspective, Taoist philosophy has a deeper understanding of the human's close dependence on the environment. Maintaining a harmonious relationship between humans and nature is a requirement for human life on this planet. From a more practical perspective, Mencius, a Chinese Confucian philosopher (372–289 BCE), has called for environmental consciousness, the conservation of forests and waters, and for diversified livelihoods as ways to enhance the quality of life as well as maintaining a harmonious relationship with the environment (Rainey, 2010).

As examined above, the three ideas of contentment, humility, and inclusivity can be incorporated into individual lives as well as social and political systems in South and Southeast Asia. This can provide an opportunity to facilitate human activities that are harmonious with nature as well as each other. This means humans become a collaborator of the environment rather than a controller.

Becoming Companions to the Earth and Friendly with Ourselves

The three value concepts presented in this chapter – contentment, humility, and inclusivity - are practical. Although these concepts are borrowed from ancient Hindu, Buddhist, Tao, and Confucius philosophies, these are not a doctrine that binds all devotees to a single system. These values inspire individual perceptions, even challenges, and questions on the part of each person. However, interconnected paths of individuals point toward a common direction by practicing these concepts. This wisdom highlights the value of a harmonious relationship with nature rather than conquering it. These values encourage a conserving lifestyle rather than a wasteful lifestyle.

Incorporating these values into individual, social, and political levels can facilitate human activities that can improve the wellbeing of people and reduce the environmental risks, including the challenges of climate change. Further, suppose individuals realize contentment, humility, and inclusivity. In that case, there is no need for separate educational activities to teach environmental protection, peaceful coexistence, and climate change.

In many catastrophes, especially those caused by natural hazards, the first responders are people who are alive and able to function. What is valuable to understand is that people practice the concepts of contentment, humility, and inclusivity when going through a catastrophe. There are many instances, people helping each other in catastrophes (O'Keefe, O'Brien and Jayawickrama, 2015). On many occasions, communities affected by crises collaborate with each other as the official assistance would arrive after many days of the emergency (Ferdinand et al., 2012). It is the ability of people to be content with what they have, humble to help everyone, and inclusivity to leave no one behind. These are important factors in their capabilities to respond to catastrophes.

The human relationship to the natural world is derived partly from the human addiction to consuming larger quantities of resources on this planet. This has resulted in successive catastrophes such as floods, droughts, landslides, and diseases (both infectious and chronic), which provides evidence of a disastrous conflict between humans and nature. As argued by Gore (1992) and shown by the data of ADRC (2019), the previous challenges of this conflict between humans and nature used to be local or regional; however, the current crises are global – extinction of species, decaying ozone layer, thinning rainforests and climate imbalances are all point toward an increased global crash between the human and nature. Human beings are losing their position as important and valuable animals on this planet. We are still claiming to be the most intelligent animal. However, the way we are using our intellect is destroying our position on this planet – the chances of becoming extinct as a species are not an illusion anymore for humans.

Based on the ancient wisdom from South and Southeast Asia, the global governance and development approaches have to facilitate contentment, humility, and inclusivity that cultivate inner wellbeing within individuals and societies. However, the current approaches are promoting greed, hatred, and violence. We have seen over the years that whenever there is an overconsumption of materials, there are increasing desires, which in return generate conflicts with the environment. However, from the ancient philosophical standpoint, when there are fewer desires within the human with contentment, humility, and inclusivity, there can be further development and flourishing of the human civilization.

Sivaraksa (1992) argues that the new generation of human beings must rethink and develop a value system based on love, society, contemplation, and self-realization. He further argues that we need to establish our economies within social and environmental contexts. The current quantitative measurements of development outcomes approaches are too simplistic to connect with nature and natural processes. There is less attention to the question of who is a human being and what a human being should do? (Sivaraksa, 1992). The explanations of contentment, humility, and inclusivity not only shed light on climate change and human activities but also examine the responses of people in ancient societies, which can be adapted to improve the contemporary human being. Understanding ancient people's responses to modern challenges is not easy. As mentioned at the beginning, the objective of this chapter is not to go back to the past, observed with romanticism, but to examine what is useful, and then to learn the lessons that are suitable to solve modern problems. The hard lessons from ancient wisdom can be learned through social, cultural, and political praxis in moving toward a more contented, humble, and inclusive society that enjoys the quality of life through a harmonious relationship with the environment.

A new examination of ancient wisdom, both from written and unwritten traditions, is essential for a deeper understanding of nature and the role of human beings on this planet. Human societies must examine individual values and behaviors through education, personal and professional practices, governance, and development systems. Any meaningful approach to change that establishes a harmonious relationship with nature must not benefit the rich at the expense of the poor or the powerful at the expense of the powerless. It is both a political and transformative process.

The existing framework that has governed the global development and human activities over the past half-century assumed that there is one model for development in the 20th and 21st centuries, which is the western model, and there are "developed" countries and "developing" countries. The idea is that the "developing" countries should follow the same processes as the "developed" countries without a critical inquiry. This approach was operationalized from economic terms, and greater dependence was placed on economic factors. This framework assumes that rapid economic growth could occur if there was central planning and control of the economy (by the state or the private sector) as a "top-down" process, emphasizing industrialization, modernization, and urbanization. Material accumulation, either in public or private forms, was expected to solve human problems but never paid attention to the increased threats humans face from the natural world.

What is becoming clear in this discussion within this chapter is that ancient people and their wisdom had an intimate knowledge of their environmental resources and the natural world. They knew when to and how to engage with the environment, what resources to access and what not to touch, what time of the year, and the sequence in which certain environmental resources to consume to get the maximum benefits. These were done through time-tested methods for cost-effectiveness and generating the least risks. In Asia, farmers knew methods to increase the fertility of topsoil by cultivating plants in a particular pattern. Since the forests were their lifeblood, people knew how to use forests for food and medicinal herbs, hunt animals within a code of ethics, and permit the environment to regenerate itself. They knew that extensive lumbering and clearing forests for modern industrial consumption would affect climate and rainfall. Only a contented, humble, and inclusive society can establish such development. Contemporary human beings must re-realize that establishing harmonious relationships within, with each other, and with the natural world is crucial for our survival and will improve our quality of life.

Toward a Conclusion

There is a benefit and an opportunity that can be created by adapting the ancient wisdom into contemporary societies in South and Southeast Asia as well as globally. While not romanticizing the ancient philosophies from South and Southeast Asia, the point of this chapter essentially is that this wisdom was a process in which many people share. It is neither elitist nor eclectic. There was an element of generational and experiential learning, which became systematized, and scientifically validated, which can be shared commonly for development. The conclusion is inescapable that to effectively deal with the climate change and adverse consequences of our activities would be to examine ancient wisdom and build on them, drawing upon the wide range of scientific and technological choices available within the context of each society and country.

Although not a proverb from Asia, from Latin America, the title of this chapter explains it all. Those who make an enemy of the earth make an enemy of themselves. If humans are to survive on this planet, let's become an ally of nature and become friendly with ourselves. In conclusion, here is an example from the Amazon rainforest as told by Henning (2002, pp. 17–18):

> In the Amazon rain forest, a Cofan Indian elder conversed with a "gringo" from the "world beyond the forest." The elder's knowledge of the rain forest surpassed that of biologists holding PhDs who came to study there. His was a practical knowledge. He is able to find and identify more than 140 plants his people used for medicinal purposes. He knows where the peach tree grows, and its fruit attracts the brilliant blue gold macaw. In addition to his knowledge, the elder has access to a great research library. Each volume is a friend. That woman over there knows all there is to know about marital discord and how to ensure a household's harmony. That old man is walking beneath the trees talks to the God of Creation. The brothers fishing along the strand are the heads of a family that has built the best boats on the river for many generations. The elder's material possessions are scant and include a machete from upriver and two cloaks from the city. His wife owns several spoons and a metal pot. Aware of the village school, run by a Jesuit priest, the visitor asks, "Do any of your children go on to high school in the city? Do any seek a university education?"
>
> The old man shakes his head. There is too much to be learned at home. "It takes many years to learn to be a Cofan," he says. His gaze drifts across the river then return to the gringo. "My heart aches for you," he says. "For me?" replies the gringo. "Why?"
>
> "Because you are so poor. We in the forest have all we want. You gringos want for so much you do not have."

References

ADRC. (2019). *Natural Disaster Data Book 2019: An Analytical Overview*. Kobe: Asian Disaster Reduction Centre.

Attfield, R. (2014). *Environmental Ethics: An Overview for the Twenty-First Century* (2nd ed.). Cambridge: Polity Press.

Bingham, P. M. (2000). Human evolution and human history: A complete theory. *Evolutionary Anthropology: Issues, News, and Reviews*, 9(6), 248–257. https://doi.org/10.1002/1520-6505(2000)9:6

Blakney, R. B. (1956). *The Way of Life, Lao tzu: A New Translation of the Tao Te Ching*. New York: A Mentor Book.

Britannica, The editors of encyclopaedia. *Licchavi*. Encyclopedia Britannica, 11 Aug. 2010, https://www.britannica.com/topic/Licchavi. Accessed 18 May 2021.

Buddhadasa, B. (1989). *Handbook for Mankind: Principles for Buddhism*. Bangkok: Supanit Press.

Buddharakkhita, A. (1989). *Mettā: The Philosophy and Practice of Universal Love*, Kandy: Buddhist Publication Society.

Bush, K. D., and Saltarelli, D. (2000). *The Two Faces of Education in Ethnic Conflict: Towards a Peacebuilding Education for Children, Innocenti Insight*. United Nations Publications.

Carrozza, C. (2015). Democratizing expertise and environmental governance: Different approaches to the politics of science and their relevance for policy analysis. *Journal of Environmental Policy & Planning*, 17(1), 108–126. https://doi.org/10.1080/1523908X.2014.914894

Ferdinand, I., O'Brien, G., O'Keefe, P., & Jayawickrama, J. (2012). The double bind of poverty and community disaster risk reduction: A case study from the Caribbean. *International journal of disaster risk reduction*, 2, 84–94. https://doi.org/10.1016/j.ijdrr.2012.09.003

Funtowicz, S. O., and Ravetz, J. R. (1993). Science for the post-normal age. *Futures*, 25(7), 739–755. https://doi.org/10.1016/0016-3287(93)90022-L

Gore, A. (1992). *Earth in the Balance: Ecology and the Human Spirit* (1st ed.). Boston: Houghton Mifflin Company.

Henning, D. H. (2002). *A Manual for Buddhism and Deep Ecology*. (Special Edition). Bangkok: The World Buddhist University.

Horton, P., & Horton, B. P. (2019). Re-defining sustainability: Living in harmony with life on Earth. *One Earth*, 1(1), 86–94.

Illich, I. (1971). *Deschooling Society*, New York: Harper & Row.

Illich, I., (1973). *Tools of Conviviality*, New York: Harper & Row.

Illich, I., (1978). *The right to Useful Unemployment and Its Professional Enemies*. London: M. Boyars; Don Mills.

Jhawar, S. R. (2004). *Building a Noble World*. Chicago: Noble World Foundation.

Nandakumar, P. (1971). *Kamba Ramayanam. A Condensed Version in English Verse and Prose*. New Delhi: Sahitya Akademi.

Ngomane, N. M. (2019). *Everyday Ubuntu: Living Better Together, the African Way*. New York: Random House.

O'Brien, G. and O'Keefe, P. (2014). *Managing Adaptation to Climate Risk: Beyond Fragmented Responses*. London: Routledge.

O'Keefe, P., O'Brien, G., & Jayawickrama, J. (2015). Disastrous disasters: A polemic on capitalism, climate change, and humanitarianism. (pp. 33–44). In. Collins, A. Jones, S. Manyena, B., & Jayawickrama, J. (eds.). *Hazards, Risks and Disasters in Society*. Amsterdam: Elsevier Inc.

Radhakrishnan, S. (1950). *The Dhammapada: With Introductory Essays, Pali Text, English Translation and Notes*. London: Oxford University Press.

Radhakrishnan, S., (1963). *Bhagavad Gita*, London: George Allen and Unwin Ltd.

Rainey, L. D. (2010). *Confucius & Confucianism: The Essentials*. Malden: Wiley-Blackwell.

Singh, H. K. J., & Mani, M. (2012). KS Maniam, Jhumpa Lahiri, Shirley Lim: A reflection of culture and identity. *International Journal of Applied Linguistics and English Literature*, 1(3), 68–75. http://dx.doi.org/10.7575/ijalel.v.1n.3p.68

Sivaraksa, S. (1992). *Seeds of Peace: A Buddhist Vision for Renewing Society*. Berkley: Parallax Press.

Smith, N. (2000). Afterword: Who rules this sausage factory? *Antipode*, 32(3), 330–339. https://doi.org/10.1111/1467-8330.00138

Smith, N. and O'Keefe, P. (1980). Geography, Marx and the concept of nature. *Antipode* 12 (2), 30–39. https://doi.org/10.1111/j.1467-8330.1980.tb00647.x

Sri, M. (2016). *Jewel in the Latus: Deeper Aspects of Hinduism.* Karnataka: Magenta Press and Publications.

Thera, S. (2008). *Kālāma Sutta. The Buddha's Charter of Free Inquiry.* The Wheel Publication, (8).

UN ESCAP. (2019). Population and development indicators for Asia and the Pacific, 2019. Unescap.org. https://www.unescap.org/sites/default/files/Population%20Data%20Sheet%202019.pdf. Accessed 25 May 2021.

Wignaraja, P. (1991). *Participatory Development: Learning from South Asia.* United Nations University Press.

World Health Organization. (2017). *Depression and Other Common Mental Disorders: Global Health Estimates* (No. WHO/MSD/MER/2017.2). World Health Organization.

Yang, F. (2019). Taoist wisdom on individualized teaching and learning—Reinterpretation through the perspective of Tao Te Ching. *Educational Philosophy and Theory*, 51(1), 117–127. https://doi.org/10.1080/00131857.2018.1464438

Zelinsky, W. (1975). The Demigod's dilemma. *Annals of the Association of American Geographers*, 65(2), 123–142. https://doi.org/10.1111/j.1467-8306.1975.tb01026.x

Zhihua, Y. (2012). Taoist philosophy on environmental protection (pp. 279–292). In Zhongjian Mou (ed.). *Taoism.* The Netherlands: Brill.

3 Post-Cyclone Livelihood Strategies and Security Status of Coastal Households in Bangladesh

An Empirical Study

Shitangsu Kumar Paul and Jayant K. Routray

Introduction

The Bangladesh coast is susceptible to cyclones and storm surges. In terms of the likelihood of occurrence, at least one major tropical cyclone strikes the country each year (Mooley, 1980; Haque, 1997), together with powerful storm surges that assert hundreds of thousands of lives and make the Bangladesh coast more unsafe than any other regions of the world (Murty & Neralla, 1992). The basic physical and meteorological conditions necessary to generate tropical cyclone exists in the Bay of Bengal and the Bangladesh coast. These pre-requisite conditions are the re-curvature phenomenon of the tropical cyclone in the Bay of Bengal, shallow continental shelf, funnel shape at the head of the Bay of Bengal, sea-level *Orography* on the Bangladesh coast, and high density of population (Murty & El-Sabh, 1992). Therefore, this unique natural setting of the country modifies and regulates the climatic conditions and makes the country more vulnerable to natural disasters such as cyclones accompanied by storm surges and coastal flooding (Ali, 1996, 1999; Paul & Routray, 2010b).

On the night of 15 November 2007, Cyclone Sidr, a Category 4 storm, made *landfall* across the southwestern coast and traveled through the heart of the country from the southwest to the northeast (Figure 3.1). With winds recorded as strong as 248 kilometers per hour, it triggered storm surges as high as 7.5 meters in some coastal districts. Due to cyclone Sidr, over 55,000 people sustained physical injuries (GoB 2008a), an estimated 1.87 million livestock and poultry perished, and crops on 2.51 million acres suffered complete or partial damage. In addition, the cyclone washed away private food stockpiles and storage and destroyed 1.4 million fruit trees. The Joint Damage Loss and Needs Assessment Mission, led by the World Bank, estimated the total cost of the damage caused by Cyclone Sidr is $1.7 billion, a figure representing about 3% of the annual gross national product of Bangladesh (GoB, 2008a). Cyclone Sidr severely affected over 7.46 million people's livelihoods (GoB, 2008a). According to a rapid emergency assessment completed by UN officials, 2.6 million people were found to require immediate food

DOI: 10.4324/9781003216476-4

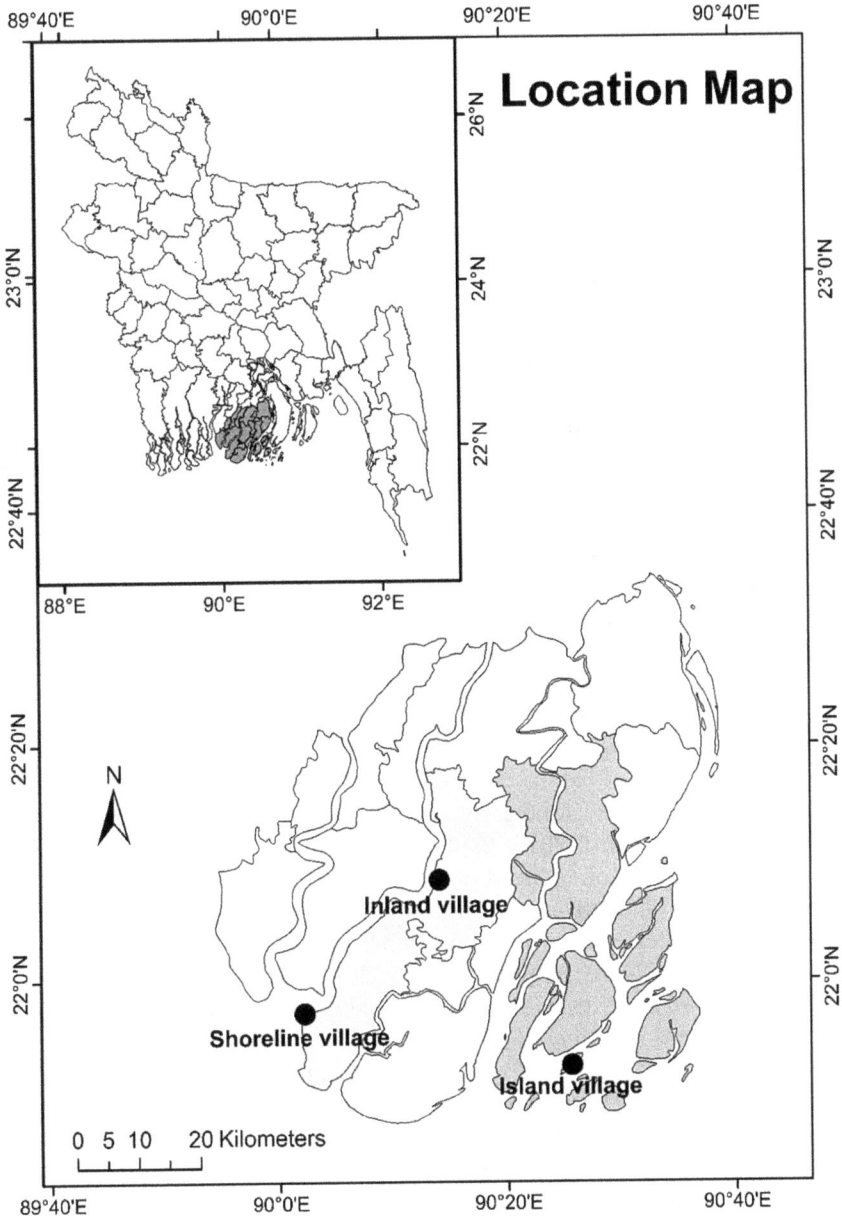

Figure 3.1 Location of study villages.

Source: Authors.

assistance across the affected areas (UN, 2007). Besides, increased salinity caused by storm surges during the cyclone and sand deposition has further hampered agricultural productivity. The officials highlighted the large-scale loss of standing crops, family food stocks, and livestock. These losses were compounded by the virtual collapse of already meager-wage-earning opportunities together with the price hike of essential food commodities in the aftermath of cyclone Sidr affected areas and made the livelihoods of cyclone victims insecure (GoB, 2008b).

To develop the conceptual framework of the study 'asset vulnerability framework' (Moser *et al.*, 1994), household's 'access to resources' (Blaikie *et al.*, 1994), Sustainable Livelihood Framework (DFID, 1999), and Household Livelihood Security Model (Sanderson, 2000) are reviewed and conceptualized, and more emphasis is given on sustainable livelihood framework and household livelihood security model to outline the role of different livelihood capitals which mediate livelihood strategies in a disaster situation and also in usual time to secure livelihood. The basic livelihood framework in Carney (1998) stresses the need to maintain 'outcome focus' thinking about how development activity impacts people's livelihoods. This framework has mainly been used for basic needs programs by CARE International, for poverty elimination by the UK Department for International Development (DFID), and for designing several fisheries, livestock, forestry, rural development, and poverty alleviation investment projects and programs, as well as for monitoring stakeholder participation by the Food and Agricultural Organization (FAO) of the United Nations. However, it has not been used in the disaster management field. Various livelihood variables related to disaster and livelihood security were not yet specifically defined, nor were their measurement methods developed (Shivakoti & Shrestha, 2005).

Apart from the livelihoods, in the present trends of disaster research, various aspects of Cyclone Sidr have received the attention of hazard researchers, such as cyclone early warning, non-response, and evacuation behavior (Paul & Dutt, 2010; Paul & Routray, 2012; Paul, 2014; Hossain *et al.*, 2014); causes of reduced deaths, injuries, and illness pattern (Paul, 2009; Paul, 2010; Paul *et al.*, 2011, Paul, 2016b); coping response and adjustment strategies to cyclone and other disasters (Mallick *et al.*, 2009; Paul & Routray, 2010a; Paul & Routray, 2010b, Paul *et al.*, 2013; Paul & Hossain, 2014); impacts of cyclone on coastal physical infrastructures and social conditions (Islam *et al.*, 2011; Paul, 2012); post-Sidr nutritional status of women and children (Paul, 2010; Paul *et al.*, 2012); and the household food security status of marginal farmers and so on (Uddin, 2012; Paul, 2013; Paul, 2016a; Paul, 2019). However, post-cyclone livelihood capital status and strategies among residents of the affected areas have not been the focus of any study. To address this research gap, this chapter will assess the livelihood capital status, identify livelihood groups, livelihood adversity and crisis, and livelihood strategies and security status of coastal households in the post-Sidr period. The findings of this study would provide useful guidelines for further interventions and targeting the insecure livelihood groups and locations with more effective post-cyclone livelihood rehabilitation-related programs.

Study Area and Methodology

Selection of Study Villages

Three villages from two of the four most severely Sidr-affected districts in Bangladesh were selected for this study (Figure 3.1). The selected villages are Angulkata of the Amtoli thana, Tatulbaria of the Taltoli thana, and Charkashem of the Rangabali thana. The first two villages are located in the Barguna district and the last in the Patuakhali district. Charkashem is an offshore island; Tatulbaria is located on the shoreline of the Bay of Bengal, and Angulkata is approximately 30 kilometers away from the sea, on the bank of the river *Paira*. All three villages experienced storm surges ranging from 3.0 to 7.0 meters. Angulkata and Tatulbaria are surrounded by polders.

On the contrary, the southern part of Charkashem is covered by mangrove trees planted under a coastal afforestation program. For the purpose of making locational distinctions, Angulkata is termed as the inland village, Tatulbaria as the shoreline village, and Charkashem as the island village. Comparative aspects of livelihoods and vulnerability to cyclones in three villages are presented in Table 3.1.

Data Collection and Analysis

Of the 788 households in three selected villages, a sample size of 331 was determined using an assumed 95% confidence level for the household

Table 3.1 Comparative aspects of livelihoods and vulnerability in three villages

Variables	Study villages		
	Inland	Shoreline	Island
Primary livelihood activities	Agriculture and allied	Fishing and allied	Fishing and allied
Secondary livelihood activities	Fishing and allied	Agriculture and allied	Agriculture and allied
Vulnerability of livelihoods to cyclones and storm surge	Moderate vulnerability to the cyclone with 2–3 meters of associated storm surge	Very high vulnerability to the cyclone with 3–4 meters of associated storm surge	Very high vulnerability to the cyclone with more than 4 meters of associated storm surge
Level of risk of livelihoods	Moderate	Very high	Very high

Source: Household Survey and Focus Group Discussion.

Note: Level of vulnerability and risk is assessed based on a five-point Likert scale.

questionnaire survey (Yamane, 1967). The number of samples selected from Angulkata and Tatulbaria was proportional to their total household size. Due to its small size, all 47 households in Charkashem were selected for this study. The individual household was the study unit, and the households were selected from Angulkata and Tatulbaria using a simple random sampling procedure. Cyclone and livelihood-related information were collected from the head of the selected households and local leaders through informal dialogue. One focus group discussion (FGD) was also conducted in each study village. Informal discussions and FGDs were carried out to gain additional insights regarding post-cyclone livelihood strategies at the household level. Both descriptive and inferential statistics are used to analyze data. In addition, a five-point Likert scale is used to assess the level of risk, vulnerability, and livelihood capital of respondents. Besides, application of the analysis of variation (ANOVA) is also used to identify statistically significant differences among the villages and livelihood groups in terms of livelihood capitals.

Measuring Livelihood Capitals

Following DFID and CARE International's livelihood framework, an analytical model is developed by identifying the relevant indicators of five livelihood capitals, and indicators are defined considering the coastal livelihoods of Bangladesh (Carney, 1998; Carney *et al.*, 1999). A five-point Likert scale is used to measure five types of livelihood capitals, considering the discrete values of 0.00, 0.25, 0.50, 0.75, and 1.

Human Capital Index (HCI)

Indicators considered to measure human capital index value are household head's education level, ability to work in adverse conditions, having disaster-related training, level of experience gained from training, solving own problems, representative of the group, exposing ideas in the group meeting, helping other to solve the problem, motivating others for community activities, facilitating community and GO-NGO initiatives, and solving conflict within the community. The below formula is used to calculate the HCI value.

$$HCI = \left(\sum HCI_1 / N + \sum HCI_2 / N ... \sum HCI_{11} / N \right) / 11$$

where HCI = Human Capital Index, HCI_1, HCI_2, ... = Human Capital Indicators, and N = Total sampled respondents.

Natural Capital Index (NCI)

The natural capital index (NCI) value is calculated by adding the average of nine selected natural capital indicators (such as access to open water bodies, forests, grazing lands, and *khas* and/or char lands, soil fertility status, the trend of soil fertility change in past ten years, the sufficiency of water for irrigation, soil salinity status, and frequency of tidal surge events) and dividing the total by 9.

$$NCI = \left(\sum NCI_1 / N + \sum NCI_2 / N \ldots \sum NCI_9 / N \right) / 9$$

where NCI = Natural Capital Index, NCI_1, NCI_2, ...= Natural Capital Indicators, and N = Total sampled respondents.

Financial Capital Index (FCI)

Indicators considered to measure the financial capital index (FCI) are available financial deposits and monetary values of liquid assets of each household, such as reserve of money in the form of cash, deposits in banks, cooperatives, and groups, remittance and pension, and liquid assets from livestock, poultry, jewelry, furniture, storage of food and cash crops, trees and other assets which can provide liquid money. After converting all forms of such assets into monetary values for each household, it is divided by the aggregate highest available financial deposit and monetary value among the study villages to get the FCI.

$$FCI = Av / Av_h$$

where FCI = Financial Capital Index, Av = Available financial deposit and monetary value of liquid assets of each household, and Av_h = Highest available financial deposit and monetary value among the study villages.

Physical Capital Index (PCI)

Physical Capital Index (PCI) is measured considering the quality of road to reach in-market, accessibility to the road, access to cyclone shelters, and access to agricultural and fishing accessories such as tractors, harvesters, boats, nets, and means of transportation such as rickshaws, van, motorbike, and bicycle.

$$PCI = \left(\sum PCI_1 / N + \sum PCI_2 / N \ldots \sum PCI_9 / N \right) / 9$$

where PCI = Physical Capital Index, PCI_1, PCI_2, ...= Physical Capital Indicators, and N = Total sampled respondents.

Social Capital Index (SCI)

Three broad groups, such as access to mass communication and individual communication, participation, and connection, and memberships in the groups, are considered to measure social capital index (SCI) values.

Mass Communication and Individual Communication Index (MCICI)

Indicators for mass communication and individual communication include household head's access to radio, television, mobile phone, newspaper, printed materials, disaster mitigation, agriculture extension, fishery, health and family planning personnel, Union Parishad, NGOs, and CBOs.

$$MCICI = \left(\sum MCICI_1 / N + \sum MCICI_2 / N ... \sum MCICI_{11} / N \right) / 11$$

where MCICI = Mass Communication and Individual Communication Index, $MCICI_1$, $MCICI_2$, ..., $MCICI_{11}$ = Mass Communication and Individual Communication Indicators, and N = Total sampled respondents.

Participation and Connection Index (PACI)

Participation and connection indicators include participation in community-based awareness campaigns, disaster mitigation activities, connection with NGOs for disaster mitigation, relationships with others in the community, and assisting others during a crisis.

$$PACI = \left(\sum PACI_1 / N + \sum PACI_2 / N ... \sum PACI_5 / N \right) / 5$$

where MICI = Participation and Connection Index, $PACI_1$, $PACI_2$, ..., $PACI_5$ = Participation and Connection Indicators, and N = Total sampled respondents.

Membership Index (MGI)

Membership is measured through involvement in voluntary, religious, microcredit, cooperative groups, and Union Parishad.

$$MGI = \left(\sum MGI_1 / N + \sum MGI_2 / N ... \sum MGI_5 / N \right) / 5$$

where MGI = Membership in Groups Index, MGI_1, MGI_2, ..., MGI_5 = Memberships in Groups Indicators, and N = Total sampled respondents.

Finally, the overall SCI is calculated based on the average of mass communication and individual communication, participation and connection, and memberships.

Assessing Livelihood Security

Household livelihood security is measured considering the variables of food security, nutritional security, income security, shelter-water and sanitation security, and health and education security. In this regard, food security is measured based on perception and satisfaction with consumed food and direct calorie intake. The quantity of food in kilogram and calorie information of consumed food is used to calculate per capita calorie intake per day per person. Likewise, nutritional security is measured by considering the body mass index (BMI) of reproductive women and the stunting, wasting, and underweight of less than five years of children. Economic security is measured considering the annual average income, expenditure, and savings status of a household. However, economic status is compared with the national average. Shelter-water and sanitation status are some other indicators of livelihood security. Specifically, the roofing materials of the house, drinking water, and toilet facilities are used to assess the shelter-water and sanitation status of a household. Similarly, child immunization, child death during birth, and annual healthcare expenditure are used to assess the health security status of a household. Besides, education security is measured considering the adult literacy and seven years and above literacy rate of households. Such statuses are compared with the national average to compare the livelihood security status of households.

Results and Discussion

Performance of Livelihood Capitals and Asset Pentagon

The present study finds that overall human capital is higher in the Shoreline village with a value of 0.41 out of 1 than in Inland (0.34) and Island (0.31) villages respectively. Application of the ANOVA indicates that the selected villages significantly differ in terms of human capital. Respondents of the Shoreline village have better accomplishment of human capital than the other two. On the contrary, the Island village (0.59) has higher natural capital than the Shoreline (0.57) and the Inland (0.50) villages. It is due to the higher access of Island villagers to open access common natural resources such as sea, river, grazing lands, and *khas* lands. Higher access to such resources had attracted many migrants to settle in that village, though this is the most vulnerable location to cyclones among the three villages. Similarly, the Shoreline village also enjoys more natural capital than Inland. As the Shoreline village is located on the Bay of Bengal's seashore, villagers have higher access to fishing in the deep sea. On the other hand, the Inland village is agro-based, and limited common resources are available. Therefore, the NCI reveals a relatively lower value. However, though a statistically significant difference does not exist, the FCI value is comparatively higher in Shoreline (0.055) than in Inland (0.049) and Island (0.031) villages (Table 3.2).

Among the villages, Island accounted for lower financial capital, which reveals the generally poor economic condition of respondents. The PCI reveals a higher value for Shoreline (0.51) than Inland (0.42) and Island (0.28) villages. The higher PCI value in Shoreline reveals that respondents had higher access and ownership of physical assets such as fishing and agricultural accessories. Fishermen in the Shoreline village usually catch fish in deep-sea, thus requiring a huge amount of investment for mechanized boats and nets. On the other hand, a large number of respondents in the Island

Table 3.2 Livelihood assets index by village and livelihood groups

Livelihood capitals	Study villages			
	Inland	Shoreline	Island	ANOVA
Human	0.34	0.41	0.31	F = 11.87**
Natural	0.50	0.57	0.59	F = 39.64**
Financial	0.049	0.055	0.031	F = 2.255
Physical	0.42	0.51	0.28	F = 54.71**
Social	0.29	0.32	0.21	F = 17.69**
Livelihood Capitals	Livelihood groups			
	Farmer	Fisher	Wage laborer	ANOVA
Human	0.39	0.36	0.30	F = 12.103**
Natural	0.55	0.60	0.53	F = 21.153**
Financial	0.08	0.04	0.03	F = 14.615**
Physical	0.46	0.48	0.33	F = 51.106**
Social	0.28	0.27	0.24	F = 9.358**

$p < 0.05$, ** $p < 0.01$,

Source: Household Survey.

village are also fishermen but usually unable to invest more money for boats and nets due to poor economic conditions. Agricultural accessories are not as expensive as fishing accessories, the number of farmers is higher than the fisher, and fishermen are engaged in subsistence-level fishing. Hence, the Inland village shows less PCI value. The SCI value shows a better situation in Shoreline (0.32), followed by Inland (0.29) and Island (0.21) villages (Table 3.2). The higher performance of SCI values in the Inland village indicates better social coherence among the inland coastal communities. On the other hand, the majority of the respondents on the Island are immigrants from different inland locations; therefore, social bonding is not as tight as in Inland locations. Likewise, institutional activities are relatively limited in the Island village due to poor communication networks and remoteness.

On the other hand, due to the proximity to the thana headquarters, respondents in Inland and Shoreline villages enjoy more government and non-government organizations' services. However, such locational advantage benefits the villagers of Shoreline and Inland, for example, access to information and creating a link with GO and NGOs. Such facilities help Inland villagers with relatively better disaster preparedness, rescue, relief, and rehabilitation.

It also reveals that accesses to livelihood capitals are significantly higher for farmers and fisher than wage laborers. This reveals a relatively weaker asset portfolio of wage laborer compared to farmers and fishers. Between farmer and fisherman, the financial and SCI value is higher for the farmer, while the natural and PCI value is higher for fishermen. This indicates a relatively better status of a farmer than a fisherman. Higher human capital unveils better physical ability, skill, and knowledge of farmers than fishers.

Similarly, having higher financial capital of fishermen reveals a higher ability to convert liquid assets and savings to gain other capital. The higher social capital of farmers demonstrates relatively better participation and connection status and social coherence of this group with various other community groups. Social coherence, a part of social capital, helps farmers to cope with a post-cyclone crisis, such as providing food to neighbors immediately after a cyclone while institutional assistances are not available and helping each other to reconstruct destroyed houses. On the other hand, the higher natural capital of fishermen reveals high access of fishing communities to common resources such as the sea and rivers. Fishing accessories require a higher investment than agriculture accessories. Therefore, physical capital for fisher groups is higher than for the farmer.

Radar diagrams are plotted based on the calculated scores of five livelihood capital indices to present in the livelihood asset pentagon (Figures 3.2 and 3.3). The shape of the pentagon represents the variations in villages' and livelihood groups' access to five livelihood capitals. The center point of the pentagon represents zero value in terms of access to assets. In contrast, deviations from the center point of the pentagon to the outer sides represent higher values of livelihood capitals. Figures 3.2 and 3.3 present the livelihood asset pentagon to depict villagers' and livelihood groups' overall access to five livelihood capitals.

Livelihood Groups and Strategies

Livelihoods in coastal Bangladesh are broadly clustered into two groups: natural-resource-based (agriculture, fishing, aquaculture, extraction of forest resources) and human-resource-based (boat building, net making, fish

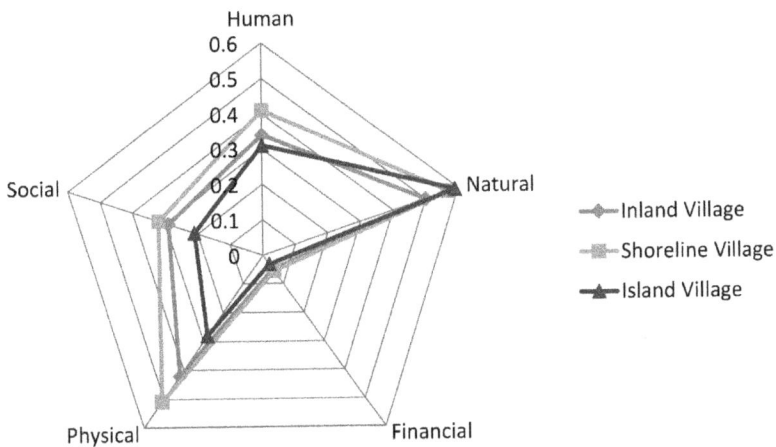

Figure 3.2 Livelihood asset pentagon by villages.

Source: Based on Table 3.2.

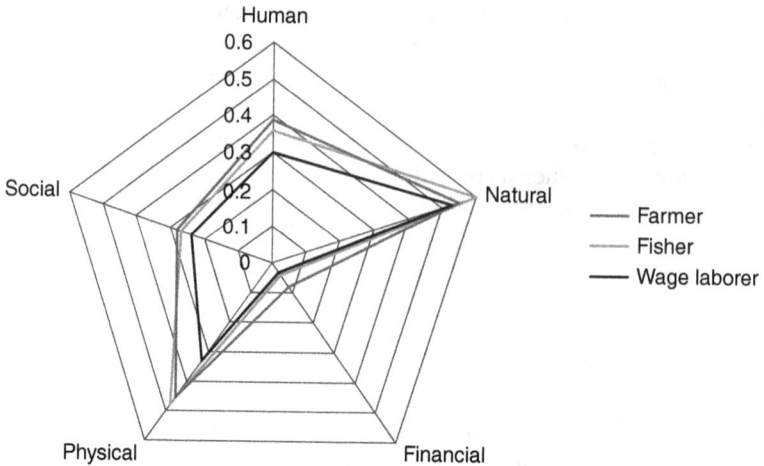

Figure 3.3 Livelihood asset pentagon by livelihood groups.
Source: Based on Table 3.2.

processing, trading). There are over seven million households in the coastal zone, out of which agricultural laborers, small farmers, fishers, and poor urban cover 71% of total households (BBS, 2001). In the present study, three major livelihood groups are identified: farmers, fishermen, and wage laborers (Table 3.3). Farmer includes small, medium, and large farmers. The present study finds that most farmers belong to the small farmer category in study villages. Likewise, the fisher group includes fishermen and fish fry collectors who usually catch fish in rivers, channels, and the sea. The wage laborer group includes agricultural and non-agricultural laborers, housekeepers, and/or maidservants. Profiles of these livelihood groups in the study villages are presented in Table 3.3.

Table 3.3 Major livelihood groups in study villages

Livelihood group	Study villages							
	Inland		Shoreline		Island		All	
	N	%	N	%	N	%	N	%
Farmer	68	34.34	20	22.22	12	27.91	100	30.21
Fisher	42	21.21	50	55.56	28	65.12	120	36.25
Wage Laborer	88	44.44	20	22.22	3	6.98	111	33.53
Total	198	100.00	90	100.00	43	100.00	331	100.00

Source: Household Survey.

Fisher Livelihood Group

Fisher households are 14% of the total coastal rural households of Bangladesh. Fishers are extremely poor, and about 70% fall into the small farmers' category. Fish fry collectors are the subcategory of the fisher livelihood group. About half a million people in the coastal zone are fry collectors. To a large extent, fish fry collectors depend on the shrimp sector, contributing 41% of household income. The present study finds that fisher is the largest livelihood group, with 36.25% of households among the study villages. In the case of individual villages, 65.12%, 55.56%, and 21.21% of households in Island, Shoreline, and Inland villages, respectively, belong to the fisher livelihood group (Table 3.3). It reveals that the number of fisher groups is significantly higher on Island, followed by Shoreline and Inland villages.

On the other hand, in the fisher livelihood group, about 44% depends on fish fry collection for sustaining livelihoods. Fishermen usually catch fish year-round, but there are seasonal variations in the availability of fish in rivers and sea. Large-scale fishermen usually catch fish in the deep sea all year round with mechanized boats. While small-scale fishers usually catch fish from May to November. Mid-November to mid-April is the lean period. They usually engage in wage laboring, repairing boats and nets, and agricultural activities. Major risks identified by this group are cyclones, storm surge, *dadon*,[1] and *mahajani*[2] system of borrowing. In the fisher group, fish fry collectors are the most subsistence sub-group that usually collect fish fry from mid-December to June. The majority of the fry collectors are younger women and children. A few fishermen who do not have their own boat and net also engage in fry collection. In general, fry collectors usually earn 80–120 BDT (1–1.5$) per day, while the amount of earning is lower for women. This fry collector group is mostly found in Island and Shoreline villages. In an Island village, both males and females collect fish fry, while in the Shoreline village, women's participation in fry collection is relatively limited. Major risks that hinder the livelihoods of poor fry collectors are cyclones, storm surges, *dadon* and *mahajani* systems, poverty, and child labor.

Wage Laborer Livelihood Group

In coastal Bangladesh, the number of agricultural laborer households is 1.81 million, which is 33.2% of total coastal households. Among the wage laborers, 55% belong to the small farmer, and 43% are landless (BBS, 2001). The present study finds that one-third of total households are wage laborers category. The number of wage laborers is higher in the Inland (44.44%) village than in Shoreline (22.22%) and Island (6.98%) villages (Table 3.3). The wage labor livelihood group is higher in Inland because of the village's agricultural economy. The majority of the wage laborers are small farmers and landless. The natural capital of the inland village offers very limited options for poor villagers apart from working as wage labor. Whereas, in Shoreline and Island villages, vast natural resources (open water bodies and forests) offer several choices for villagers, such as independent entrepreneurship for fishing either on a small or large scale. Hence, the number of wage labor is found on a limited scale in the Shoreline village and a very limited scale in the Island village. Wage labor usually works all year round. However, there is variation in the availability of work. The average wage per day is 120 to 150 BDT (1.5–2.0$) for wage labor, irrespective of village locations. November to February is the lean period for wage labor in coastal villages because of the limited availability of work in agricultural fields and the non-availability of fish in the rivers. It is also found that major risks hindering the livelihoods of wage labor are poor transportation, low wages, uncertainties in getting a job every day, and illness.

Farmer Livelihood Group

Farmers are the third-largest livelihood group with 30.21% of total surveyed households. In the farmer livelihood group, 77.66% is small, and 23.34% is the large farmer. Earlier studies reveal that in the past few decades, many large and medium farmers have turned into the small farmer category in the coastal zone of Bangladesh because of the fragmentation of land due to increasing population and pauperization. Hence, the number of small farmers have increased, and the number of medium and large farmers have sharply decreased in the coastal zone (PDO-ICZM, 2004). The present study finds that farmer is the second-largest livelihood group in Inland, with 34.34% of households, 22.22% in Shoreline, and 27.91% in Island villages, respectively (Table 3.3). It is found that the Inland village is predominantly agro-based, and the majority of the respondents are directly or indirectly dependent on agriculture. On the contrary, Island and Shoreline villages' agriculture is not dominant like Inland villages. Therefore, the proportion of farmer livelihood groups is relatively limited. Besides, farming activity depends on available cultivable land, and cropping pattern depends on land suitability for single, double, or multiple uses. This study also finds that the Island village is suitable for single cropping, the shoreline village is both single and double cropping, and the Inland village is double and triple cropping. In single-cropped land, farmers used to cultivate transplanted Aman during the Kharif season (April to November). The rest of the year is the lean period, and they usually engage in other types of income-earning activities such as fishing, fish fry collecting, livestock rearing, etc. Soil salinity, lack of irrigation facilities, cyclones, and induced surges are major risks for single-crop farming. The period for double cropping is April to November, and the cropping pattern is Rabi-Fellow-T. Aman. January to March is a lean period. Triple-cropped lands follow the cropping pattern of Rabi-T. Aus-T. Aman. The period of cropping is March to May (pre-Kharif) for T-Aus, June to November (Kharif) for T. Aman, and November to February (*Rabi*) for winter vegetables, pulses, and other Rabi crops. Major risks for double cropping are water scarcity for irrigation, tidal surge, and soil salinity. High soil salinity and lack of irrigation facilities are the key reasons for single cropping in Island and Shoreline villages. In Inland villages, though lands are triple-cropped, scarcity of irrigation facilities and moderate soil salinity in the winter hinder *Rabi* crop cultivation. Based on field survey and FGDs, major risks and uncertainties of farming identified are cyclones, storm surges, floods, mortgage of land, and the loan from *mahajans*.

Livelihood Adversity and Crisis

Almost every year, coastal populations experience damage to assets, crops, poultry, livestock, houses, and livelihoods due to cyclones, coastal flooding, tidal surge, coastal erosion, etc. Cyclone is the most common form of hazard in coastal Bangladesh and strikes on average every two to three years. The Bay of Bengal surrounds the southern part of Bangladesh, and any cyclone formed in the sea is always accompanied by storm surges and inundates entire coastal belts. Therefore, any surge that strikes the land always causes huge impacts on the coastal population's livelihoods. The cyclones cause devastation of agricultural production, damage houses and other infrastructure, and result in deaths. Coastal erosion is a key problem, as most of the lands are part of the active delta, which are recently inhabited and cultivated, worn out due to abnormal water movement and coastal flooding.

The present study finds that 33% of the respondents have identified cyclones as a major impediment in all three villages. Secondly, storm surge is identified as a major hazard by 30.5 and 27.2% of respondents in Shoreline and Island villages and 28.4% of respondents in Inland villages (Table 3.4). Because cyclone is always accompanied by storm surge in Shoreline and Island villages, the low strength of cyclone in Inland villages remains relatively safer. Saline water intrusion is identified as a major problem by the respondents of the Island village. During the dry season, soil salinity emerges as a severe problem for farmers, which is further aggravated due to shrimp cultivation. In addition, some other problems such as loss of fruit trees, agricultural production, fish production, and illness of inhabitants are also identified by the villagers of all three locations. Moreover, some other adversities such as flood, coastal erosion, and sand deposition on agricultural land were identified by a few respondents. The respondent identifies the illness of inhabitants irrespective of village location, while coastal flooding and sand deposition on agricultural land is identified as problems in the Inland village.

Income Diversification and Livelihood Strategies

Farmers in all the study villages are engaged in agriculture and simultaneously engaged in fishing, animal husbandry, selling crops, coconut, betel nut, dairy products, etc. Due to the subsistence nature of income-earning at the household level, most small farmers and subsistence fishers involve in wage laboring as their primary occupation. Field surveys and key informant interviews unveil that income diversification of rural households has been taking place rapidly due to the changes in demographic characteristics such as

Table 3.4 Major adversity and crisis in a year in the study villages (multiple responses)

Adversity and crisis	Study villages							
	Inland		Shoreline		Island		All	
	N	%	N	%	N	%	N	%
Cyclone	198	36.1	90	30.5	43	27.2	331	33.0
Storm surge	156	28.4	90	30.5	43	27.2	289	28.8
Salinity intrusion	0	0.0	5	1.7	42	26.6	47	4.7
Loss of trees	85	15.5	37	12.5	15	9.5	137	13.7
Loss of agricultural productions	55	10.0	23	7.8	4	2.5	82	8.2
Loss of fish production	26	4.7	28	9.5	6	3.8	60	6.0
Illness	14	2.6	22	7.5	5	3.2	41	4.1
Flood	4	0.7	0	0.0	0	0.0	4	0.4
Coastal erosion	9	1.6	0	0.0	0	0.0	9	0.9
Sand deposition on agricultural land	2	0.4	0	0.0	0	0.0	2	0.2
Total	549	100.0	295	100.0	158	100.0	1002	100.0

increasing population and migration, and cultural transformations such as occupational diversity, women's participation in economic activities, etc. Such changes are exemplified by people's increasing involvement in various income-earning activities, involvement in other's occupational domains such as farmer to seasonally fishing, and women's increasing engagement in income-earning activities apart from housework. These, in turn, change the traditional livelihood pattern of coastal areas (PDO-ICZM, 2004).

Most coastal households are heavily dependent on natural resources to earn livelihoods. Hence, occupations are seasonal in nature. Similarly, the scope to switch from one occupation to another is limited because such switching of occupations needs extra skills apart from the main occupation. Though there are constraints in switching occupations, the present study finds that household with single income-earning activity is rare in all surveyed villages. Many households depend on multiple income-earning activities. As such, fishermen remain unemployed about two-thirds of the time a year. Therefore, they engage in wage laboring or working in the agricultural field as hired labor for survival. Likewise, women are also increasingly involving themselves in different income-earning activities apart from everyday household tasks. Different combinations of activities performed by men and women in three study locations are presented in Table 3.5.

Livelihood Security Status

In the present study, livelihood security is measured based on several general categories: food security, nutritional security, income security, shelter-water and sanitation security, health security, and education security. In the case of

Table 3.5 Combination of livelihood activities at household level in study villages

Male	Female
Agriculture, farm labor	Home gardening, poultry, and cattle rearing
Agriculture, petty business	Partial involvement in agriculture, poultry, and cattle rearing, aquaculture
Farm labor, rickshaw/van pulling	Paddy harvesting and husking, net making
Non-farm labor, fishing, rickshaw/ van pulling	Poultry and goat rearing, collecting cow dung for fuel
Agriculture, sharecropping, paddy husking	Paddy harvesting and husking, puffed rice (*muri*) preparing and selling, duck keeping, egg selling
Agriculture, cattle grazing, pond aquaculture, fishing	Wage labor, earthwork (food for work/cash for work), fish fry collection
Shrimp and other pond aquaculture, farming, day labor	Fish fry collection, wage labor (maidservant), net repairing
Shrimp fry collection Wage labor, fishing net preparing and repairing	Wage labor, maidservant, cow-dung, and fuelwood collection
Fishing, fish trading, agriculture, aquaculture	Wage labor, paddy husking, puffed rice (*muri*) preparing, earthwork, *kantha*[3] making
Pond aquaculture, business, agriculture, and sharecropping	Fish fry collection, fishing, net making, and repairing

food security, in general, most people had experienced availability of food for seven to nine months in a year. The situation worsens during post-cyclone based on locational exposure, gender, income, and occupation of households. In coastal Bangladesh, household food intake is highly centered on rice consumption, and such rice consumption declines 46.82% from normal time to post-cyclone situation among the sampled households. The present study finds that it is directly linked with locational exposure, gender, income, and occupation of households. One-quarter of respondents found that the amount of food usually consumed was adequate, while only 23% found the quality of such food to be adequate. Similarly, more than 80% of respondents were worried with a high level of anxiety (on a five-point Likert scale) for future food consumption during post-cyclone, which is linked with location, gender, income, education, and occupation as well. Hence, such findings reaffirm that female-headed households, lower-income groups, illiterates and unstable occupational groups such as day laborers and maidservants, and households in most exposed remote locations are the most vulnerable to post-cyclone food insecurity.

This study also finds that more than 20.5% of households are 'absolutely' food insecure based on a daily per capita caloric intake of less than 2122 kcal, and 17.2% are hard-core food insecure with a caloric intake of 1805 kcal. A significant difference exists among the villages in terms of caloric availability, whereas the Island village reveals to be relatively more food insecure. Likewise, wage laborer livelihood groups are found more food insecure in terms of per capita calorie intake than farmers and fishers. Regression result also reveals that indicators of maternal characteristics, household demographic and socio-economic characteristics, livelihood capital, and coping strategies are significant factors influencing food security (the regression model is not presented in this chapter). However, such influencing factors significantly vary across the villages based on available resources, demographic, and socio-economic conditions. This helps identify location-specific indicators that enhance and reduce post-cyclone food security status.

Like food security, other aspects of livelihood security such as nutritional status, income level, health sanitation, drinking water facility, healthcare facilities, and educational status of household members significantly vary across the villages and livelihood groups. The Island village and wage labor livelihood groups are significantly lacking behind in all aspects of livelihood security measures. The farmer livelihood group shows relatively better status than fisher and wage laborer groups. In a few cases, such as educational security, the fisher livelihood group revealed backwardness in all aspects. From a nutritional security point of view, the number of reproductive women having underweight, wasting, and underweight children are highly prevalent on Island, while Shoreline and Inland villages reveal relatively better status. On the other hand, wasting is a severe problem in all three locations, which is higher than the national average. Likewise, the majority of the coastal population is economically insecure, and a large number of households have an annual income of less than 45,000 BDT (650$ per year), while the average savings is 27,393 BDT (less than 400$). In most cases, savings for lower-income households are the security deposits to the NGOs, which are inaccessible during needs. About 50% of total household expenditure is for food consumption, and for poorer households, such expenditure is more than 60%. Besides, households borrow money from multiple sources. While they borrow money from one source to repay others, thus creating dependency. The Island village and wage laborer livelihood groups reveal economic insecurity in all aspects of economic security indicators.

The housing of fisher and wage laborer livelihood groups, and as a location, the Island village reveals dilapidated condition. Other issues such as sanitation and drinking water facility reveal significant improvement in all three locations. Sanitation status is relatively better in the Shoreline village because of NGO activities than in Island and Inland villages, while the Island reveals a poor sanitation situation that needs to be considered. Drinking water facility is homogenous across the villages and livelihood groups, though poorer households spend more time in collecting water because of not having tube well in their dwelling units. Child immunization and deaths of children are a great concern in Island villages. However, incidences of diseases are mostly similar across the villages, though healthcare facilities and affordability to pay for healthcare vary significantly across the villages. The Island village reveals a disadvantageous status than Shoreline and Inland villages.

Similarly, the educational attainment of household heads and household members reveals the backwardness of Island villages to be more than Inland

and Shoreline villages. However, such a differential status of various livelihood security indicators reveals relatively livelihood insecurity of Islanders, while the Shoreline village reveals relatively better status, followed by the Inland village (Table 3.6). Among the livelihood groups, farmers are relatively more secure than wage laborers and fishers.

Table 3.6 Comparative aspects of livelihood security indicators in three villages

Indicators	Study villages			
	Inland	Shoreline	Island	Remark
Food insecurity (Food insecure households in terms of >2122 kcal/day/person)	46.5%	31.1%	53.5%	National average is 40% (FAO, 2008)
Nutritional insecurity Underweight reproductive women (BMI > 18.5)	30.56%	22.09%	31.71%	National average is 50% (UNICEF, Undated)
Wasted children (below five years)	20.93%	19.15%	28%	National average is 17.4% (BHDS, 2007)
Underweight children (below five years: weight-for-age)	40.70%	36.17%	44%	National average is 41% (BHDS, 2007)
Stunted children (below five years)	45.35%	44.68%	48%	National average is 40% (BHDS, 2007)
Economic insecurity Average annual income	48,605 BDT	85,722 BDT	52,717 BDT	The national average is 55,888 TK (757.7 USD) (2010)
Average annual expenditure	46,481 BDT	75,816 BDT	49,720 BDT	Data not available
Average annual savings	17,259 BDT	46,736 BDT	30,172 BDT	Data not available
Shelter, water, and sanitation Security Roofing material of the house	CI Sheet 88.89%	CI Sheet 93.33%	Straw and Polythene 58.05%	Data not available
Average room/HH	1.63 room	1.67 room	1.21 room	Data not available
Average house size/HH	15.18 meters	18.72 meters	9.91 meters	Data not available
Households have toilet facilities.	79.80%	91.11%	55.81%	Patuakhali district, Barguna district, and the national average is 92.15%, 93.28%, and 78.4%, respectively

(Continued)

Table 3.6 (Continued)

Indicators	Study villages			
	Inland	Shoreline	Island	Remark
Drinking water (tube well)	100%	100%	100%	Patuakhali district, Barguna district, and national average are 51.46%, 61.56%, and 95.9%, respectively
The average distance of tube well from the dwelling unit	249 meters	170 meters	227 meters	Data not available
Health security Child immunization	96.19%	98.89%	76.74%	National average is 95% (UNICEF, 2012)
Child deaths during childhood	19.19%	23.33%	37.21%	Data not available
Annual average healthcare expenditure	3450 BDT	5968 BDT	3101 BDT	Data not available
Educational security Illiteracy of household head	65.7%	52.2%	79.1%	Data not available
Illiteracy of household members	39.07%	31.08%	77.25%	National average is 49.6%
Seven years and above literacy	60.03%	68.32%	22.75%	Patuakhali district, Barguna district, and national average are 51.6%, 55.3%, and 48.8%, respectively

Conclusion

The present study identifies three major livelihood groups: farmer, fisher, and wage laborer. The distribution of livelihood capitals among the three villages reveals significant differences. The Island village lacks behind the Inland and Shoreline villages in terms of livelihood capital, though the natural capital for Island is relatively higher. The scarcity of other capitals hinders the proper utilization of the potential of the natural capital of Islanders. Social capital for Island is significantly lower than other two villages, which unveils relatively lesser social coherence of Islanders, which is most important to survive in a disaster situation. As most of the inhabitants of the Island village are immigrants from inland locations, thus social bonding is relatively less. Likewise, among the livelihood groups, wage laborers have less livelihood capital than farmers and fishermen. The majority of the households, irrespective of their village locations, have identified cyclones as major adversity that significantly destroys their livelihoods. Most of the coastal households are not dependent only on single-income-earning activities. Households in study villages diversify income sources wherever possible and, most importantly, face a crisis in livelihoods. Such diversification of income-earning activities unveils the transformation process of traditional typical coastal livelihoods. In general, the livelihoods security of Islanders and Shoreline villagers are more susceptible to cyclones and storm surges. On the other hand, as a livelihood group, wage laborers and fish fry collectors are most vulnerable to any hazard events and remain insecure in the post-cyclone period. Based on the findings, this study reveals that the livelihood security of cyclone survivors of the Island village and wage laborer and fish fry collector livelihood groups are more susceptible to cyclones due to the lack of livelihood capital than the other two villages and livelihood groups considered in the present study. Therefore, the present study advocates for targeting vulnerable locations, such as the Island village, and vulnerable livelihood groups, such as wage laborers and fish fry collectors, for disaster risk reduction that could more adequately address the post-cyclone livelihood insecurity of the inhabitants of coastal Bangladesh. Besides, creating post-cyclone income-earning activities such as 'cash for work' and 'food for work' for vulnerable livelihood groups and locations could be fruitful to reduce post-cyclone hardships for shorter periods. Apart from such activities, post-cyclone livelihood rebuilding, such as credit support for boats, nets, livestock, and agriculture, could provide benefits for longer periods. Moreover, emphasis should be given to coordinated disaster risk reduction and rebuilding of livelihoods, such as assistance for producing food rather than providing food. In other words, supports should be expanded and continued for income-generating activities among the coastal rural poor in disaster-prone areas, reducing the prevalence of post-cyclone livelihood insecurity for a longer period.

Notes

1 *Dadon* is an advance payment of money for delivering catch fish under stipulated terms.
2 The term *Mahajan* is applied to people hereditarily engaged in money lending.
3 *Kantha* is a type of embroidery popular in Bangladesh. Women typically use old saris and cloths and layer them together and stitch to make them like a light blanket.

References

Ali, A. (1996). Vulnerability of Bangladesh to climate change and sea level rise through tropical cyclones and storm surges. *Water, Air, & Soil Pollution*, 92(1–2), 171–179. https://doi.org/10.1007/BF00175563

Ali, A. (1999). Climate change impacts and adaptation assessment in Bangladesh, *Climate Research*, 12(2–3), 109–116. https://www.jstor.org/stable/24866005

BBS. (2001). *Preliminary Report of Household Income and Expenditure Survey-2000*. Dhaka: Ministry of Planning.

Blaikie, P., Cannon, T., Davis, I., & Wisner, B. (1994). *At Risk: Natural Hazards, People's Vulnerability and Disasters*. London: Routledge.

Carney, D. (1998). Implementing the sustainable rural livelihoods approach. In: Carney, D. (ed): *Sustainable rural Livelihoods-What Contribution Can We Make?* Papers presented at the Department for International Development's Natural Resources Advisers' Conference, July 1998, 3–23. London: Department for International Development.

Carney, D., Drinkinwater, M., Rusinow, T., Neefjs, K., Wanmali, S., & Singh, N. (1999). *Livelihoods Approaches Compared: A Brief Comparison of Livelihoods Approaches of the UK Department of International Development (DFID), CARE, Oxfam and United Nations Development Program (UNDP)*. London: DFID.

DFID. (1999). Sustainable Livelihoods Guidance Sheets, Section 2, The Livelihoods Framework. Department for International Development, London.

FAO. (2008). Special Report FAO/WFP Crop and Food Supply Assessment Mission to Bangladesh, Retrieved from http://www.fao.org/docrep/011/ai472e/ai472e00.htm on 1/21/2010

GoB. (2008a). *Cyclone Sidr in Bangladesh: Damage, Loss and Needs Assessment for Disaster Recovery and Reconstruction*. Dhaka: Government of Bangladesh.

GoB. (2008b). *Super Cyclone Sidr 2007: Impacts and strategies for interventions*. Dhaka: Government of Bangladesh.

Haque, C.E. (1997). Atmospheric hazards preparedness in Bangladesh: A study of warning, adjustments and recovery from the April 1991 cyclone. *Natural Hazards*, 16(2–3), 181–202.

Hossain, M. N., Paul, S. K., Roy, C., & Hasan, M. M. (2014). Factors influencing human vulnerability to cyclones and storm surges in the coastal Bangladesh, *Journal of Geo-Environment*, 11, 1–29.

Islam, A.K.M.S., Bala, S.K., Hussain, M.A., Hossain, M.A., & Rahman, M. (2011). Performance of coastal structures during Cyclone Sidr. *Natural Hazards Review*, 12(2), 111–116. http://dx.doi.org/10.1061/(ASCE)NH.1527-6996.0000031

Mallick, B., Witte, S.M., Sarkar, R., Mahboob, A.S., & Vogt, J. (2009). Local adaptation strategies of a coastal community during Cyclone Sidr and their vulnerability analysis for sustainable disaster mitigation planning in Bangladesh. *Journal of Bangladesh Institute of Planners*, 2, 158–168.

Mooley, D.A. (1980). Severe cyclonic storms in the Bay of Bengal, 1877–1977. *Monthly Weather Review*, 108(10), 1647–1655.

Moser, C., Gauhurts, M., & Gonhan, H. (1994). *Urban Poverty Research Source Book: Sub-city Level Research*. Washington, DC: World Bank.

Murty, T.S., & El-Sabh, M. (1992). Mitigating the effects of storm surges generated by tropical cyclones - a proposal. *Natural Hazards*, 6(3), 251–273. https://doi.org/10.1007/BF00129511

Murty, T.S., & Neralla, V. R. (1992). On the recurvature of tropical cyclones and the storm surge problem in Bangladesh. *Natural Hazards*, 6(3), 275–279. https://doi.org/10.1007/BF00129512

Paul, B.K. (2009). Why relatively fewer people died? The case of Bangladesh's Cyclone Sidr. *Natural Hazards*, 50(2), 289–304. https://doi.org/10.1007/s11069-008-9340-5

Paul, B.K. (2010). Human injuries caused by Bangladesh's Cyclone Sidr: an empirical study. *Natural hazards*, 54(2), 483–495. https://doi.org/10.1007/s11069-009-9480-2

Paul, B.K., & Dutt, S. (2010). Hazard warnings and responses to evacuation orders: The Case of Bangladesh's Cyclone Sidr. *The Geographical Review*, 100(3), 336–355. https://doi.org/10.1111/j.1931-0846.2010.00040.x

Paul, B.K., Rahman, M.K., & Rakshit, B.C. (2011). Post-Cyclone Sidr illness patterns in coastal Bangladesh: an empirical study. *Natural hazards*, 56(3), 841–852. https://doi.org/10.1007/s11069-010-9595-5

Paul, S.K. (2010). Prevalence of nutritional insecurity in the central coast of Bangladesh: Evidences from Cyclone Sidr'. *The Journal of Geo-Environment*, 10, 22–35.

Paul, S.K. (2012). Vulnerability to tropical cyclone in the Southern Bangladesh: Impacts and determinants. *Oriental Geographer*, 53 (1–2), 19–40.

Paul, S. K. (2013). Post-cyclone livelihood status and strategies in coastal Bangladesh. *Rajshahi University Journal of Life & Earth and Agricultural Sciences*, 41, 1–20.

Paul, S.K. (2014). Determinants of evacuation response to cyclone warning in coastal areas of Bangladesh: A comparative study. *Oriental Geographer*, 55(1–2), 57–84.

Paul, S. K. (2016a). Consumption pattern and status of food security in coastal Bangladesh: A study on post-cyclone period. *The Journal of Geo-Environment*, 13, 1–20.

Paul, S. K. (2016b). Death, injury and illness caused by Bangladesh's cyclone Sidr: an empirical study. *Plan Plus*, 7(1). Retrieved from https://planplusjournal.com/index.php/planplus/article/view/80

Paul, S. K. (2019). Post cyclone household food security in coastal Bangladesh. *People at Risk: Disaster and Despair*, 185–209.

Paul, S. K., & Hossain, M. N. (2014) Adaptation of cyclone victims in the central coast of Bangladesh, *Jahangirnagar Bishwavidyalay Bhugol O Paribesh Samikkhan*, 33, 27–47.

Paul, S. K., Hossain, M. N., & Ray, S. K. (2013). Monga in northern region of Bangladesh: a study on people's survival strategies and coping capacities. *Rajshahi University Journal of Life & Earth and Agricultural Sciences*, 41, 41–56.

Paul, S.K., Paul, B.K., & Routray, J.K. (2012). Post-Cyclone Sidr nutritional status of women and children in coastal Bangladesh: an empirical study. *Natural Hazards*, 64(1), 19–36. https://doi.org/10.1007/s11069-012-0223-4

Paul S.K., & Routray J.K. (2010a). Flood proneness and coping strategies: The experiences of two villages in Bangladesh. *Disasters*, 34(2), 489–508.

Paul S.K., & Routray J.K. (2010b). Household response to cyclone and induced surge in coastal Bangladesh: Coping strategies and explanatory variables. *Natural Hazards*, 57(2), 477–499. https://doi.org/10.1007/s11069-010-9631-5

Paul, S.K., & Routray, J.K. (2012). An analysis of the causes of non-responses to cyclone warnings and the use of indigenous knowledge for cyclone forecasting in Bangladesh. In: Leal Filho W (Ed). *Climate Change and Disaster Risk Management*, 15–39. Berlin: Springer.

PDO-ICZMP. (2004). Living in the Coast-People and Livelihoods. PDO-ICZMP (Program Development Office for Integrated Coastal Zone Management Plan) Water Resources Planning Organization (WARPO), Ministry of Water Resources, Government of the People's Republic of Bangladesh. p. 11.

Sanderson, D. (2000). Cities, disaster and livelihood. *Environment and Urbanization*, 12(2), 93–102. https://doi.org/10.1177%2F095624780001200208

Shivakoti, G., & Shrestha, S. (2005). Analysis of livelihood asset pentagon to assess the performance of irrigation systems (part 1 - analytical framework). *Water International*, 30(3), 356–362. https://doi.org/10.1080/02508060508691876

Uddin, M.E. (2012). Household food security status of marginal farmers in selected storm surge prone coastal area of Bangladesh, *The Agriculturists*, 10(1), 98–103. https://doi.org/10.3329/agric.v10i1.11070

UN. (2007). Final report on Cyclone Sidr, United Nations rapid initial assessment report with a focus on 9 worst affected districts, 22 November 2007, http://www.lcgbangladesh.org/derweb/cyclone/cyclone_assessment/2007-11-22_UN%20 Rapid%20Assessment%20Report_Bangladesh.pdf, Final Report, United Nations.

UNICEF (Undated). Child and Maternal Nutrition in Bangladesh, retrieved from http://www.unicef.org/bangladesh/Child_and_Maternal_Nutrition.pdf, viewed on 3 November, 2010.

Yamane, T. (1967). *Statistics: An Introductory Analysis*, 2nd ed., New York: Harper and Row.

4 Community Perceptions of Climate Change Governance Practices

A Case Study of Hatiya Subdistrict, Bangladesh

Md. Faruk Hossain, Salim Momtaz and Anita Chalmers

Introduction

Climate change is a multifaceted global problem, and an effective response requires the involvement of numerous stakeholders and institutions (Regmi & Bhadari, 2012). It affects all countries, particularly the global south (Boyd and Ghosh, 2013; Bulkeley, 2010). Climate change governance constitutes a wide range of activities of coordination that expedite climate change adaptation (CCA) and mitigation efforts (Fröhlich & Knieling, 2013). It works at multiple levels managing complex relations between the actors associated with it (Marquardt, 2017; McCarney et al., 2011). Climate change governance is often a significant challenge at national, regional, and local levels in unstable political systems and where problematic bureaucratic infrastructure exists (Barua et al., 2014). Patt (2017) notes that the main obstacles to effective climate change governance are not financial; rather, they are associated with growing knowledge, networks, and institutions. Owing to the lack of effective climate change governance, the outcomes expected from climate risk events-related projects are not realized (Bhuiyan, 2015).

Bangladesh is one of the most disaster-prone countries in the world (Ministry of Environment Forest and Climate Change [MoEFCC], 2009), owing to its unique geographic location, the dominance of floodplains, low elevation, high population density, extensive poverty, and the low capacity of people to withstand shock (Huntjens & Zhang, 2016; Islam, 2010; Islam, 2012; Parvin et al., 2008; Rai, Huq, & Huq, 2014). The country is facing unprecedented challenges because of the adverse effects of climate change (Jaeger & Michaelowa, 2015; Mahmud & Prowse, 2012, Momtaz & Shameem, 2015). The Intergovernmental Panel on Climate Change (IPCC) has predicted that by 2050 and beyond, climate change (in association with sea-level rise) will be responsible for the submergence of 17–20 percent of Bangladesh's landmass (Islam, 2010; Moniruzzaman, 2019). This will create a risk for 27 million people by 2050 (Olsson et al., 2014). Moreover, the World Bank (2000) has projected that by 2080, 43 million people will be affected by climate change in Bangladesh, despite the country's minimal contributions to climate change. The Bangladesh government has formulated numerous

DOI: 10.4324/9781003216476-5

policies, plans, and programs to reduce the vulnerability of affected people (Pervin, 2013). This country has implemented a significant number of community-level CCA projects over the years. Many of these projects have been supported by local governments, non-governmental organizations (NGOs), and community members (Rahman & Huang, 2019).

Climate change governance involves various processes relating to climate change plans and programs and their effective and successful implementation for climate-vulnerable people. Its prevailing climate governance system is a complex relationship of various stakeholders, including governmental organizations (GOs) and NGOs. These organizations operate at different levels concerning their respective influences. The present governance system limits people's capacity to manage and adjust to climate-based shocks and stresses (Alam et al., 2011). Evidence suggests that community participation in local environmental policy and planning processes results in greater accountability and effectiveness (Abels, 2007), leading to improved CCA outcomes (Keskitalo & Kulyasova, 2016).

This study focuses on five major criteria to evaluate the climate change governance system of Bangladesh: accountability, transparency, participation, government effectiveness, and resources. Accountability is a major criterion of good governance for both actors and institutions and indicates the performance of governance mechanisms (Biermann et al., 2010; Scobie, 2017). Transparency is a major mechanism for securing accountability and legitimacy in climate governance systems (Biermann et al., 2010; Gupta & Mason, 2016). Government effectiveness is often seen as a result-based approach to good governance in dealing with climate risk-related issues (Balsiger & Van Deveer, 2012; Biermann et al., 2010). Participation expedites results from climate change governance at the local, national, regional, and international levels (Jodoin, Ducky, & Lofts, 2015). Resources are important to implement climate risk events-related plans and programs and are necessary for climate adaptation and mitigation (Allred, Schneider, & Reeder, 2016). Based on existing literature on environmental governance, these criteria are considered useful to evaluate climate change governance practices in Bangladesh.

Theoretical Framework

In general, governance is described as the set of decisions, actors, processes, institutional structures, and mechanisms, incorporating the division of authority and underlying norms involved in determining a course of action (Pelling, 2011). Blair (2017) argues that governments mainly coordinate governance processes but that inputs from various relevant stakeholders must be incorporated, particularly at the local level, where results are most often reflected. This concept can be applied to climate change governance also. Climate change governance can reduce the effects of climate-induced disasters and related risks (Forino et al., 2015). Climate change governance is a diverse and complex issue that incorporates a number of states, markets, and social actors, and requires a relationship to be developed among them (Forino et al., 2015). Climate change issues require integrated actions at multiple levels of governance systems (Rabe, 2007; Schreurs, 2010). Various actors and their interrelationships operate within a theoretical framework for CCA and disaster risk reduction (DRR) integration. Local-level governance approaches address climate change issues through encouragement, awareness, institutions, and capacity development (Bulkeley & Kern, 2006).

Climate vulnerability and resilience each have a major role in developing adaptive capacity and risk perception at the community level (Fatti & Patel, 2013). Community-based knowledge of climate trends and adaptive strategies can be essential where information on climate risk events-related effects is lacking (Ayers et al., 2014). Community members, as with other actors (such as government and NGOs) are indispensable; they provide significant support for the governance processes and improve CCA outcomes (Keskitalo & Kulyasova, 2016).

Materials and Methods

Research Design

A mixed-method approach was applied for the present study, including key informant interviews and focus group discussions (FGDs) using a semi-structured questionnaire to assemble different types of data in a variety of ways to produce a better interpretation (Creswell, 2013; Onwuegbuzie & Collins, 2007).

The head of the households was incorporated in the semi-structured interviews to get their views at the community level. Participants from both FGDs and key informant interviews were interviewed on the climate change governance practices at the community levels to endorse the views of the household surveys (Figure 4.1).

Study Sites

The coastal belt of Bangladesh is highly vulnerable to sea-level rise due to climate risk events-related impacts in the Bay of Bengal for its low-lying deltaic setting (Nishat et al., 2009). The length of the coastline of Bangladesh is around 710 kilometers, and 23 percent of the country is under the coastal zone. These areas are critically vulnerable to climate change-induced disasters (Islam, 2010). This study was carried out in Hatiya *Upazila*[1] (subdistrict) under the Noakhali district, which is located in the south-central coastal zone of Bangladesh (Figure 4.2). The total area of Hatiya Upazila is 1,508.23 square kilometers. The location of this Upazila is 22°07' and 22°35' north latitude and 90°56' and 91°11' east longitude. It is bounded by Noakhali Sadar and Ramgati Upazilas on the north, the Bay of Bengal on the south and the east, and the Manipura Upazila of Bhola district on the west (BBS, 2015). Hatiya Upazila is located in the innermost part of the Bay of Bengal, and the eastern estuary of the Meghna River converges (Parvin, Takahashi, & Shaw, 2008).

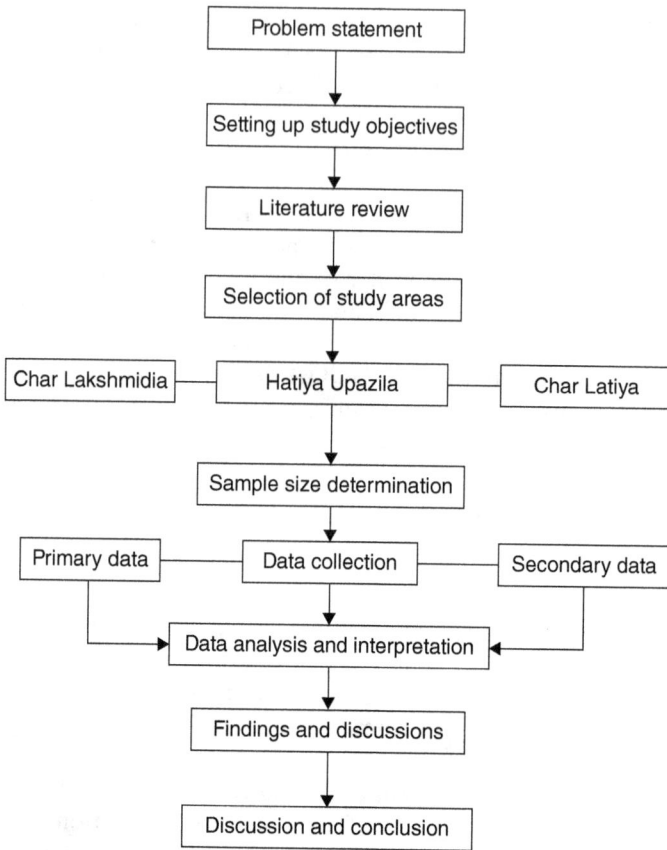

Figure 4.1 Research Design Flow Chart.

Hatiya Upazila, owing to the local conditions, has been affected by devastating cyclones and storm surges in 1960, 1961, 1970, 1985, 1991, 1994, 1997, 1998, and 2007 (Parvin et al., 2008; Union Council, 2009). Among these, 1970, 1991, and 1994 cyclones affected the Char[2] Iswar Union,[3] the surveyed union, severely. This union faces various climate risk events such as cyclones, riverbank erosion, and saline water intrusion. This Upazila, due to its geographic location and poor socio-economic condition of people, is known as one of the most climate risk events vulnerable Upazilas of Bangladesh (GoB, 2017).

Char Ishwar Union is located in the southern part of Hatiya Island (Figure 4.2). This union is located on the bank of the Meghna River. The total area of this union is 26.20 square kilometers with a total population of 42,221 (Union Council, 2019). The density of population of this union is 312 per square kilometer (GoB, 2015). The number of total villages under this union is 15, and two of them, Char Latiya and Char Lakshmidia, were selected for this study. The number of households of Char Latiya and Char Lakshmidia is 577 and 941, respectively (Table 4.1).

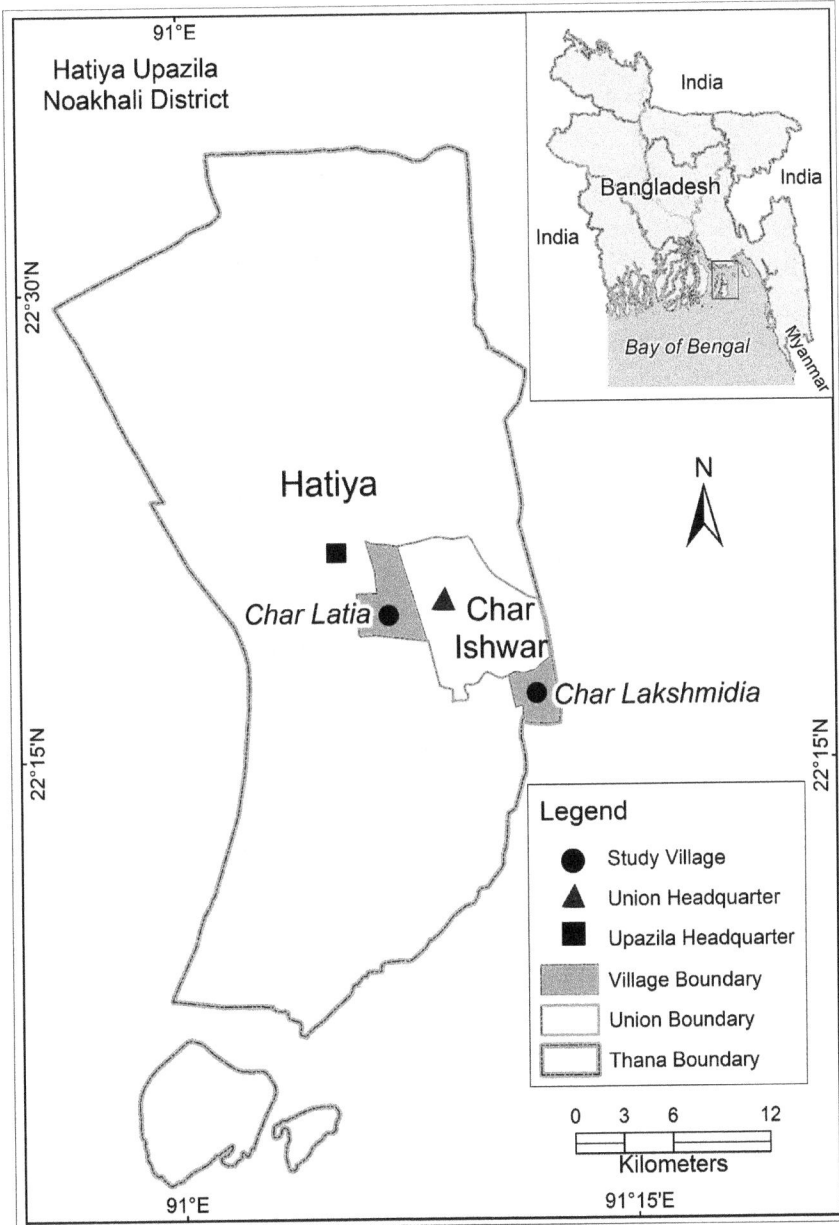

Figure 4.2 Administrative Map of Hatiya *Upazila* (subdistrict).

Source: Adopted from Banglapedia (2014).

Table 4.1 Details of the Study Villages

Name of district	Name of Upazila/ subdistrict	Name of union	Name of villages	Number of households
Noakhali	Hatiya	Char Ishwar	Char Latiya	577
			Char Lakshmidia	941
Total				1,518

Source: GoB (2015).

Sample Size and Respondent Characteristics

Some 181 household heads were interviewed from two study villages using a semi-structured questionnaire (Figure 4.1). These household heads were affected directly or indirectly by climate risk events-related impacts. These two villages are located close to the Meghna River, and the landform of these villages is plain land. Once these villages were protected by the embankment, but due to river erosion, parts of these villages are open to the river now. The questionnaire for heads of the household survey was pretested in both villages to ensure their reliability. Various stakeholders are involved in climate change governance practices at the community level, such as local government departments, local NGOs, local elected political leaders, community leaders, civil society representatives, journalists, vulnerable community people, etc. Nine (9) key informants (KI) interviews were conducted by using a checklist of questions, including two local government officials, two non-government officials, two elected representatives, two community leaders, and one journalist to understand the climate change governance practices at the community level. The interviews were for 40–60 minutes. These KIs were chosen because of their involvement in various climate risk events-related issues in the study area. Additionally, four (4) FGDs were arranged in the residence of the head of the households, and these lasted from 60 to 90 minutes. The majority of the participants were males (see Table 4.2).

For the present study, data were collected from one of the climate risk events-related vulnerable areas, the Hatiya subdistrict of Bangladesh (Figure 4.2), by using a random sampling technique.

Data analysis

Data analysis transforms data and information into findings (Patton, 2002). Data and information were organized and classified based on different issues, including governance criteria mentioned above: accountability, transparency, participation, government effectiveness, and resources. Subsequently, various issues were discussed and compared on the basis of community aspects to see their patterns and relationships to elaborate on the findings and to present the arguments. Quantitative and qualitative data were analyzed by adopting

Table 4.2 Head of the Householders' Demographic Characteristics

Factors	Classes	Percentage (%)
Gender	Male	91
	Female	9
Age range (years)	18–24	2
	25–34	22
	35–44	33
	45–54	19
	55–64	15
	65+	9
Level of education	Illiterate	10
	Primary	52
	Secondary	29
	Technical/vocational	5
	College/university	4
Occupation	Farming	56
	Fishing	20
	Business	11
	Services	8
	Housekeeping	5
House types	Pucca	3
	Semi-pucca	15
	Kutcha	82

descriptive and frequency data analysis tools of SPSS (version 24) and Nvivo 12, respectively.

The weighted average index (WAI) was the main statistical tool used, following the scaling technique (Khongsatjaviwat & Routray, 2015). The WAI for each indicator was calculated by adding up the response numbers for each indicator, multiplied by a weighted value between 0 and 1, and dividing the sum by the number of total responses (Pakzad et al., 2016). This gave an overall weighted average score for each particular indicator. The formula for calculation of WAI is as follows (Ha & Thang, 2017):

$$WAI = \Sigma S_i F_i / N$$

where WAI is the weighted average index ($0 \leq WAI \leq 1$), S_i is the scale value assigned at its priority, F_i is the frequency of household respondents, and N is the total number of observations. These indices were designed based on social scale; the value of each index was kept from 0 to 1. The type of each index is described as follows.

Perception Index

A perception index was applied to evaluate the level of heads of households' perceptions of climate change governance practices in the study sites. It includes five levels (see Table 4.3).

Table 4.3 Perception Index Levels

Categories	Very good	Good	Medium	Poor	Very poor
Scale	1	0.75	0.5	0.25	0

Satisfaction Index

The formula for calculating satisfaction index is as follows:

$$\text{WAI} = \left(1.00 \cdot f_1 + 0.75 \cdot f_2 + 0.50 \cdot f_3 + 0.25 \cdot f_4 + 0 \cdot f_5\right) / \text{N}$$

WAI is the weighted average index ($0 \leq \text{WAI} \leq 1$); f_1 is the frequency of first scale choice; f_2 is the frequency of second scale choice; f_3 is the frequency of third scale choice; f_4 is the frequency of fourth scale choice; and f_5 is the frequency of fifth scale choice. The overall assessment (OA) is calculated from the average WAI value (Nooriafshar, Williams, & Maraseni, 2004).

In the case of qualitative data, NVivo12 software facilitates the management and analysis of qualitative data (Bryman, 2012). This software manages data and ideas, queries data, visualizes data, and provides reports from data (Bazeley, 2007). Through this software, nodes and strings were developed depending on the narrative text, such as interviews and notes from the field study. Various systematic analyses and logical interpretations of data through analysis produce a better understating of the themes and issues in the relevant research. After the analysis of qualitative and quantitative data following various methods, the results were compared and contrasted for different aspects of the variables.

Results and Analysis

Climate Change Governance Practices at the Local Level

Effective climate change governance practices are important to achieve better results from climate risk events-related projects at the local level. Climate change governance practices are different at the local (subdistrict) level than those at the national level. The majority of the activities related to climate change are managed and implemented by the Upazila (subdistrict) administration headed by the Upazila Nirbahi Officer (UNO). At the local level, many issues are considered based on the local context. Five criteria—accountability, transparency, participation, government effectiveness, and resources—were chosen in this study to evaluate climate change governance practices. These criteria are keys to providing effective responses to climate risk events-related issues that meet the expectations of local communities. Respondents were asked to provide their levels of agreement and disagreement against several statements related to these criteria.

Accountability

Accountability is often lacking in various climate change-related projects at the local level. Various climate change-related projects are mainly implemented by local construction firms, resulting in them focusing on their own interests rather than maintaining quality in their work. Thus, people were not getting the expected benefits because of the absence of proper implementation strategies in various government projects.

Table 4.4 illustrates the WAI and OA results regarding the accountability levels of climate change-related activities. Local people sometimes visited these projects in their areas to see the work progress, and they deemed the OA value weak (WAI = 0.31). Moreover, locals' impressions emerged from visits to GO and NGO offices to observe work status. The OA values were weak; the WAI values were 0.26 and 0.38, respectively. Similarly, one indicator (e.g. 'government officials' visit') was also weak (WAI = 0.31), while another (e.g. 'NGO officials' visit') was strong (WAI = 0.48). This shows a significant difference in WAI (0.31 versus 0.48) between the GO and NGO official visits at the local level (see Figure 4.3). NGO officials visited their climate project sites more compared with governmental officials. Moreover, local people perceived that the existing climate change governance practices in terms of accountability were unlikely to bring significant change to locals' lives and livelihoods.

Transparency

In the beginning, local administration receives various work orders from the union councils under its jurisdiction relating to climate risk events-induced disasters. Then the Upazila administration selects works on a priority basis and takes necessary measures to move forward for the implementation of those works from Upazila to union and village level.

Table 4.4 Perceptions of Respondents on Accountability in Climate Change-Related Activities

Issues	WAI	OA
Locals' visits to climate projects[5]	0.31	W
Locals' visits to governmental organization (GO) offices	0.26	W
Locals' visits to non-governmental organization (NGO) offices	0.38	W
GO officials' visits to climate projects	0.31	W
NGO officials' visits to climate projects	0.48	M

Note: very weak (VW): 0.01–0.2; weak (W): 0.21–0.4, medium (M): 0.41–0.6; strong (S): 0.61–0.8; very strong (VS): 0.81–1. WAI: weighted average index, OA: overall assessment.

Overall WAI for accountability

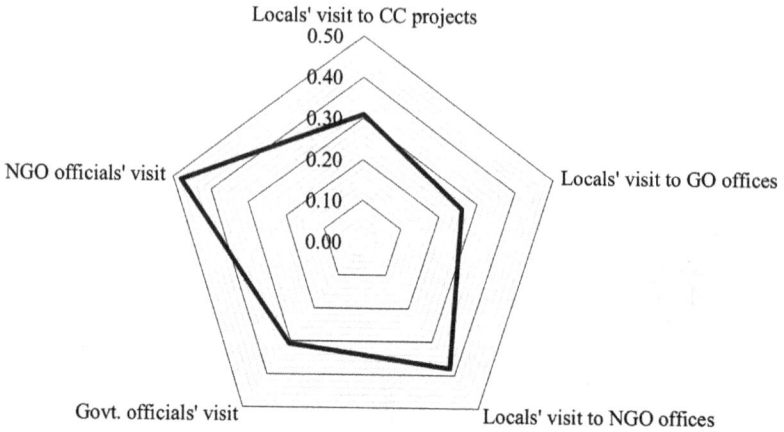

Figure 4.3 Respondents' Perceptions of Accountability.
Source: Author's fieldwork (2019).

A government officer of Hatiya Upazila explains in October 2019:

> Every government program has some specific formats, and local government authorities follow those formats to implement various climate change-related projects and programs.

In the Colonial Period (1757–1947), district-level local government departments demanded various resources from the central government, and the central government would usually take executive decisions regarding these demands. The situation has not changed much in present times in terms of how decisions are made on resource allocation. There is a lack of transparency and less inclusiveness in the activities of the local administration on the climate change issue. The respondents were asked to indicate their perceptions of transparency. To assess this, three issues were considered: support for climate victims, access to information, and timeliness (Barua, Fransen, & Wood, 2014). Based on a Likert scale (Bañas et al., 2020; Cen-López & Aguilar-Perera, 2020), more than half (55 percent) of the respondents perceived that the 'level of effort to help the climate victims' issues was 'bad' (see Figure 4.4). Around one-fourth of respondents nominated 'medium,' and 22 percent noted it as 'good.' In the case of 'access to information,' around half (44 percent) of respondents nominated it as 'bad,' while 38 percent indicated it was 'good.' When the respondents were questioned about the 'timeliness' of climate change-related activities, more than half (57 percent) noted it was 'bad.' In contrast, only one-fourth (25 percent) of respondents perceived it as 'good.' More than half of the respondents noted that transparency was not

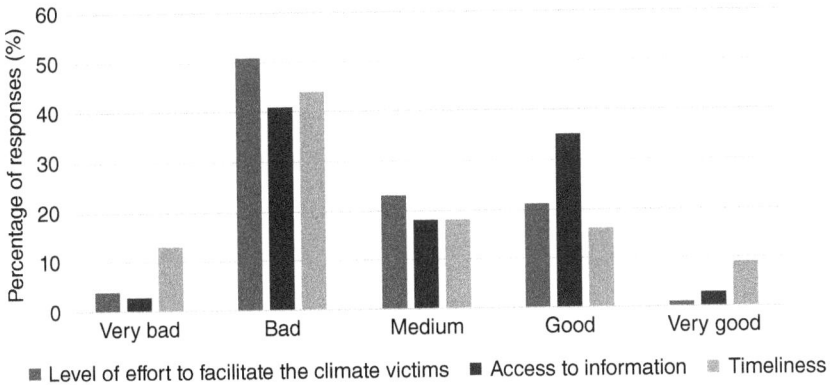

Figure 4.4 Respondents' Perceptions of Transparency.
Source: Authors' fieldwork (2019).

adhered to properly at the local level. A local journalist of Hatiya explains this to the authors in the study site in October 2019:

> Actually, we don't see transparency in managing climate change-related projects and programs in our areas. The local construction firms whom the local governments or central government nominates do whatever they want, and they don't maintain the quality of work. The allocated funds aren't used properly due to the misdeeds of various government departments and political leaders.

Participation

Participation of relevant local stakeholders in project implementation brings better outcomes. Following the government format, the Upazila administration incorporates stakeholders from different levels to implement climate risk events-related works effectively. However, the participation of community representatives is often ineffective. The local government organizations (GOs) are often incapable of dealing with diverse participants and stakeholders. The UNO, the head of local administration in Hatiya Upazila, described this participation issue in October 2019:

> We actually try to incorporate various stakeholders such as government officials, NGO officials, local political leaders, community-based organization (CBO) representatives, and others in various climate risk events-related programs. However, we can't do accordingly in many cases due to a lack of interest in participation. Of course, with the available resources, we can get better results by implementing various projects properly.

We always incorporate local-level representatives such as Union Council chairmen members, representing some of the specific communities. We also invite a few NGOs that are working at the grassroots level. Apart from this, we also invite a few farmers. So, we try to listen to various groups and their problems and feedback to overcome the issues.

Field survey results show that around half (48 percent) of the respondents perceived local participation as weak (very weak + weak)[4] (see Table 4.5). More than one-third (37 percent) of respondents thought local participation was 'medium,' and only 15 percent indicated it was 'strong.' More than half (54 percent) of the respondents mentioned the involvement of stakeholders in climate change-related activities was 'very weak.' Moreover, more than one-third (37 percent) of respondents perceived it as 'very strong.' The WAI values for the participation of locals and various stakeholders are 0.36 and 0.45, respectively (see Table 4.5). So, respondents demonstrated that stakeholders, apart from locals, participated more in climate change-related activities.

Therefore, community member participation in climate change governance practices at the local level does not operate at the expected level (WAI 0.36). Nevertheless, the OA relating to the involvement of various stakeholders in climate change-related activities (i.e., planning and implementation) is comparatively good (WAI 0.45) (see Table 4.5).

Table 4.5 Perceived Views of Locals on Participation in Climate Change-Related Activities

Statements	Responses						
	Very weak (%)	*Weak (%)*	*Medium (%)*	*Strong (%)*	*Very strong (%)*	*WAI*	*OA*
Participation of community people in climate change govern-ance practices in terms of engagement and contribution	28	20	37	12	3	0.36	W
Involvement of stakeholders in climate change-related activities, such as in planning and implementation	8	46	9	33	4	0.45	M
Total respondents	181						

Source: Author's fieldwork (2019).

Note: very weak (VW): 0.01–0.2; weak (W): 0.21–0.4, medium (M): 0.41–0.6; strong (S): 0.61–0.8; very strong (VS): 0.81–1. WAI: weighted average index, OA: overall assessment.

Government Effectiveness

Government effectiveness represents the institutional capacity for policy formulation and implementation, the quality of public services, the degree of independence from political interventions, managing resources, addressing technical issues, and the level of government commitment to policy (Barua et al., 2014; Kaufmann, Kraay, & Mastruzzi, 2010; Kaufmann et al., 2011). Government effectiveness is an important aspect affecting governance in a particular area. The local administration usually follows national-level policies and implementation strategies on climate change when implementing projects. They mainly follow instructions from the central government to ensure implementation at the field level. They also follow some generic responsibilities that address climate change, such as organizing campaigns and awareness-raising programs for the community (GoB, 2012). Respondents were asked to state their perceptions of government effectiveness in the study area. More than half (54 percent) of the respondents (strongly disagree + disagree) revealed that projects relating to climate change did not address the community's problems. However, one-third (33 percent) of respondents perceived that climate change-related activities were affected by political pressure (see Table 4.6).

Table 4.6 Perceived Views of Locals on Government Effectiveness in Climate Change-Related Activities

Issues	Responses						
	Strongly disagree (%)	Disagree (%)	Neither agree nor disagree (%)	Agree (%)	Strongly agree (%)	WAI	OA
Climate change-related projects addressed the problems of community people	11	43	13	23	10	0.45	M
Climate change projects were affected by political pressure	7	32	28	31	12	0.57	M
Proper policy and implementation of climate projects	10	46	15	25	4	0.42	M
The reliability of government commitment to policies	14	48	20	15	3	0.36	W

Source: Authors' fieldwork (2019).

Note: very weak (VW): 0.01–0.2; weak (W): 0.21–0.4, medium (M): 0.41–0.6; strong (S): 0.61–0.8; very strong (VS): 0.81–1. WAI: weighted average index, OA: overall assessment.

However, approximately half (43 percent) of the respondents indicated that climate risk events-related projects were adopted in the study area, while 39 percent disagreed with this. More than half (56 percent) of the respondents noted that climate risk events-related projects were implemented properly in the area by adhering to policies and implementation strategies, whereas one-third (33 percent) disagreed with this. Further, 62 percent of the respondents disagreed about 'the reliability of the government's commitment to policies, while a few (18 percent) supported it.

Resources

The UNO, the chief executive officer at the subdistrict level, mentioned that the Upazila administration was trying to support a large number of climate victims living in the area with limited government allocation. The UNO of Hatiya Upazila explains the scarcity of resources in October 2019:

> The government's allocations for managing the effects of climate change were inadequate and insufficient for the recipients. The Upazila administration couldn't implement various climate change-related projects due to insufficient funds. We receive funds from government departments, donors, and NGOs. These funds are thinly distributed to the Union Council authorities also.

The Union Council authorities could not perform their task properly because of corruption, partisan politics, and nepotism. As a result, vulnerable people would often not receive benefits from these types of government allocations. The UNO also demonstrated resource scarcity at the local level in October 2019:

> If the Upazila administration is allocated 100 blankets to provide the disaster-affected people, usually more than 1,000 people gather to get these blankets. So, one out of ten can receive a blanket. The people who aren't enlisted for the blankets and couldn't obtain blankets become angry and blame the government departments for this.

Local communities were asked about the availability of resources for work on climate change issues. Around half (49 percent) of the respondents, neither agreed nor disagreed: they rated the issue at a 'neutral' level (see Figure 4.5). Among the respondents, 35 percent disagreed, whereas 16 percent agreed that the available resources could be used properly to improve the existing situation. One of the respondents in the FGD session explains in Hatiya in October 2019:

> We come to know from various news media that a good amount of funds are being allocated to Bangladesh, but we don't see these funds in our areas. So we are in doubt about the destination and effectiveness of these funds.

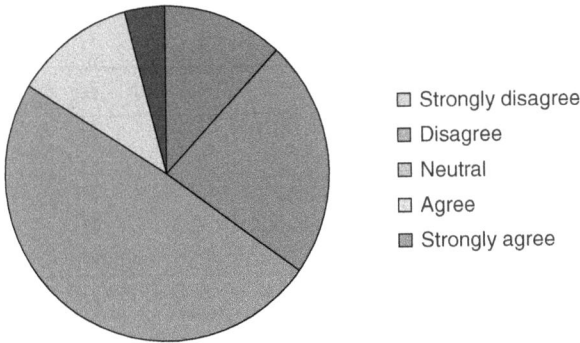

Figure 4.5 Respondents' Perceptions of Resource Availability.
Source: Authors' fieldwork (2019).

One of the elected representatives of Hatiya opines on October 2019 in the following way:

> We aren't getting the required support to tackle the climate change issue. We need more funds to tackle the impacts of climate change at the community level in this area. If we can make the best use of available resources in climate change activities, it can bring better outcomes for the communities.

Allocated resources are not enough to implement projects and programs relating to climate risk events at the local level. Moreover, available funds are not being used properly due to poor performance of local administrations, political interference, and misuse of funds at various levels of implementation.

Overall Climate Change Governance Status at the Local Level

Figure 4.6 shows the overall WAI results on climate change governance in the study area. For transparency, the OA result was rated as weak (WAI = 0.35), derived from the data in Figure 4.5 (i.e., transparency was not maintained

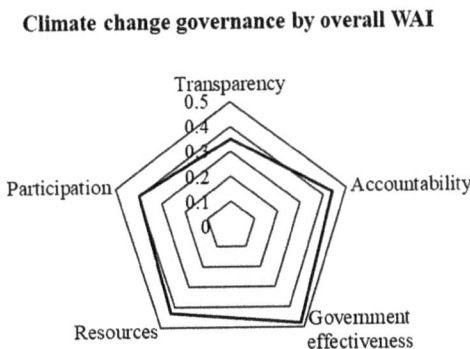

Figure 4.6 Climate Change Governance Status at the Local Level by Overall WAI.

Table 4.7 Climate Change Governance Status
by WAI and OA Results

Criteria	Total WAI	OA
Transparency	0.35	W
Accountability	0.44	M
Government effectiveness	0.45	M
Resources	0.43	M
Participation	0.40	W

Note: very weak (VW): 0.01–0.2; weak (W): 0.21–0.4,
medium (M): 0.41–0.6; strong (S): 0.61–0.8; very
strong (VS): 0.81–1. WAI: weighted average index,
OA: overall assessment.

properly in various climate change activities). Conversely, with respect to accountability, government effectiveness, and resources, these three criteria were comparatively better in the study area. The WAI results are 0.44, 0.45, and 0.43, respectively (see Table 4.7). WAI values for accountability, government effectiveness, and resources are calculated from Tables 4.3, 4.5, and Figure 4.5, respectively. According to the respondents, these three criteria improved gradually and made a significant positive impact on locals' lives and livelihoods.

Being a part of various activities relating to climate change, different stakeholders did not participate in local communities as much as expected. The OA result shows that participation at the community level was weak. The WAI value for participation was 0.40 (calculated from Table 4.5). Therefore, the overall WAI results demonstrate that the obtained values for 'accountability,' 'government effectiveness,' and 'resources' are better than those for 'transparency' and 'participation' (see Figure 4.6).

A community leader of Hatiya Island explains the concerns regarding climate change governance practices at the community level in October 2019:

> We are keen to provide relief during the disaster and post-disaster period. If we go through climate risk events-related adaptation practices in the planning stage, we can overcome many climate risk events-related challenges at the community level. However, these practices aren't started yet.

Perceptions about the Effectiveness of Climate Change Governance

In all, 98 percent of respondents agreed that they were affected by climate change-related events such as floods, cyclones, storm surges, saline water intrusion, etc. They were asked about their perceptions of the effectiveness of climate change governance. They were provided with 25 statements and asked to rank their levels of agreement or disagreement (see Table 4.8).

Government effectiveness is an important aspect affecting governance in a particular area. When implementing projects, the local administration usually follows national-level policies and implementation strategies on climate risk

Table 4.8 Effectiveness of Climate Change Governance as Perceived by Respondents

	Issues	*Responses*				
		Strongly agree (%)	*Agree (%)*	*Neither agree nor disagree (%)*	*Disagree (%)*	*Strongly disagree (%)*
1	Presence of institutional setting for climate change governance	7	52	3	29	9
2	Presence of adequate plans and programs	4	58	2	26	10
3	Policy and plans are in favor of people	5	48	4	31	12
4	Participation of various stakeholders is ensured	3	53	3	31	10
5	Climate risk events-related projects ensure the involvement of locals	3	57	3	29	8
6	Climate risk events-related projects involve locals in decision-making processes	4	56	2	25	13
7	Existing climate risk events-related projects are supportive for local livelihoods	3	50	7	26	14
8	Female members' participation is encouraged in various projects	5	70	1	21	3
9	Role plays in the overall development	7	65	4	17	7
10	Climate risk events-related projects are enhancing public awareness	6	79	3	11	1
11	Climate change governance has changed over time	4	62	10	13	11
12	Improved climate change governance is expected	20	68	2	7	3

(Continued)

Table 4.8 (Continued)

	Issues	Responses				
		Strongly agree (%)	*Agree (%)*	*Neither agree nor disagree (%)*	*Disagree (%)*	*Strongly disagree (%)*
13	Personal rights and responsibilities in climate governance practices are not well defined	5	28	8	34	25
14	Adequate fund allocation for climate projects	16	9	0	30	45
15	Proper use of climate risk events-related funds	4	7	0	36	53
16	Access to information on climate change projects	2	43	3	37	15
17	Timely completion of climate projects	2	13	3	41	41
18	Regular monitoring of climate projects	3	19	2	41	35
19	The climate change governance system is effective	5	37	3	23	32
20	Rules and policies are well informed	3	41	6	43	7
21	Employment opportunities for locals have not been created	10	25	3	27	35
22	Vulnerability and uncertainty of livelihood have not decreased	10	26	4	22	38
23	Confrontations are observed among government and non-government organizations in climate change-related projects	2	43	21	18	16
24	Stakeholder capacity building	4	35	15	33	13
25	Institutional capacity building	0	27	13	46	14

Source: Author's fieldwork (2019).

events. They mainly follow instructions from the central government to ensure implementation at the field level. They also follow some generic responsibilities that address climate risk events, such as organizing campaigns and awareness-raising programs for the community people. The results from the field survey showed that at least 50 percent of responses supported the effectiveness of climate change governance in 12 out of 25 issues (see Table 4.8).

This study demonstrated the generally low level of local community participation in climate change governance. Overall, community perceptions and views were frequently not considered when planning and implementing various climate risk events-related projects at the local level. Community participants claimed that they did not receive the required support in climate risk events-related issues from local administrations because of the limited capacity to solve climate risk events-related issues, even though they were affected significantly by climate risk events. Several initiatives relating to climate risk events were observed nationally, but climate change governance does not function well at the local level.

Challenges for Local Climate Change Governance

At the local level, challenges are observed beyond the community's capacity to respond to (Keskitalo & Kulyasova, 2016). Climate change governance is important at the local level to meet the challenges of climate risk events-related issues. In Hatiya, the Upazila administration has formed committees to work on disaster events from the Upazila to the union level. They have had some experiences tackling climate risk events, but they have not been a great focus of their actions. The present study identified various challenges in local climate governance practices. According to Sippel and Jenssen (2009), these can be classified as 'economic,' 'informational,' 'institutional,' and 'political/cultural' (see Table 4.9).

The local administration has a low level of understanding of climate risk events-related issues. However, local government officials were assigned to conduct activities relating to the climate change issue. Local political pressure groups, government departments, and other beneficiaries are not making an adequate effort to properly implement these projects. Public awareness of climate risk events issues is also important at the local level; however, local people lack awareness about how to deal with specific climate change issues. Moreover, they cannot ensure their rights at the community level in various climate change-related projects.

Consequently, they do not see the expected outcomes from projects implemented by GOs and NGOs. Moreover, community perceptions and views have not always been considered when planning and implementing projects related to climate risk events at the local level. Suppose programs relating to climate risk events are implemented properly with adequate consideration of local knowledge. In that case, they could bring about significant changes to the community resilience required to face climate risk events-related effects. Therefore, the proper implementation of various climate risk events-related projects by GOs and NGOs is important to enhance local community resilience.

Table 4.9 Challenges for Local Climate Change Governance

Class types	Challenges
Economic	Inadequate funds for climate change-related projects
	Climate change issues receive less focus
Informational	Lack of trained people
	Lack of expertise
	Public participation is not always maintained
	Minimal monitoring and evaluation
Institutional	Little access to information
	Dependency on national government
	Lack of effective local governance system
	Lack of coordination
	Less effort to incorporate into a structured
	Lack of cooperation
	Weak regulatory framework
Political/Cultural	Less control over utilities
	Lack of climate change policy for the local level
	Local political pressure
	Less prioritized issue

Source: Author's fieldwork (2019).

Discussion and Conclusion

Various local organizations in the form of public, private, and civic bodies make efforts to assess the effects of climate change at the community level (Rahman & Huang, 2019). The communities of Hatiya Island are highly vulnerable to climate risk events, more so than the mainland. This area has been affected by severe climate risk events in the past, and the number of these events has increased, causing enormous damage to lives and property. Thus, it is vital to ensure effective climate change governance practices to properly implement projects and programs addressing climate risk events at the local level.

The local government departments emphasize climate change governance criteria while operating climate risk events-related activities. However, they cannot manage projects properly in many cases due to a shortage of funds and various limitations such as the lack of skilled staff and required political support at the local level (Leon, 2016). In addition, accountability is not well developed among stakeholders at the local level. Transparency in climate risk events-related projects is necessary because this helps maintain accountability and participation (Biermann et al., 2010; Gupta & Mason, 2016). This issue is followed in many projects and programs at the local level. However, more possibilities must be considered at various levels for the successful completion of climate risk events-related projects.

Sometimes, local government departments do not integrate the knowledge of community people; this is vital for the successful completion of locally led climate projects. At times, they do not examine the local context, culture, dynamics, or needs to ensure the effectiveness of climate risk events-related projects. According to the KIs local-level organizations cannot differentiate between climate finance and development finance when funds are channeled into various projects; this creates challenges for other relevant stakeholders.

NGOs, like other stakeholders, seek to reduce insecurity and protect and support people who may be vulnerable to climate change (Bannerman et al., 2011). NGOs are performing better than government departments at the local level. Therefore, the capacity of local-level GOs must be strengthened to properly implement climate change-related projects. Improved implementation will lead to greater benefits for the communities from these projects. Despite the various common challenges at the local level (Keskitalo & Kulyasova, 2016), climate change governance practices are considered 'moderate,' based on the perceptions of locals. However, a combined effort, especially by GOs and NGOs at the community level, may substantially reduce climate risk events-related effects on lives and livelihoods. All relevant stakeholders must come forward and make collective efforts to tackle climate risk events in this area for this to happen.

Although community people are adopting various measures to overcome the climate risk events-related challenges, these measures are not bringing the expected outcomes. In general, it was found that community perceptions and views were often not considered when planning and implementing various climate risk events-related projects at the local level. Climate risk events are gaining importance for stakeholders at the community level because of increased exposure to climate risk events-induced disasters: daily life is becoming more difficult. Community members face the adverse effects of climate risk events because they are not yet prepared to tackle the issues properly. However, the community people perceive climate risk events-related issues better at present. They can undertake measures themselves, reducing adverse effects through effective climate change governance practices at the community level. The ultimate victims of climate risk events are community people. So, by using the findings of these types of research, the capacity of the community people of other parts of Bangladesh can be scaled up.

Acknowledgments

This book chapter is based on the first author's Ph.D. research in the Sustainable Resource Management program under the School of Environmental and Life sciences at the University of Newcastle, Australia. This research was conducted under the University of Newcastle's UNIPRS and UNRSC scholarships program.

Notes

1 The Upazila is an important administrative unit in Bangladesh comprising several 'unions,' which in turn are composed of several villages (Nishat et al., 2009).
2 Char is a tract of land surrounded by the waters of an ocean, sea, lake, or stream. Chars in Bangladesh are observed as a 'by-product' of the rivers' hydro-morphological activities (Chowdhury & Chowdhury, 2014).

3 The union parishad (council) is the lowest level of local government. It usually covers a few villages.
4 Very weak and weak are considered a 'weak' category to efficiently demonstrate the evaluation results.
5 Climate change-related projects are conducted by GOs and NGOs and provide mainly adaptation benefits to the most climate-vulnerable people of Bangladesh. These projects are financed by the central government or development partners.

References

Abels, G. (2007). Citizen involvement in public policy-making: Does it improve democratic legitimacy and accountability? The case of PTA. *Interdisciplinary Information Sciences, 13*(1), 103–116.
Alam, K., Shamsuddoha, M., Tanner, T., Sultana, M., Huq, M. J., & Kabir, S. S. (2011). The political economy of climate resilient development planning in Bangladesh. *Institute of Development Studies Bulletin, 42*(3), 52–60.
Allred, S., Schneider, R., & Reeder, J. (2016). The role of natural resource professionals in addressing climate change. *Climate, 4*(38), 1–17. doi:10.3390/cli4030038.
Ayers, J., Huq, S., Wright, H., Faisal, A. M., & Hussain, S. T. (2014). Mainstreaming climate change adaptation into development in Bangladesh. *Climate and Development, 6*(4), 293–305. doi:10.1080/17565529.2014.977761.
Balsiger, J., & Van Deveer, S. D. (2012). Navigating regional environmental governance introduction. *Global Environmental Politics, 12*(3), 1–17.
Bangladesh Bureau of Statistics. (2015). Bangladesh population and housing census 2011: Community report (Zila: Noakhali). Retrieved August 14, 2021 from www.bbs.gov.bd
Banglapedia. (2014). Hatiya Upazila. In S. Islam (Ed.), *Banlapedia: National encyclopedia of Bangladesh*. Dhaka, Bangladesh: Asiatic Society of Bangladesh.
Bannerman, M., Rashid, M. H. O., & Rejve, K. (2011). NGO-government partnerships for disaster preparedness in Bangladesh. Retrieved 20 August, 2021 from https://odihpn.org/magazine/ngo%c2%96government-partnerships-for-disaster-preparedness-in-bangladesh/
Barua, P., Fransen, T., & Wood, D. (2014). Climate policy implementaion tracking framework. Retrieved August 16, 2021 from http://wri.org/publication/climate-policy-tracking
Bazeley, P. (2007). Perspectives: Qualitative computing and NVivo. In P. Bazeley (Ed.), *Qualitative data analysis with NVivo*. London: Sage.
Bhuiyan, S. (2015). Adapting to climate change in Bangladesh. *South Asia Research, 35*(3), 349–367. doi:10.1177/0262728015598702.
Biermann, F., Betsill, M. M., Gupta, J., Kanie, N., Lebel, L., Liverman, D., Schroeder, H., Siebenhüner, B., and Zondervan, R. (2010). Earth system governance: A research framework. *International Environmental Agreements: Politics, Law and Economics, 10*(4), 277–298.
Blair, A. A. C. (2017). *Climate change threats and adaptation responses in small island developing states: A comparative analysis of Antigua & Barbuda and Vanuatu.* (PhD), University of Newcastle, Australia.
Boyd, E., & Ghosh, A. (2013). Innovations for enabling urban climate governance: Evidence from Mumbai. *Environment and Planning C: Government and Policy, 31*(5), 926–945. doi:10.1068/c12172.

Bryman, A. (2012). *Social research methods* (4th ed.). New York, USA: Oxford University Press.

Bulkeley, H. (2010). Cities and the governing of climate change. *Annual Review of Environment and Resources*, 35(1), 229–253. doi:10.1146/annurev-environ-072809-101747

Bulkeley, H., & Kern, K. (2006). Local government and the governing of climate change in Germany and the UK. *Urban Studies*, 43(12), 2237–2259

Chowdhury, S. Q., & Chowdhury, M. H. (2014). Char. In S. Islam (Ed.), *Banglapedia: The national encyclopedia of Bangladesh*. Dhaka: Asiatic Society of Bangladesh.

Creswell, J. W. (2013). *Research design: Qualitative, quantitative and mixed methods approaches*. California, USA: SAGE Publications Inc.

Fatti, C. E., & Patel, Z. (2013). Perceptions and responses to urban flood risk: Implications for climate governance in the south. *Applied Geography*, *36*, 13–22. doi:10.1016/j.apgeog.2012.06.011.

Forino, G., von Meding, J., & Brewer, G. J. (2015). A conceptual governance framework for climate change adaptation and disaster risk reduction integration. *International Journal of Disaster Risk Science*, 6(4), 372–384. doi:10.1007/s13753-015-0076-z.

Fröhlich, J., & Knieling, J. (2013). Conceptualising climate change governance. In J. K. A. W. L. Filho (Ed.), *Climate change governance* (pp. 9–26). Verlag Berlin Heidelberg: Springer.

GoB. (2012). *Bangladesh climate public expenditure and institutional review*. Dhaka, Bangladesh: Government of the People's Republic of Bangladesh.

GoB. (2015). *Bangladesh population and housing census 2011: Community Report (Zila: Noakhali)*. Dhaka, Bangladesh: Bangladesh Bureau of Statistics (BBS).

GoB. (2017). *Proceedings of upazila inception workshop on integrating community based adaptation Noakhali*. Dhaka, Bangladesh. Retrieved August 10, 2021 from www.bd.undp.org/.../Inception%20workshop%20Report,%20Hatiya,%20Noakhali%2

Gupta, A., & Mason, M. (2016). Disclosing or obscuring? The politics of transparency in global climate governance. *Current Opinion in Environmental Sustainability*, 18, 82–90. doi:10.1016/j.cosust.2015.11.004.

Ha, H. D., & Thang, T. N. (2017). Fishery communities' perception of climate change effects on local livelihoods in Tam Giang Lagoon, Vietnam. In Shivakoti, G., Thang, T., Dung, N., Hulse, D., & Sharma, S. (Eds.), *Redefining diversity & dynamics of natural resources management in Asia, Volume 3* (pp. 111–124). Netherlands: Elsevier.

Huntjens, P., & Zhang, T. (2016). *Climate justice: Equitable and inclusive governance of climate change*. Retrieved August 11, 2021 from http://www.thehagueinstitute forglobaljustice.org/wp-content/uploads/2016/04/Climate-Justice-April-2016.pdf

Islam, M. A. (2012). Innovative approach of ecosystem-based adaptation in coastal Bangladesh. In N. Uy, & R. Shaw (Eds.)., *Ecosystem-based adaptation Community, Environment and Disaster Risk Management, Volume 12* (pp. 85–106). Bingley: Emerald Group Publishing Limited. doi:10.1108/S2040-7262(2012)000001201.

Islam, N. (2010). Climate change and the challenge to social policies. *Global Social Policy*, *10*(1), 18–20.

Jodoin, S., Ducky, S., & Lofts, K. (2015). Public participation and climate governance: An introduction. *Review of European Community & International Environmental Law*, 24(2), 117–122.

Kaufmann, D., Kraay, A., & Mastruzzi, M. (2010). The worldwide governance indicators: Methodology and analytical issues. Retrieved 20 August, 2021 from www. govindicators.org

Kaufmann, D., Kraay, A., & Mastruzzi, M. (2011). The worldwide governance indicators: Methodology and analytical issues. *Hague Journal on the Rule of Law*, 3(02), 220–246. doi:10.1017/s1876404511200046.

Keskitalo, E. C. H., & Kulyasova, A. A. (2016). The role of governance in community adaptation to climate change. *Polar Research*, 28(1), 60–70. doi:10.1111/ j.1751-8369.2009.00097.x.

Khongsatjaviwat, D., & Routray, J. K. (2015). Assessment of capacities of TAOs and CBOs for local development in Thailand. In Dutt, A. K., Noble, A. G., Costa, F. J., Thakur, S. K., Thakur, R., & Sharma H. S. (Eds.), *Spatial diversity and dynamics in resources and urban development: Volume 1: Regional Resources* (pp. 501–527). Dordrecht: Springer.

Leon, A. G. (2016). Climate change governance in Megadiverse countries: The case of REDD+in Latin America. *International Journal of Climate Change: Impacts and Response*, 8(3), 61–80.

Marquardt, J. (2017). Conceptualizing power in multi-level climate governance. *Journal of Cleaner Production*, *154*, 167–175. doi:10.1016/j.jclepro.2017.03.176.

McCarney, P., Blanco, H., Carmin, J., & Colley, M. (2011). Cities and climate change. In W. D. S. C. Rosenzweig, S. A. Hammer, S. Mehrotra (Ed.), *Climate change and cities: First assessment report of the urban climate change research network* (pp. 249–269). United Kingdom: Cambridge University Press, Cambridge.

Ministry of Environment, Forest and Climate Change. (2009). *Bangladesh climate change strategy and action plan 2009*. Dhaka, Bangladesh: MoEFCC, Government of the People's Republic of Bangladesh.

Momtaz, S., & Shameem, M. (2015). *Experiencing climate change in Bangladesh: Vulnerability and adaptation in coastal regions*. Netherlands: Elsevier.

Moniruzzaman, A. N. M. (2019). How climate change is affecting livelihoods in Bangladesh? [video]. My Justice. Retrieved August 21, 2021 from https://justicehub. org/article/major-general-anm-muniruzzaman-climate-change-is-affecting-livelihoods-in-bangladesh/

Nishat, A., Hussein, S. G., Matin, M. A., Molla, A. R., & Tellam, I. (2009). Adapting to climate variability and change in Bangladesh. In T. Devisscher, O'Brien, O'Keefe, & I. Tellam (Eds.), *The adaptation continuum groundwork for the future* (pp. 15–30). The Netherlands: ETC Foundation.

Nooriafshar, M., Williams, R., & Maraseni, T. (2004). The use of virtual reality in education. Paper presented at the American Society of Business and Behavioral Sciences, 2004 Seventh Annual International Conference, Cairns, Australia.

Olsson, L., Opoondo, M., Tschakert, P., Agrawal, A., Eriksen, S. H., Ma, L., Ma, S., Perch, L., & Zakieldeen, S. A. (2014). Livelihoods and poverty. Retrieved August 22, 2021 from https://www.ipcc.ch/pdf/assessment-report/ar5/wg2/WGIIAR5-Chap13_FINAL.pdf

Onwuegbuzie, A. J., & Collins, K. M. T. (2007). A typology of mixed methods sampling designs in social science research. *The Qualitative Report*, *12*(2), 281–316.

Pakzad, P., Osmond, P., & Corkery, L. (2016). Developing key sustainability indicators for assessing green infrastructure performance. Paper presented at the International High- Performance Built Environment Conference—A Sustainable Built Environment Conference, Sydney, 17–18 December 2016, Series (SBE16).

Parvin, G. A., Takahashi, F., & Shaw, R. (2008). Coastal hazards and community-coping methods in Bangladesh. *Journal of Coastal Conservation*, 12(4), 181–193. doi:10.1007/s.

Patt, A. (2017). Beyond the tragedy of the commons: Reframing effective climate change governance. *Energy Research & Social Science*, 34, 1–3. doi:10.1016/j.erss.2017.05.023.

Patton, M. Q. (2002). *Qualitative research and evaluation methods* (3rd ed.). Thousand Oaks: Sage.

Pelling, M. (2011). *Adaptation to climate change: From resilience to transformation*. London: Routledge.

Pervin, M. (2013). Mainstreaming climate change resilience into development planning in Bangladesh. Retrieved August 22, 2021 from http://pubs.iied.org/10045IIED

Rabe, B. G. (2007). Beyond Kyoto: Climate change policy in multilevel governance systems. *Governance: An International Journal of Policy, Administration, and Institutions*, 20(3), 423–444. doi:10.1111/j.1468-0491.2007.00365.x.

Rahman, M. M., & Huang, D. (2019). Climate governance initiatives: Snapshots from Bangladesh. *Journal of Geoscience and Environment Protection*, 07(09), 131–147. doi:10.4236/gep.2019.79010.

Rai, N., Huq, S., & Huq, M. J. (2014). Climate resilient planning in Bangladesh: A review of progress and early experiences of moving from planning to implementation. *Development in Practice*, 24(4), 527–543. doi:10.1080/09614524.2014.908822.

Regmi, B. R., & Bhadari, D. (2012). Climate change governance and funding dilemma in Nepal. *TMC Academic Journal*, 7(1), 40–55.

Schreurs, M. A. (2010). Multi-level governance and global climate change in east Asia. *Asian Economic Policy Review*, 5(1), 88–105. doi:10.1111/j.1748-3131.2010.01150.x

Scobie, M. (2017). Accountability in climate change governance and Caribbean SIDS. *Environment, Development and Sustainability*, 20(2), 769–787. doi:10.1007/s10668-017-9909-9.

Sippel, M., & Jenssen, T. (2009). What about local climate governance? A review of promise and problems. *Social Science Research Network (November) [online]*. Retrieved August 22, 2021 from https://mpra.ub.uni-muenchen.de/20987/

Union Council. (2009). *Workplan of union disaster preparedness*. Retrieved August 20, 2021 from Char Iswar Union Council, Hatiya, Noakhali, Bangladesh.

Union Council. (2019). *Union disaster management committee emergency plan (in Benglai)*. Retrieved August 19, 2021 from Hatiya, Noakhali, Bangladesh.

World Bank. (2000). Bangladesh: Climate change and sustainable development. Retrieved August 12, 2021 from http://documents.worldbank.org/curated/en/906951468743377163/pdf/multi0page.pdf

5 Gender, Climate Justice, and (Un)Sustainable Development in Indonesia

Ann R. Tickamyer and Siti Kusujiarti

Introduction

Indonesia faces multiple challenges, both geophysical and socio-political, from impending climate change. As a tropical island nation located in the Pacific Ring of Fire, it is particularly vulnerable to the disasters, extreme weather events, and ecological and environmental depredations that rising temperatures bring (Asian Development Bank [ADB], 2021). As an emerging economy that has made significant progress in achieving development goals of increasing economic growth, reducing poverty, and nurturing a new democratic polity, climate change threatens its economic and political stability. Finally, goals of equality, empowerment, and social justice for women and cultural and ethnic minorities are jeopardized by the uneven causes and consequences of anthropogenic climate change that make these groups particularly vulnerable. In this chapter, we argue that three pillars of development – social inclusion and equality, economic development, and environmental sustainability (Sachs, 2012) – are inextricably linked generally and specifically in Indonesia, with its future depending on the successful integration of all three. We demonstrate these connections by showing how gender justice and women's empowerment are critical for development efforts to plan, mitigate, and adapt to climate change across diverse landscapes of risk – riskscapes – within the archipelago. Conversely, where equal rights for women and other groups lag, sustainable development will suffer.

The example of Indonesia is important in its own right and as a case study and exemplar for other nations. As a large, geographically, and socially diverse nation, the example permits the comparison of differential risks, vulnerabilities, and outcomes experienced in different locales. Unsustainable development practices impede progress toward both gender and climate justice. This chapter builds on past studies of the ways gender analysis is critical to understanding and implementing development policies and programs that favor sustainability and address climate change in Indonesia and beyond.

DOI: 10.4324/9781003216476-6

Three Pillars of Development

The United Nations' Sustainable Development Goals (SDGs) adopted in 2015 clearly link social justice, economic development, and environmental protection and sustainability in 17 statements that map an agenda for the future of the planet and its inhabitants (United Nations Department of Social and Economic Affairs, n.d.). SDGs follow the conclusion of the Millenium Development Goals (MDGs) in 2015, lauded as a success, but limited and highly variable in their actual accomplishment (Sachs, 2012). SDGs elaborate, extend, and interrelate many of the initial goals, implicitly and explicitly demonstrating their connections. While multiple goals deal with the three pillars of human health and well-being: economic development, environmental sustainability, and social inclusion, three specific goals speak directly to the issue at hand. Goal 5 specifies "Achieve gender equality and empower all women and girls"; Goal 8 states, "Promote sustained, inclusive and sustainable economic growth, full and productive employment, and decent work for all"; and Goal 13 specifies "Take urgent action to combat climate change and its impacts." Each of these goals has relevance for the Republic of Indonesia in its efforts to emerge from the shadows of poverty, colonialism, authoritarian rule, and environmental destruction. As an emerging economic power with the world's fourth-largest population and a history of leadership among nonaligned nations, it has immense potential for global prominence and leadership. Economic growth has steadily reduced poverty and increased per capita income; the transition to democratic rule with the fall of the New Order regime has increased representation and participation, and women's increasing educational attainment and activism have challenged highly paternalistic and patriarchal ideologies and practices embedded in both the state and religion. Government agencies have made detailed plans to monitor progress toward SDGs and have ongoing efforts to collect relevant data necessary to monitor progress (Reagan, 2018).

However, the nation is also saddled with the legacies of its past and ongoing challenges to overcome them. The positive trends are put into question by countervailing tendencies that make progress uneven at best. Poverty and inequality remain widespread and are increasing with the COVID-19 epidemic and its economic consequences (OECD, 2021). The majority of workers are outside the formal economy and experience great precarity, with approximately 60% of the labor force classified as informal by the International Labor Organization and women are disproportionately represented (ILO, n.d.). Democracy is undermined by extreme corruption that is little better than the situation that toppled the authoritarian New Order regime in 1998. Fragmentation of the state from devolution and decentralization following the overthrow of the New Order results in the highly uneven implementation of national goals and policies (Aspinall, 2018).

Women remain disadvantaged in the household, community, and society. Despite the introduction of gender quotas in parliamentary elections and promises to implement gender mainstreaming across policy platforms and programs, they are underrepresented in politics and political arenas (Tickamyer et al., 2014; Wornell et al., 2015). Patriarchal practices hold sway in civic society, family, and household, and efforts to reform the marriage law to a more egalitarian form have dragged on for decades. The spread of conservative forms of Islam has introduced new dilemmas and problems for women's representation and participation in public spheres (Robinson, 2018; Tickamyer & Kusujiarti, 2012).

Looming over all potential for progress are the threats to health and safety from environmental destruction brought about by misuse of natural resources, and increasingly, by climate change (Djalante, 2018). The Indonesian glass is half empty and half full. Progress on all three pillars generally, and SDGs specifically, is threatened by the failures alluded to above, both natural and anthropogenic disasters, and above all, by the relentless march of climate change.

Environment, Disaster, and Climate Change

As an island nation located between the Indian and Pacific Oceans in the highly unstable "Ring of Fire," Indonesia is particularly vulnerable to both quick and slow-onset natural disasters and climate change. The archipelago is composed of well over 17,000 islands with many thousands of miles of coastline and low-lying areas vulnerable to storms, sea-level rise, and flooding. The magnitude of death and destruction from the 2004 Indian Ocean earthquake and tsunami (Frankenberg et al., 2011) brought international attention to the devastation in the region, but these events are not unique, nor are they the only sources of disaster. Numerous forms of disaster, both natural and human origin, including earthquakes, volcano eruptions, landslides, tsunamis, cyclones, floods, deforestation, fires, oil and chemical spills, and mudflows, are standard for the nation. One estimate states that Indonesia averaged 289 disasters per year over a recent 30-year period with approximately 8,000 deaths annually (gfdrr.org, n.d.). The most recent report from 2020 exceeds this average with 409 deaths and disappearances and 6,451,439 people affected and evacuated due to disasters (UN Sustainable Development Goals, 2021). The estimated 160 million deaths and many more casualties in Aceh on the island of Sumatra from the tsunami and earthquake, while extreme, do not define the extent of the nation's vulnerability.[1]

Climate change promises to increase the numbers and intensity of disastrous events with mounting evidence that they are reciprocal influences: the disaster is both product and driver of climate change (Hore et al., 2018, National Academies of Sciences, Engineering, and Medicine, 2016; Schipper & Pelling, 2006). No place better illustrates this as well as the complex interconnections between human and natural disasters (Freudenberg et al., 2008) than the capital city, Jakarta on the island of Java. Sprawling, coastal, and densely populated, Jakarta is now estimated to be 40% below sea level from a combination of rising oceans and subsidence from excessive pumping of groundwater to meet the needs of an ever-growing population (NASA Earth

Observatory, 2021). The resulting devastating annual flooding and extreme air pollution from uncontrollable traffic and congestion make the city an ongoing disaster site with many casualties that only promises to get worse. The extent and intractability of the problem have resulted in the government announcing plans to relocate the capital to another island.

The problems that Jakarta illustrates are repeated across the archipelago manyfold in one form or another. Economic growth and development have catapulted Indonesia from a poor, developing nation into the category of upper middle income in the global economy. While the gains are real, impressive, and with significant benefits for a growing middle class, growth has been largely predicated on exploitation and mismanagement of natural resources that have resulted in highly uneven development in who benefits and with what costs. Since 1999, poverty has been reduced from over 24% to just under 10% but then has risen again substantially from the impacts of the COVID pandemic. Throughout, there have been persistent high levels of inequality and precarity (OECD, 2021). The World Bank estimates that climate change could cost GDP between 2.5 and 5%, with the biggest impacts on the poor (World Bank, n.d.). Extreme environmental damage to one of the premier reservoirs of tropical forests and biodiversity on the planet, massive pollution of air and water, and widespread destruction of habitat and biodiversity are endemic (Margono et al., 2014; Wells et al., 1999). Deforestation makes it one of the highest emitters of greenhouse gases in the world (OECD, 2021). Reining in misuse and mismanagement remains ephemeral and aspirational. The overthrow of the 32-year Suharto led New Order government in 1998 was fueled by disgust with the corruption, crony capitalism, and nepotism that characterized both politics and the economy with many negative impacts on the environment, yet these remain rampant today with money politics dominating the political landscape (Aspinall, 2018). Growth, prosperity, stability, and sustainability are severely threatened by both internal structural problems and the forces of climate change.

Social Construction of Impacts and Outcomes: Gendered Riskscapes

Regardless of how they are classified, both disaster and climate change are ultimately human-sourced, although, for disasters, it is common to distinguish between "natural" and "technological" sources. The distinction is spurious with all disasters, both physically based and socially constructed in terms of hazards, risk, and vulnerability (Freudenberg et al., 2008). Similarly, climate change is now indisputably linked to human action with specific sources and impacts varying from local to global. There is great variation both socially and spatially in risk, vulnerability, and management of disaster and climate change reflecting physical and geographic differences, cultural arrangements, and social hierarchies (Hore et al., 2018; Watts & Bohle, 1993; Wisner et al., 2004). Gender is foremost among the factors that influence impacts and outcomes. Women have been documented to be among those most at risk and also central to recovery, mitigation, and adaptation (Enarson

& Meyreles, 2004; Enarson et al., 2018; Demetriades & Esplen, 2008; Fulu 2007; Kusumasari, 2015; Nelson et al., 2002).

Other factors that influence disaster and climate change risk and outcomes are social capital, social conflict, and governance structures, all closely intertwined with gender. Social capital increases resilience (Meyer, 2018), conflict diminishes it (Freudenburg, 1997), and governance inclusivity, planning, proactivity, legitimacy, and absence of corruption determine how well recovery proceeds and who is included (Mercer, 2010; Thomalla et al., 2006). While relatively few studies explicitly link all these factors, and often there is a lack of sex-disaggregated data to test the connections, existing accounts suggest that gender is heavily implicated with women's empowerment, agency, and broad participation in both pre and post-disaster work, reducing vulnerability and creating more positive outcomes. Conversely, their exclusion and subordination have more negative consequences (Tickamyer & Kusujiarti, 2020).

All factors can be tied together with the concept of "riskscape," the "landscapes of risk that exist in relation to practice" (Müller-Mahn et al. 2018, p. 197) that define the social, spatial, and temporal intersections of risk, vulnerability, recovery, and resilience. They underscore differences among women and between women and men, depending on geography, period, and culture. "Riskscapes are shaped by the practices generated by gender relations and forms of individual and community social capital, community conflict and governance and strength of civic society" (Tickamyer & Kusujiarti, 2020, p. 239).

In Indonesia, the links between gender, development, disaster, and climate change are clearly illustrated in differing riskscapes. A disadvantage in crisis reflects disadvantage in social standing, which mirrors disparities in women's status across the nation. The great diversity of social, cultural, and political arrangements is reflected in variations in women's status and position in social hierarchies across geographic locations and landscapes over time. Differences in women's status and access to societal resources are tied to both historical differences and current practices and often entail deep contradictions between rhetoric and reality (Tickamyer & Kusujiarti, 2012). Even where gender arrangements are relatively equitable, women lag men in most realms, although overall, they have made steady progress in literacy, educational attainment, and labor force participation.

The New Order, during its 32-year regime from 1965 to 1998, made concerted efforts to homogenize women's roles as primarily domestic while simultaneously conscripting their labor for national development. The contradiction meant that many women negotiated an uneasy relationship between increased agency and a demeaning ideology. Women were policed by mandatory participation in state-sponsored hierarchically organized "involuntary voluntary" women's organizations, the Family Welfare and Empowerment Program or PKK (*Pemberdayaan Kesejahteraan Keluarga*) and *Dharma Wanita* that both dictated behavior and delivered social services in local communities (Tickamyer & Kusujiarti, 2012, pp. 161–191). These and the expansive growth of women's religious organizations affiliated with national religious movements continued to mobilize women's mass participation after regime change and democratization, although on a more genuinely voluntary basis (Robinson, 2018). They also maintain strong influences on gender roles and practices, with many of the contradictions and dilemmas intact.

The national emphasis on women's domesticity as their *Kodrat* or inherent nature during the New Order did not erase the many differences in women's status across the archipelago (Tickamyer & Kusujiarti, 2012). Rather, it heightened women's disadvantage in different riskscapes. At the same time, mandatory participation in organizations such as the PKK provided opportunities for civic engagement, development of social capital, and leadership skills as an inadvertent and contradictory byproduct of their involvement that moved the needle closer to empowerment for women while also influencing riskscapes.

Indonesian riskscapes demonstrate that gender equality and women's empowerment have major effects on impacts and outcomes of disaster and climate change. Where women have greater status, mobility, and economic, political, and civic opportunities, in other words, where they have more agency and are more empowered, they are better able to weather disasters and cope with climate change. Where they lack power and agency, they are more vulnerable, more at risk, and more likely to fare worse in disaster and climate change (Tickamyer & Kusujiarti, 2020).

Variations in women's vulnerability and outcomes are illustrated by accounts of major disaster events in Indonesia across different riskscapes. These include the 2004 tsunami and earthquake in Aceh on Sumatra, the 2006 earthquake in Bantul, Yogyakarta, on Java, and the 2010 eruption of Mount Merapi in 2010 also in Yogyakarta (Kusujiarti, 2017; Kusujiarti et al., 2015; Tickamyer & Kusujiarti, 2020) and by other studies of disaster and climate change conducted in Indonesia in recent years (Kusumasari, 2015; Kusumasari & Alam, 2012; Lavigne et al., 2008; Lisna et al., 2011; Mulyasari & Shaw, 2013; Pratiwi et al., 2017). The three primary examples, tsunami, earthquake, and volcano illustrate the range of disaster riskscapes and the variation in women's social standing in these areas with corresponding influence on disaster vulnerability, recovery, and resilience. We have written about it extensively elsewhere (Tickamyer & Kusujiarti, 2020). Here, we summarize gendered riskscape characteristics and differences and their implications.

Indonesian Gendered Riskscapes

The three disasters are generally classified as natural disasters, but each demonstrates the social components of creating risk and vulnerability as well as resilience and recovery. They vary substantially in scale and thus are not strictly comparable. Aceh is by far the most destructive of the disasters and, in fact, one of the largest globally in recent history (and one that created devastation throughout South and Southeast Asia). The earthquake in Bantul and the eruption of Merapi in different parts of Yogyakarta were much more circumscribed but also devastating within their spheres.

Aceh: In the disaster in Aceh, women were disproportionately vulnerable and victimized by the events directly and in the long aftermath and recovery process (Kusujiarti, 2017; Kusujiarti et al., 2015). Women are thought to comprise the majority of deaths and casualties, victims of a history of social conflict and women's subordination. Aceh is the site of a longstanding civil war, with hostilities only ending in the aftermath of the disaster (Burke, 2008). The many years of brutal armed conflict between GAM (*Gerakan Aceh Merdeka*), the Free Aceh Movement, and the government created a corrosive community of violence, distrust, and exclusion. Women's mobility and public involvement were severely curtailed, and gender violence was rampant both before and after the tsunami. The aftermath of the disaster led to peace negotiations between Aceh and the central government, largely due to the need to focus on recovery; however, the process was exclusionary, and the outcome further disadvantaged women. Women were excluded from the negotiations, and little effort was made to consider their situation or interests. The primary terms of the settlement, the implementation of self-governance within the Republic, control of natural resources, and the official establishment of a restrictive version of Shari'ah law further disadvantaged women. Shari'ah, as interpreted by Acehnese religious and political authorities, severely limited women's activities and participation in public affairs. During the long years of recovery and reconstruction (formally declared completed by the Indonesian government in 2014), women continued to be ignored, restricted, or actively constrained from civic engagement (Arianty, 2018). Additionally, the onset of new environmental problems from flooding, deforestation, and extension of palm oil plantations during the recovery period further disadvantaged women and increased their labor burdens (Lisna et al., 2011). As a riskscape, Aceh exemplifies a disaster site characterized by its legacy of conflict and distrust and its ongoing practices of social exclusion, patriarchy, and top-down governance and planning. The social landscape has magnified the risks and vulnerabilities experienced by women.

Bantul: In many ways, the earthquake that struck offshore of the Bantul district in the Special Province of Yogyakarta, Java, in 2006, presents the antithesis of Aceh. While not strictly comparable because of the difference in scale, nevertheless, the magnitude of the damage was severe, with over 6,000 deaths, many thousands more injured and left homeless, and massive damage to infrastructure, housing, industry, and livelihoods. Although damage was widespread, two-thirds of the casualties occurred in the district of Bantul. Loss of jobs was especially severe in small-scale services and manufacturing, mainstays of the economy for many residents (Suzetta et al., 2006). Unfortunately, official numbers are not available, but similar to Aceh, it is thought that women were disproportionately victimized (Pristiyanto, 2008). Unlike Aceh, however, the population was resilient, and recovery proceeded quickly and without serious conflict.

In an effort to determine why it was so different, major components of its social and cultural arrangements can be identified as contributors to successful recovery and reconstruction. Among the significant factors are the absence of conflict, good governance reflected in strong, popular, local leadership that welcomed public input and participation in planning, and Javanese cultural beliefs and practices that emphasize mutual self-help and assistance (*gotong royong*) and consensus-based decision-making (*musyawarah*). Additionally, women in this part of Indonesia typically are economically active and participate in civic affairs, occasionally in leadership positions. The legacy of a particularly strong and active PKK and other social activities meant that many women have had opportunities to develop social networks, social capital, and leadership experience that could be mobilized for disaster response. As a social service organization, the PKK was active in recovery efforts, as were other civic and voluntary organizations. Beyond recovery, in some communities, residents joined together to develop more sustainable and self-sustaining food provision and agricultural practices, fully involving both women and men. In general, the experience of Bantul suggests the importance of more egalitarian and participatory approaches to disaster relief and recovery that also hold lessons for how to approach climate change.

Merapi: Indonesia's location in the Pacific "Ring of Fire" makes active volcanoes a standard part of its geography, with numerous eruptions of varying sizes and degrees of damage. Mount Merapi, located on the border of two provinces, Yogyakarta and Central Java, is the most active and the most consistently destructive. Despite the danger of eruption, its environs are densely populated, primarily by agriculturalists who value the fertility of the soil and have strong cultural attachments to the location and its indigenous belief system that addresses risk and response with faith in ancestral spirits (De Coster, 2002; Lavigne et al., 2008). A major eruption in 2010 resulted in the loss of approximately 350 lives, including that of the hereditary spiritual leader (*juru kunci*), believed to be the key to divine protection through access to the volcano spirit. Damage included many more injuries, massive loss of property, and the dislocation of upward of 350,000 people. Evacuation, both voluntary and mandated, created intense trauma for residents who were attached both materially and spiritually to their land. Loss of homes and property was compounded by the loss of livelihoods and community and deep psychosocial distress related to environmental damage (Warsini et al., 2014, 2016). The wholesale destruction of livestock and former farmland that was no longer suitable for planting made it necessary to find other sources of employment, often with little success, resulting in dependence on assistance from the government, family, and private donors. Even when work was available, the large influx of labor drove down wages and made Merapi refugees vulnerable to exploitation. Men were most likely to find employment in sand mining, women in tourism, often in jobs catering to "disaster tourism" (Kenny, 2005). Some multi-national corporations relocated to the area to take advantage of cheap labor. Temporary housing arrangements created by the government forced evacuees to move multiple times and live with strangers in close quarters, disrupting existing social capital, solidarity, and community ties. Conflict increased as well as violence against women, domestic disputes, and sexual harassment. Resettlement to more permanent housing involved a 3–5-year wait and still could not recreate lost communities. A decade later, survivors still showed symptoms of posttraumatic stress (Wati et al., 2020).

Comparison: The three disaster sites present suggestive contrasts in both impacts and consequences that help extract factors that are important for reducing vulnerability, risk, and damage and for increasing recovery and resilience. All three cases depict catastrophic natural disasters with impacts that are amplified or diminished by combined local history, cultural practices, and social structure. All three demonstrate the importance of gender relations and women's empowerment in either exacerbating or mitigating impacts. The differences between them illustrate the complex intersections of geophysical and socio-cultural factors that define different riskscapes and their consequences.

The most extreme case, Aceh, exemplifies all the factors with negative effects. These include a history of women's exclusion and subordination, social conflict, community corrosiveness, and poor governance and leadership. Women remain severely disempowered in Acehnese society, and the result is evident in both vulnerability and response to disaster and in daily life. The practice of a highly conservative and patriarchal form of Islam that is restrictive to women's agency heightens their vulnerability. It seems few lessons have been learned from past experience, and the result is ongoing risk and vulnerability, especially for women and other marginalized groups.

At the other extreme, although nowhere near attaining gender equality, Bantul comes much closer to demonstrating the positive effects of women's relative empowerment, social capital, and organizational experience, along with the absence of conflict and a more open and participatory government. It also suggests a link between inclusiveness and gender equality and willingness to take proactive measures to reduce risk and advance measures protective of the environment. In the aftermath of the disaster, local governments and groups with citizen input have attempted to improve environmental sustainability through new initiatives and practices (Tickamyer & Kusujiarti, 2020).

Finally, the Merapi case falls somewhere in between. Although located in the same province as Bantul with shared roots in Javanese history and culture as well as Muslim and New Order gender ideology and practices, women were more resistant or less mindful of their discipline and much less active in civic and service organizations than in Bantul, possibly because they were more isolated than their counterparts in the more urbanized district. The absence of conflict prior to the eruption of Merapi did not prevent its appearance post-disaster, exacerbated by stress and inept assistance and recovery practices. Unfortunately, few studies have collected gender-disaggregated data or attempted to examine gender differences in women's experiences compared to men, leaving many questions about their risk and recovery. Nevertheless, women's relative lack of agency coupled with the other social factors identified here suggests their important role in risk and recovery.

Linking Gender Justice and Environmental Justice

In addition to these iconic cases, there are increasing numbers of examples of women's efforts and activism in Indonesian environmental movements that indicate the importance of women's empowerment to protect themselves, their families and communities, and the environments and natural resources on which they depend (Hendrastiti, 2019; Hendrastiti and Kusdinar, 2019; Hendrastiti & Kusujiarti, 2020) As a form of postcolonial feminism, women mobilize against powerful national and multi-national corporate and state interests whose destructive practices in mining, deforestation, and agroforestry threaten their access and stewardship of the environment, natural resources, and ecosystem services. In Bengkulu in the Western part of Indonesia and Sumba in the East and many locations in between, there are examples of women seeking to assert their rights to voice and agency in environmental activism, linking gender justice and environmental justice with varying degrees of success and failure (Hendrastiti & Irianto, 2020). Hendrastiti and her colleagues argue that women mobilize to protect their families and, in the process, assert their rights to be heard and participate as citizens. Legal rights and actual practice on the ground, however, often do not correspond, and it requires organization and activism to assert and seek justice. Whether the goal is to stop pollution from mining or to maintain access to forest products that are part of food security, sovereignty, and livelihoods, they pursue a combination of environmental and legal objectives.

Conclusion

In Indonesia, as in virtually every corner of the globe, gender justice and environmental justice are intimately linked and are necessary for the pursuit of sustainable development. The emphasis in the SDGs and numerous other platforms on three pillars: inclusivity and equality, economic development, and environmental sustainability, require women's empowerment. The three specific goals noted initially – "Achieve gender equality and empower all women and girls"; "Promote sustained, inclusive and sustainable economic growth, full and productive employment, and decent work for all"; and "Take urgent action to combat climate change and its impacts" – have a long way to go before becoming a reality. Progress on all fronts is uneven and mixed with both real achievements and signs of progress and disappointing failures to live up to promises and expectations. Indonesia exemplifies the significance of these goals and their connections, as well as the difficulties in achieving them. The heterogeneity built into a nation as geographically, demographically, culturally, and politically diverse across the different riskscapes that characterize Indonesia combined with its extreme vulnerability to disaster and climate change ensures that the road to sustainability will continue to be rough with both setbacks and accomplishments.

What is the case in Indonesia will also be relevant in diverse locations and contexts. More research on the connections between goals and practices of gender justice, environmental justice, and sustainability within and across riskscapes and national boundaries is critical for developing good policy and programs to achieve all three goals in widely varying social and cultural contexts. Urgent action is needed on all fronts, beginning but not ending with gender justice.

Note

1 Estimates of casualties in Aceh apart from the rest of the tsunami disaster field vary substantially from the relatively conservative figure cited here to as many as a quarter of a million deaths. The extreme level of destruction in a region experiencing ongoing conflict means that an exact figure will never be established.

References

Arianty, D. (2018). Women's responses to Islamic Law in Aceh. In R. Hefner (Ed.), *Routledge Handbook of Contemporary Indonesia* (pp. 346–353). London and NY: Routledge Taylor & Francis Group.

Asian Development Bank (ADB). (2021). *Climate Risk Country Profile: Indonesia.* Retrieved November 22, 2021 from https://www.adb.org/sites/default/files/publication/700411/climate-risk-country-profile-indonesia.pdf

Aspinall, E. (2018). Democratization: Travails and achievements. In R. Hefner (Ed.), *Routledge Handbook of Contemporary Indonesia* (pp. 83–94). London and NY: Routledge Taylor & Francis Group.

Burke, A. (2008). Peacebuilding and rebuilding at ground level: Practical constraints and policy objectives in Aceh. *Conflict, Security & Development* 8 (1),47–69. https://doi.org/10.1080/14678800801977088

De Coster, B. (2002). *Perception des Risques Naturels par les Populations sur les Flancs du Volcan Merapi. Java-Centre, Indonésie.* DVD Film 35'+ report.

Demetriades, J. and Esplen, E. (2008). The gender dimensions of poverty and climate change adaptation. *IDS Bulletin* 39, 4. https://doi.org/10.1111/j.1759-5436.2008.tb00473.x

Djalante, R. (2018). Review article: A systematic literature review of research trends and authorships on natural hazards, disasters, risk reduction and climate change in Indonesia. *Natural Hazards and Earth System Sciences* 18, 1785–1810. https://doi.org/10.5194/nhess-18-1785-2018

Enarson, E., Fothergill, A. and Peek, L. (2018). Gender and disaster: Foundations and new directions for research and practice. In H. Rodriguez, W. Donner, and J. Trainor (Eds.), *Handbook of Disaster Research* (pp. 205–223). NY: Springer. https://doi.org/10.1007/978-3-319-63254-4_11

Enarson, E. and Meyreles, L. (2004). International perspectives on gender and disaster: Differences and possibilities. *International Journal of Sociology and Social Policy* 24 (10/11), 49–93. https://doi.org/10.1111/j.1759-5436.2008.tb00473.x

Frankenberg, E., Gillespie, T., Preston, S., Sikoki, B., and Thomas, D. (2011). Mortality, the family and the Indian ocean tsunami. *The Economic Journal* 121 (554), F162–F182. https://doi.org/10.1111/j.1468-0297.2011.02446.x

Freudenberg, W. Gramling, R., Laska, S., and Erikson, K. (2008). Organizing hazards, engineering disasters? Improving the recognition of political-economic factors in the creation of disasters. *Social Forces* 87 (2), 1015–1038, https://doi.org/10.1353/sof.0.0126

Freudenburg, W. (1997). Contamination, corrosion and the social order: An overview. *Current Sociology* 45, 19–39. https://journals.sagepub.com/doi/10.1177/001139 297045003002

Fulu, E. (2007). Gender, vulnerability and the experts: Responding to the Maldives tsunami. *Development and Change* 38, 843–864. https://doi.org/10.1111/j.1467-7660. 2007.00436.x

Global Facility for Disaster Risk and Recovery (GFDRR.org). (n.d.). Retrieved September 18, 2019 from https://www.gfdrr.org/en/indonesia

Hendrastiti, T. K. (2019). Oral story of women's anti-mining group in Sumba: A narrative of subaltern movement for food sovereignty. *Jurnal Perempuan* 24 (1), 1–12. https://doi.org/10.34309/jp.v24i1.291

Hendrastiti, T. K. and Irianto, S. (2020). Persistent courage of the local women resistance toward undemocratic policies. *ETNOSIA: Jurnal Etnografi Indonesia* 5(2), 184–199. https://doi.org/10.31947/etnosia.v5i2.9863

Hendrastiti, T. K., and Kusdinar, P. (2019). Involvement of women village leaders in developing dialogues on forest conflict resolution. *Jurnal Perempuan* 24 (4), 259–270. https://doi.org/10.34309/jp.v24i4.381

Hendrastiti, T. K. and Kusujiarti, S. (2020). Ginger torch flower (Unji): The identity of women's agency in the national park. *International Journal of Advanced Science and Technology* 29 (3), 5690–5699. http://sersc.org/journals/index.php/IJAST/article/view/6194

Hore, K., Kelman, I., Mercer, J. and Gaillard, J. C. (2018). Climate change and disasters. In H. Rodríguez, W. Donner, and J. Trainor (Eds.), *Handbook of Disaster Research, Handbooks of Sociology and Social Research* (pp. 145–159). NY: Springer. https://doi.org/10.1007/978-3-319-63254-4_8

International Labour Organization (ILO). (n.d.). *Informal economy in Indonesia and Timor-Leste*. Retrieved September 22, 2021 from https://www.ilo.org/jakarta/areasofwork/informal-economy/lang--en/index.htm

Kenny, S. (2005). Reconstruction in Aceh: Building whose capacity? *Community Development Journal* 42 (2), 206–221. https://doi.org/10.1093/cdj/bsi098

Kusujiarti, S. (2017). Tsunami, civil society and Shari'ah Law in Aceh, Indonesia: Intersection of disaster, decentralization and gender relations. In T. Reilly (Ed.), *The Governance of Local Communities: Local Perspectives and Challenges* (pp. 221–233). New York: Nova Publishers.

Kusujiarti, S., Miano, E. W., Pryor, A. L. and Ryan, B. R. (2015). Unveiling the mysteries of Aceh, Indonesia: Local and global intersections of women's agency. *Journal of International Women's Studies* 16, 186–202. https://vc.bridgew.edu/jiws/vol16/iss3/13

Kusumasari, B. (2015). Women adaptive capacity in post disaster recovery in Indonesia. *Asian Social Science* 11 (12), 911–2025. http://dx.doi.org/10.5539/ass.v11n12p281

Kusumasari, B. and Alam, Q. (2012). Local wisdom based disaster recovery model in Indonesia. *Disaster Prevention and Management* 21, 351–369. https://doi.org/10.1108/09653561211234525/full/html

Lavigne, F., De Coster, B., Juvin, N., Flohic, F., Gaillard, J.-C., Texier, P., Morin, J., and Sartohadi, J. (2008). Peoples' behaviour in the face of volcanic hazards: Perspectives from Javanese communities. *Indonesia, Journal of Volcanology and Geothermal Research* 172, 273–287. https://doi.org/10.1016/j.jvolgeores.2007.12.013

Lisna, E., Safrida, and Kusujiarti, S. (2011). Strategies for Strengthening Gender Sensitive Approach in Disaster Management in Aceh Province. A Case Study on

Flood Disaster in Pidie Jaya and Aceh Jaya Districts. Paper presented at the *Annual International Workshop and Expo on Sumatera Tsunami Disaster and Recovery in Conjunction with 4th South China Sea Tsunami Workshop (AIWEST-DR)*. Banda Aceh, Indonesia. November.

Margono, B. A., Potapov, P. V., Turubanova, S., Stolle, F., and Hansen, M. C. (2014). Primary forest cover loss in Indonesia over 2000–2012. *Nature Climate Change* 4, 730– 735. https://doi.org/10.1038/nclimate2277

Mercer, J. (2010). Disaster risk reduction or climate change adaptation: Are we reinventing the wheel? *Journal of International Development* 22, 247–264. https://doi.org/10.1002/jid.1677

Meyer, M. (2018). Social capital in disaster research. In H. Rodríguez, W. Donner and J. Trainor (Eds.), *Handbook of Disaster Research, Handbooks of Sociology and Social Research* (pp. 263–286). New York: Springer. https://doi.org/10.1007/978-3-319-63254-4_14

Müller-Mahn, D., Everts, J. and Stephan, C. (2018). Riskscapes revisited – Exploring the relationship between risk, space and practice. *Erdkunde* 72, 197–213. https://doi.org/10.3112/erdkunde.2018.02.09

Mulyasari, F. and Shaw, R. (2013.) Role of women as risk communicators to enhance disaster resilience of Bandung, Indonesia. *Natural Hazards* 69, 2137–2160. 10.1007/s11069-013-0798-4

NASA Earth Observatory. (2021). *As Jakarta Grows So Do the Water Issues.* Retrieved September 22, 2021 from https://earthobservatory.nasa.gov/images/148303/as-jakarta-grows-so-do-the-water-issues

National Academies of Sciences, Engineering, and Medicine. (2016.) *Attribution of Extreme Weather Events in the Context of Climate Change.* Washington, DC: National Academies Press.

Nelson, V., Meadows, K., Cannon, T., Morton, J., and Martin, A. (2002). Uncertain predictions, invisible impacts, and the need to mainstream gender in climate change adaptations. *Gender and Development* 10 (2), 51–59. https://doi.org/10.1080/13552070215911

OECD. (2021). *Economic Survey of Indonesia.* Retrieved from https://www.oecd.org/economy/indonesia-economic-snapshot/. March 18.

Pratiwi, N. A., Y. Y. D. Rahmawati, and I. Setiono. (2017). Gender equality in climate change adaptation: A case of Cirebon, Indonesia. *The Indonesian Journal of Planning and Development* 2 (2), 84–86. https://doi.org/10.14710/ijpd.2.2.74-86

Pristiyanto, D. (2008). *National Plan for Disaster Risk Reduction 2006–2009*, published by the Ministry of Development Planning and Indonesian National Board for Disaster Management. 2006. Jakarta. Pembentukan BPBD Berdasarkan Permendagri. 46/2008 dan Perka BNPB 3/2008.

Reagan, H. (2018). *Overview of Sdgs Implementation in Indonesia & Using Susenas in Measuring Wash Sdgs Indicators. International Workshop on Sustainable Development Goal (SDG) Indicators*, 26–28 June 2018, Beijing, China. http://www.stats.gov.cn/english/pdf/202011/P020201102587178396531.pdf

Robinson, K. (2018). Gender and politics post-New Order. In R. Hefner (Ed.), *Routledge Handbook of Contemporary Indonesia* (pp. 309–321). London and NY: Routledge Taylor & Francis Group.

Sachs, J. (2012). From millenium development goals to sustainable development goals. *Lancet* 379, 2206–11. https://doi.org/10.1016/S0140-6736(12)60685-0

Schipper, L., and Pelling, M. (2006). Disaster risk, climate change and international development: Scope for and challenges to, integration. *Disasters* 30, 19–38. https://doi.org/10.1111/j.1467-9523.2006.00304.x

Suzetta, H. Paskah, A., and Cua, E. (2006). *Preliminary Damage and Loss Assessment: Yogyakarta and Central Java Natural Disaster: A Joint Report of BAPPENAS, the Provincial and Local Governments of D.I. Yogyakarta, the Provincial and Local Governments of Central Java, and International Partners.* The 15th Meeting of the Consultative Group of Indonesia (CGI). Jakarta, Indonesia. Retrieved November 22, 2021 from https://think-asia.org/bitstream/handle/11540/2453/damage-assessment-indonesia-earthquake.pdf?sequence=1

Thomalla, F., Downing, T., Spanger-Siegfried, E., Han, G. and Rockström, J. (2006). Reducing hazard vulnerability: Towards a common approach between disaster risk reduction and climate adaptation. *Disasters* 30, 39–48. https://doi.org/10.1111/j.1467-9523.2006.00305.x

Tickamyer, A. R. and Kusujiarti, S. (2012). *Power, Change and Gender Relations in Rural Java: A Tale of Two Villages.* Athens: Ohio University Press.

Tickamyer, A. R. and Kusujiarti, S. (2020). Riskscapes of gender, disaster and climate change in Indonesia. *Cambridge Journal of Regions, Economy and Society* 13, 233–251. https://doi.org/10.1093/cjres/rsaa006

Tickamyer, A. R., Kusujiarti, S. and Wornell, E. J. (2014). Gender justice, climate change and sustainable development in Indonesia. In Z. Zhu and J. Li (Eds.), *Environment and Sustainable Development in Asia, Vol. 4, Globalization, Development and Security in Asia* (pp. 745–746). Singapore: World Scientific Publishing.

UN Sustainable Development Goals. (2021). *Indonesia. Voluntary National Review.* Accessed September 27, 2021 from https://sustainabledevelopment.un.org/memberstates/indonesia

United Nations Department of Social and Economic Affairs. (n.d.) *Sustainable Development.* Accessed September 20, 2021 from https://sdgs.un.org/

Warsini, S., Mills, J., West, C., and Usher, K. (2014). The psychosocial impact of the environmental damage caused by the MT Merapi eruption on survivors in Indonesia. *EcoHealth* 11, 491–501. https://doi.org/10.1007/s10393-014-0937-8

Warsini, S., Mills, J., West, C., and Usher, K. (2016). Living through a volcanic eruption: Understanding the experience of survivors as a phenomenological existential phenomenon. *International Journal of Mental Health Nursing* 25, 206–213. https://doi.org/10.1111/inm.12212

Wati, D. E., Mustikasari, M., and Panjaitan, R. U. (2020). Post traumatic stress disorder description in victims of natural post eruption of Merapi one decade. *Jurnal Ilmu Keperawatan Jiwa*, 3 (2), 101–112. https://doi.org/10.32584/jikj.v3i2.522

Watts, M. J., and Bohle, H. G. (1993). The space of vulnerability: The causal structure of hunger and Famine. *Progress in Human Geography* 17 (1), 43–67. https://doi.org/10.1177/030913259301700103

Wells, M., Guggenheim, S., Khan, A., Wardojo, W. and Jepson, P. (1999). *Investing in Biodiversity: A Review of Indonesia's Integrated Conservation and Development Projects.* World Bank Publications. doi:10.1596/0-8213-4419-6.

Wisner, B., Blaikie, P., Cannon, T., and Davis, I. (2004). *At Risk: Natural Hazards, People's Vulnerability and Disasters* (2nd ed.). London: Routledge.

World Bank. (n.d.). *Indonesia Climate Change Overview: Country Summary.* Retrieved September 25, 2021 from https://climateknowledgeportal.worldbank.org/country/indonesia

Wornell, E. J., Tickamyer, A. R., and Kusujiarti, S. (2015). Gender mainstreaming principles in Indonesia's REDD+ program: A document analysis. *Journal of Sustainable Development* 8 (8), 159–170. http://dx.doi.org/10.5539/jsd.v8n8p159

Part II

Responding to climate change

Sociopolitical underpinnings of multiple
actors and institutions

6 The Resilience of Fisheries Households to Climate Shock in Tam Giang – Cau Hai Lagoon, Vietnam

Ha Dung Hoang, Salim Momtaz,
Maria Schreider, Chung Van Nguyen and
Tien Dung Nguyen

Introduction

The Tam Giang – Cau Hai (TG–CH) Lagoon is located along the coast of Thua Thien Hue Province, Central Vietnam, between the coordinates 160014'00"–160042'00" North and 107022'00"–107057'00" East. The lagoon system is around 70 km long and is separated from the sea by an extensive system of dunes, running parallel to the coastal area of Thua Thien Hue Province. Thanks to its abundant natural resources, the TG–CH Lagoon plays a significant role in the social-economic development of Thua Thien Hue Province in general and the livelihoods of nearly half a million people living along the coast of this province in particular. Many people here are fishers, who account for 43.3% of the population in the lagoon area (Binh, Huong, Thien, & Chau, 2018). The local fishers' livelihoods mainly depend on the exploitation and use of natural resources of the lagoon and involve activities such as fishing, aquaculture, agriculture, and services (port services, seafood processing, tourism) (Figure 6.1).

The lagoon is located in a tropical monsoon climate, so its areas are frequently subjected to natural disasters such as occasional typhoons, monsoons, storm surges, sea-level rise, and El Nino phenomena (Thanh, Vu, Nguyen, & Zheng, 2013). Recent studies have confirmed that climate change in Vietnam has brought extreme weather events that are increasing in number and difficult to predict. In the TG–CH Lagoon, shocks related to climate change have affected the day-to-day lives of SSF communities (Ha & Thang, 2017). Climate change has directly and indirectly affected the natural environment, ecosystem, and local fisheries. Specifically, the adverse effects of climate shocks have caused a decline in daily fish catch and aquatic resources. The majority of fishers agreed that the climate and shocks have been changing in recent decades. Most local communities have reported strange changes in floods, fluctuations in temperature and rainfall, stronger storms, more frequent drought, and unpredictable freshening of the lagoon water (Figure 6.2). All of these changes have seriously impacted SSF in the TG–CH Lagoon.

DOI: 10.4324/9781003216476-8

Figure 6.1 Fishery activities and fishing gear in estuarine areas of the TG–CH Lagoon.

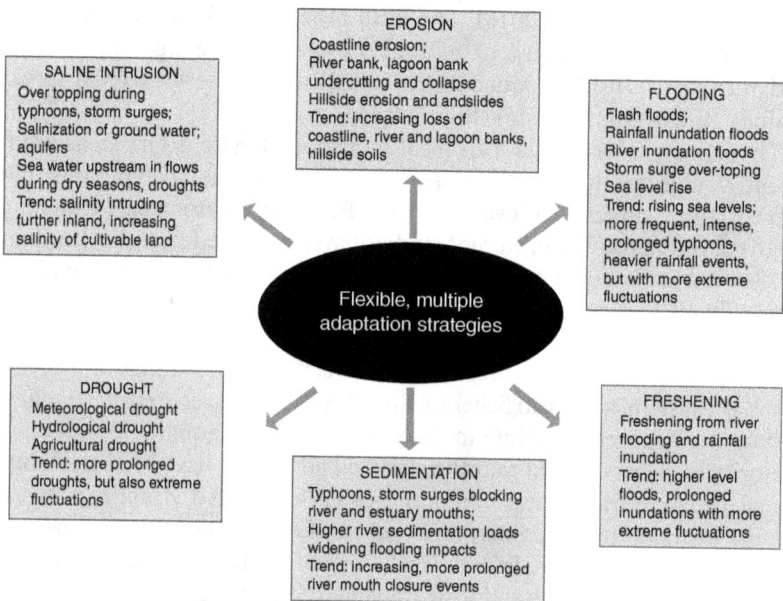

Figure 6.2 The impacts of climatic shocks and stressors on various coastal livelihoods in Central Vietnam.

Source: Truong et al. (2010).

In this context, the resilience of SSF has been recognized as a multi-level, context-specific process. Vulnerability and resilience assessment complexity makes it difficult to measure through simple indicators (Binh et al., 2018). In addition, understanding the recovery and resilience of fishing households (or fishing communities) needs more than just theories; therefore, practical assessment requires indicators specific and measurable resilience (Béné, 2013).

Thus, there is a need for a deeper understanding of the types of shocks, adaptive mechanisms, economic and social conditions that make people vulnerable, and the factors that influence their ability to respond to shocks and stressors. Therefore, further research is needed to analyze resilience rather than describing the frequency and severity of shocks or the types of coping strategies adopted by individuals and groups (Béné et al., 2016; Binh et al., 2018).

This study aims to fill a knowledge gap regarding the resilience assessment of SSF communities to climate risks and shocks in coastal areas of Vietnam. It also examines the vulnerability of fishery communities in the TG–CH Lagoon areas in the context of increased natural disaster risks.

Research Methodology

Research Approach

The overall approach to this research involved participatory research methods. Participatory research methods have been widely applied in social studies. These methods have several advantages and strengths, including (1) ensuring objectivity of information and data collected; (2) variety of collected data providing a deeper understanding of issues under investigation; (3) ensuring initiative and empowerment of information providers; (4) using data collection tools quickly and flexibly (Chambers, 2005; Huong, 2010; Momtaz & Kabir, 2018).

The present study also highlights community participation in assessing climate change and shocks to their livelihoods. This follows Momtaz and Kabir's (2018) view of the community's role in evaluating environmental impacts. The research participants were key informants and/or representatives from local government, agriculture, and rural development; Fishery Association agencies from provincial to community levels; other agencies (non-governmental organizations [NGO]s, community-based organizations [CBO]s, and private actors); and local fishers.

Key informant interviews and focus group discussions (FGDs) were used for data collection and analysis. At the same time, several participatory rural appraisal (PRA) tools such as timelines, seasonal calendars, and mapping were also applied to facilitate group discussions. This approach focused on the use of local knowledge and participation of local fishers in a quick appraisal, assessment, and analysis (Chambers, 2005) and allowed the researcher to work closely with local people paying particular attention to the culture and traditional customs of the community. Secondary data were collected and served as entry points to the participatory assessment process in fishery communities at the provincial, district, and communal levels. Also, reports from related organizations were scrutinized.

Analytical Framework

The interaction between climate change and vulnerability creates increasing pressure on people's livelihoods, especially the poor. In recent years, there has been an increase in the interest of scientists and policymakers in resilience studies. The Climate Change Conference has further propped this up in Copenhagen in December 2009 that drew greater attention of the global community to climate change effects, adaptation, and mitigation. Organizations and individuals have developed resilience assessment methods based on conceptual frameworks using the concept of resilience, methods of research, and data collection. Approaches to resilience research thus far are diverse, and there have been many different research frameworks developed to date (e.g., Douxchamps, Debevec, Giordano, and Barron (2017)).

In practice, most research frameworks indicate that different variables can be applied to analyze resilience. The frameworks also describe the linkages between social, economic, and ecological variables essential for evaluating resilience. One such framework is developed by the Resilience Measurement Working Group (RMTWG) (Smith & Frankenberger, 2017). The RMTWG framework can be applied in monitoring and evaluating programs, intervention planning, and research on how to support households to respond to shocks. This framework consists of three main components: (1) absorption, the ability to minimize negative impacts of shocks; (2) adaptive capacity, which involves the making of decisions about alternative livelihood strategies based on changing conditions; and (3) transformative capacity, expanded to facilitate resilience in the long term (Smith & Frankenberger, 2017).

Given the complexity of quantifying indicators, a single index cannot measure resilience capacity (Smith & Frankenberger, 2017). Measuring resilience requires combining several indicators into an overall outcome. Figure 6.3 shows the indicators of the three capacities used to describe the resilience of fishery households in this study. In this section, the relevance of each indicator to the resilience of fishers is presented. As can be seen, some indicators listed in Figure 6.3 were used to evaluate more than one capacity.

Finally, based on the above review, the present research applied the tools and resilience evaluation with respect to the research framework of the RMTWG. The framework shows the outcomes of the resilience strategies, including suggestions and recommendations to help fishers improve resilience and reduce the impact of climate shocks on their livelihood in the short and long term.

Selection of Study Sites

The TG–CH Lagoon is connected to the South China Sea (*Biển Đông* in Vietnamese) via two estuaries: Thuan An in the north and Tu Hien in the

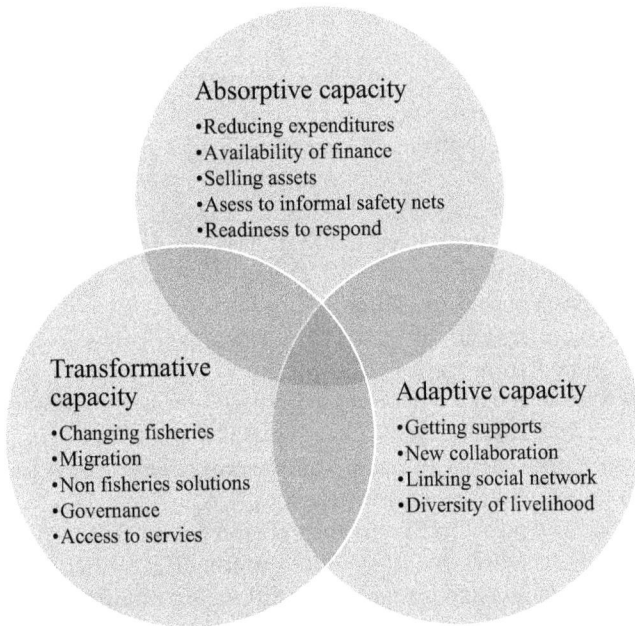

Absorptive capacity
•Reducing expenditures
•Availability of finance
•Selling assets
•Asess to informal safety nets
•Readiness to respond

Transformative capacity
•Changing fisheries
•Migration
•Non fisheries solutions
•Governance
•Access to servies

Adaptive capacity
•Getting supports
•New collaboration
•Linking social network
•Diversity of livelihood

Figure 6.3 Indicators employed to measure resilience capacity.
Source: Adapted from RMTWG and Smith and Frankenberger (2017).

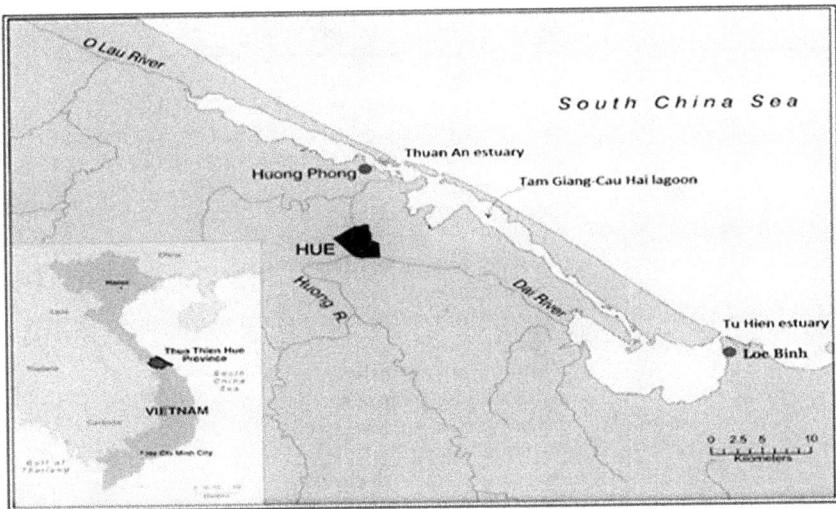

Figure 6.4 Location of TG–CH Lagoon.
Source: Adapted from Armitage and Marschke (2013).

south (Map 1). The lagoon covers an area of 21,600 ha, with a width of 1–10 km and an average depth of 1.5–2 m. The deepest point is more than 10 m, at Thuan An estuary (T. D. Thanh, 2007). The lagoon system runs through five districts' territories: Phong Dien, Quang Dien, Huong Tra, Phu Vang, and Phu Loc. It supports 133 fisheries villages in 33 lagoon communes (Binh et al., 2018).

As this research focused on the SSF of the TG–CH Lagoon, site selection was based on several criteria to ensure that the sites were representative of the overall fishery community of the TG–CH Lagoon. The selected sites meet the specific conditions outlined in Table 6.1.

Based on these criteria, as well as the researcher's own experience and knowledge gained from past studies of the TG–CH Lagoon, two coastal communes were selected for the research: Loc Binh Commune located in the southern part and Huong Phong Commune in the central area of the TG–CH Lagoon. Within each commune, a number of fishery villages were chosen in which to implement the research activities.

Huong Phong and Loc Binh communes are respectively located in the Thuan An and Tu Hien estuaries of the TG–CH Lagoon, Thua Thien Hue Province (Figure 6.5). Loc Binh has a terrain running along the mountainside for 14 km with the other side bordered by the Cau Hai Lagoon, while the Huong Phong Commune is bordered on both sides by the Huong and Bo rivers facing the Tam Giang Lagoon. Agriculture and SSF are the main occupations in both

Table 6.1 Criteria for selecting the study sites

	Criteria	Indicator
1	Natural conditions	Having lagoon water surface
		Experiencing several types of climate shocks
2	Social aspects	Fishery communities have lived there for a long time
		Fishers have a diversity of cultures, traditional customs, and religions
		Both poor and non-poor fishery households[2]
3	Livelihoods	Mainly dependent on lagoon resources
		Long-standing traditional fisheries or newly established fisheries
4	Occupations	Diversity of occupations (agriculture, fishery, and others)
		Mobile and fixed fishing
		Aquaculture (intensive and extensive)
		Upper and lower tide ponds
		Onshore fishing
		Polyculture and fish cages
		Net enclosure aquaculture and upper tide ponds
5	Data	Available data, information, and documents from various sources (FAs, local governments, local organizations)
		Local communities are accessible for information collection

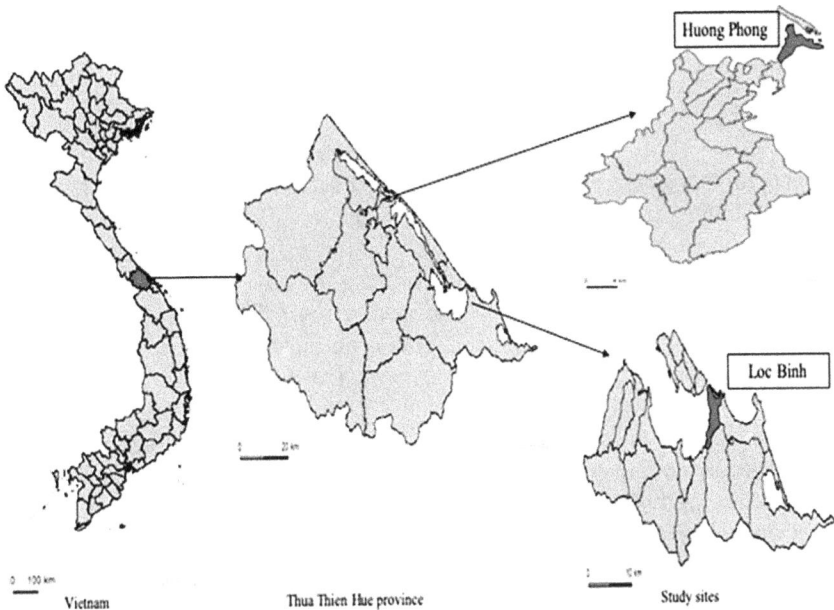

Figure 6.5 Location map of study sites.
Source: Modified from Global Administrative Areas (GADM, 2018).

communes, followed by small businesses and public services. SSF is directly dependent on the exploitation of the lagoon's natural resources. Because of their location close to the estuaries, Huong Phong and Loc Binh are subject to extreme annual weather events that mostly affect fishers.

In the Loc Binh and Huong Phong communes, SSF has been a very common occupation in all communities in these estuarine areas. Traditionally, based on differences in fishing techniques and locations, local fishers were divided into two groups based on the gear they employed: fixed or mobile. While mobile fishers conduct small-scale activities requiring little financial capital by using compact mobile fishing gear such as Chinese traps, gillnets, and push nets, the fixed-gear group uses fishing gear attached to specific locations in the lagoon, including bush traps, fixed lift nets, fish corrals, and bottom nets. In both communes, aquaculture is ubiquitous, including lower tidal and upper tidal earth ponds involving almost 75% of total surveyed households.

Household Selection for Survey

A random sampling method was used to collect samples for this study. This was to ensure that selected households were representative of all fishery households in the study sites (Ha & Thang, 2017). A sample of fishery households in the two communes was randomly selected for interview. The sampling was based on the fisher population. The required sample size of households for conducting the questionnaire survey was determined using the sample size formula of Yamane (1967):

$$n = N / \left(1 + N^* e^2\right)$$

where, in this study, n = sample size (number of interviewed households), N = total number of fishery households in the study sites; and e = error level.

Households were randomly selected for the survey from the 215 fishery households in the two communes by applying the above formula to achieve a 97% level of confidence with a 3% level of error. Of the 181 fishery households, 110 were in Loc Binh and 71 in Huong Phong. Six group discussions were organized in the two communes with the participation of 30 fishers. Nine key informants were interviewed in-depth. Finally, one seminar with the participation of six representatives of stakeholders was held in the field to collect feedback and comments about the primary research findings.

To select potential participants in the household survey, key informant interviews, and group discussions, the researcher collaborated with heads of local FAs to separate fishery households into three groups—mobile fishing, fixed fishing, and aquaculture—based on the name lists of households received from communal governments. The researcher then visited or contacted potential participants to invite them to participate in the research activities.

Data Collection Methods

Primary Data Collection

This research collected primary data through many sources that emphasized the participation of fishery communities. The study applied PRA tools to collect primary data, including surveys, survey questionnaires, and in-depth understanding of the subject through interviews in the community and discussion groups.

Questionnaires, checklists, and other PRA tools have been translated into English and Vietnamese. The Vietnamese version had been consulted with research experts and English translators. The aim is to ensure consistent accuracy and understanding of every question and its purposes, including using local words to make people understand. Translations of these documents were presented to local authorities and communities to provide a common understanding and secure agreement to participate in the research activities.

Secondary Data Collection

Secondary data such as social-economic conditions, natural conditions at the study sites, and regulations on the utilization and conservation of fisheries resources were collected from available materials such as workshop documents; reports; statistical books; newspapers and magazines from the library

Table 6.2 Summary of data collection sources in the study sites

Source	Scope	Purpose
Household survey	181 households randomly selected from 215 fishery households in two coastal communes (110 Huong Phong and 71 Loc Binh households); income and livelihood activities; various questions on historical shock impact Used a semi-structured questionnaire	Description of shock effects Analysis of changes in livelihood, resource dependency, as well as resilience and response strategies
Key informant interview	Nine key informants Used a checklist of questions	Analysis of policy implementation and intervention. Find out the response strategies
Focus group discussion	Three focus group discussions were held in each commune; each group had 3–5 people representing two aquaculture groups, two mobile fishing groups, and two fixed fishing groups	Analysis of the history of climate shocks and classification of shock groups
Seminar with stakeholders	The study findings were presented at a seminar with the participation of representatives of local communes, district and province, and FAs	Raise perceptions and concerns with related stakeholders Receive comments and feedback on the research methodology, field trip activities, results of the research, and recommendations

of the Hue University; local governments; FA offices; and other stakeholders. Secondary data were also collected from the Internet. The data collected to describe historical and present climate shocks and resilience are summarized in Table 6.2.

Data Analysis

Qualitative Analysis

Qualitative data analysis was used to describe the levels of perception of local fishers and the resilience process in facing shocks. PRA tools including maps and timelines were used to identify the duration, frequency, and location of shocks, while seasonal calendars were used to find out the seasonal aquaculture and fisheries schedules. The resilience of fishery households was analyzed through changes in a fishery household's livelihoods before and after disturbances.

Quantitative Analysis

ASSESSING FISHERY HOUSEHOLDS' ABILITY

Household ability to handle shocks refers to the specific fishery and non-fishery efforts and strategies that fishers employ to reduce or minimize adverse effects. In the context of increasing climate risks and shocks in the TG–CH Lagoon, fishery households' ability to handle shocks can be defined as a set of activities or strategies that fishers used to survive during a natural disaster and climate shocks. Surveyed households at the two research sites were asked: 'When facing climate shocks, how did you feel you could handle them?' They gave responses via a five-level scale for evaluation, where 1 = very easily and 5 = could not handle it at all. The assessment's result enables fishery households to precisely understand their current situation, as well as adaptive strategies that can be applied to their families.

RESILIENCE ANALYSIS

The resilience analysis involved a response strategy analysis for selected households, drawing on resilience assessment techniques recently field-tested in fishery communities at the study sites. The primary data for these resilience assessments were collected through a series of household interviews, complemented by community-level focus groups. Descriptions of the response strategies were used for the resilience assessment of households. Béné (2013) argued that resilience costs are an indication of the resilience level of a system (household or community in the context of this research): 'The lower the resilience costs, the more resilient the system is (to a given shock)' (Béné, 2013; Béné et al., 2016). The literature review indicates that resilience can be measured by analyzing the response strategies applied by fishery households and communities to overcome and recover after a specific disturbance.

The Resilience of Fisheries Households in the Tam Giang – Cau Hai Lagoon

Climate Shocks at Study Sites

At both study sites, most respondents perceived climate change as manifested in the fluctuation of precipitation and temperature over recent years. To obtain more details of their perception of shocks, the fishers were asked to list all the climate shocks that had affected them in the previous 5 years. In general, the local fishers mentioned seven types of climate shock in both communes. These shocks were both short and long-term, including storms, floods, extreme cold, drought, freshening, water pollution, and other shocks (Table 6.3).

In general, shocks because of climate change and natural disasters had a significant impact on the life and livelihood of fishers in the two communes of Loc Binh and Huong Phong. Fishery households, especially the group of

Table 6.3 Household ability to handle shocks (response scale 1–5)

Climate shock	Ability to handle climate shocks		
	N	Mean value	Standard deviation
Storm	152	3.67	0.926
Flood	178	3.85	0.813
Extreme cold	120	3.54	0.897
Drought	81	3.49	0.896
Freshening	135	4.03	0.938
Water pollution	144	4.20	0.790
Others	21	3.81	1.078

Source: Household survey (2018).

mobile fishing households, were the most directly and severely affected (Ha & Thang, 2017). Therefore, the fishers had to strengthen their ability to deal with shocks that occurred. Resilience and adaptation measures were the basis for the fishery communities recovering from shocks.

The Ability of the Fishery Households to Handle Shocks

Assessing fishery households' ability to deal with shocks is very important (Béné et al., 2016). Table 6.3 presents households' ability to cope with climate shocks in the current study. The combined results show that although storms and floods were shocks that had a severe impact on the livelihoods of the fishers, fishers suggested that they were able to cope with shocks, although with some difficulty.

Table 6.3 shows that the fishers rated their ability to cope with storms at 3.67 and with floods at 3.85, corresponding to the ability to cope with 'a bit of difficulty.' This is consistent with the results of the group discussions; the participants said that they had been living in the lagoon for a long time and were quite familiar with natural disasters such as floods and storms. These shocks occurred regularly every year and caused severe impacts on fishers' livelihoods and lives. Although local fishers did not have enough resources, such as financial and physical capital, to deal with storms and floods, they had enough experience to minimize the damage caused by these shocks.

Conversely, the ability of fishers to respond to shocks related to water sources, such as freshening and water pollution, was limited. Recent water-related shocks had been caused by uncontrolled aquaculture development and fluctuations in weather conditions such as temperature and rainfall and the lagoon environment. Such shocks had recently caused difficulties for fishers involved in both fishing and aquaculture. Most fishers have argued that it was difficult to cope with water-related shocks, as they appeared abnormal and occurred on a large scale across the entire TG–CH Lagoon. The survey results show that freshening (4.03) and water pollution (4.20) were shocks that fishers felt they could not cope with.

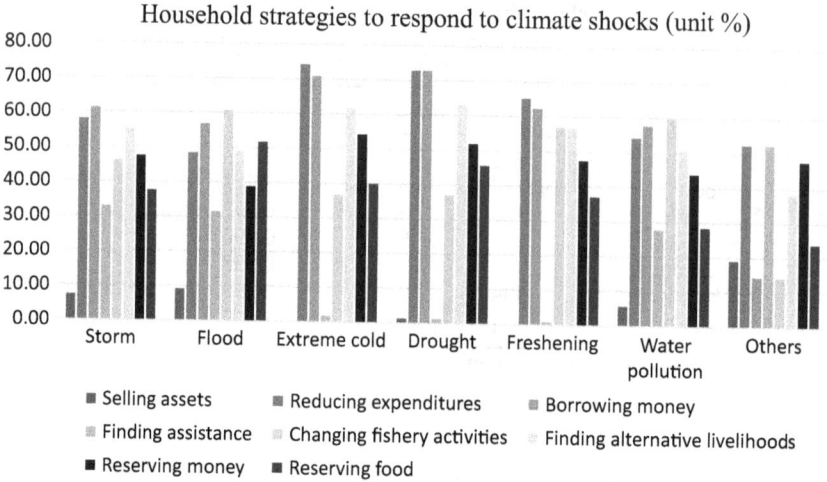

Household strategies to respond to climate shocks (unit %)

Legend:
- Selling assets
- Reducing expenditures
- Borrowing money
- Finding assistance
- Changing fishery activities
- Finding alternative livelihoods
- Reserving money
- Reserving food

Figure 6.6 Response strategies for dealing with each type of climate shock (grouped by shock).

Source: Household survey (2018).

The results of group discussions reveal that it took a long time for fishery households to recover from aftershocks. The recovery time for the mobile fishing group was longer than that for the fixed fishing and aquaculture groups. This is understandable because the income of mobile fishing households depended on only one fishing activity. If shocks occurred, it was difficult for them to respond; thus, their recovery was slower than for other groups in their community.

Figure 6.6 shows that fishers applied many different solutions to deal with shocks occurring in their locality. In particular, to deal with storms and floods, fishery households have applied relatively uniform solutions.

Response Strategies of Fishery Households

Since small-scale fishers in this study were vulnerable in a range of ways from other groups in their communities under climate variability and change, they developed coping strategies. Some employed coping practices related to specific climate-induced vulnerability, as pointed out by local fishers in household interviews, FGDs, and key informant interviews, as discussed below and summarised in Table 6.4.

The results from the household survey at the two study sites of Loc Binh and Huong Phong show that to cope with the negative impacts of climate shocks, fishers implemented the following four types of strategic solutions:

Table 6.4 Summary of response strategies employed by fishery households

Strategy category	Type of response	Detail/Indicator
Short-term coping strategies	Borrowing money	- Fishers borrowed money from friends, relatives, local microcredit, banks, FAs (in Loc Binh only), local organizations, and so on.
	Reducing expenditure	- Reduced spending on fishery activities such as fuel, new fishing gear, aquatic fingerlings, and aquatic feed.
		- Reduced spending on recreational activities and daily habits such as cigarettes, alcohol, and lottery tickets.
		- Reduced food expenditure, including regular food purchases and expensive foodstuffs such as beef and pork.
		- Reduced general expenses for the family such as clothing, electronic devices, and motorbikes.
		- Reduced treatment costs such as medicines and medical examinations at medical facilities.
		- Reduced spending on education for children.
		- Reduced spending on religious activities such as worship.
	Reserving money	- Deposited money in banks.
		- Sent money to a local credit fund.
		- Deposited money in microcredit institutions.
	Reserving food	- Reserved rice and other foodstuffs.
		- Stored food for flood and storm seasons.
	Selling assets	- Valuable assets such as motorbikes, televisions, and jewelry.
		- Houses and land.
		- Aquaculture ponds, fish corrals, and fishing boats.

(Continued)

Table 6.4 (Continued)

Strategy category	Type of response	Detail/Indicator
Fishery-related strategies	Changing fishery activities	- Changed fishing gear. - Changed targeted species. - Changed fishing grounds. - Changed the aquaculture calendar. - Reduced the number of fishing days. - Changed the number of crew members.
Livelihood diversification	Finding alternative livelihoods	- Local employment. - Trading. - Seafood processing. - Poultry and cattle breeding. - Rice cultivation. - Marine fishing.
	Migration	- Seasonal labor migration. - Inside or outside the community.
	Exit the fishery	- Started a new job/livelihood.
Social network-based strategies	Finding assistance	- Other fishers. - Local FAs. - Local governments. - NGOs. - Local Women's Union.
	Seek support from friends and peers	- Friends. - Relatives.

Household Response Typology

The matrix in Figure 6.7 shows that, in general, *short-term coping strategies* were the most frequent responses put in place by fishery households. Among these strategies, *borrowing money* and *reduction of expenditures* were commonly adopted, while asset selling was notable for being rarely done in either the Loc Binh or Huong Phong communes.

Several other essential observations emerge beyond the generic pattern of short-term coping strategies. First, social network-based approaches (seeking

Loc Binh	Money 1	Expenses	Diversification	Money 2	Food	Fisheries	Assistance	Assets
%	75.74	73.77	65.08	57.87	42.30	40.16	17.21	4.59

Huong Phong	Fisheries	Food	Diversification	Assistance	Expenses	Money 1	Money 2	Assets
%	78.28	32.58	25.79	25.34	22.17	20.81	14.48	5.43

Both	Money 1	Expenses	Diversification	Fisheries	Money 2	Food	Assistance	Assets
%	61.13	60.05	54.63	50.30	46.33	39.71	19.37	4.81

Code for responses:

Short-term coping strategies: Money 1 = borrowing money; Expenses = reducing general expenditures; Money 2 = reserving money; Food = reserving food; Assets = selling family assets;

Fishery-related strategies: Fishery = change fishery activities;

Livelihood diversification: Diversification = finding alternative livelihoods;

Social network-based strategies: Assistance = seek support from friends and peers;

Figure 6.7 Climate shock–response matrices for the two communes. Values are the percentage of total responses ranked from the most (left) to the least (right) often implemented. The color codes represent the four categories described in Table 6.4.

Source: Household survey (2018).

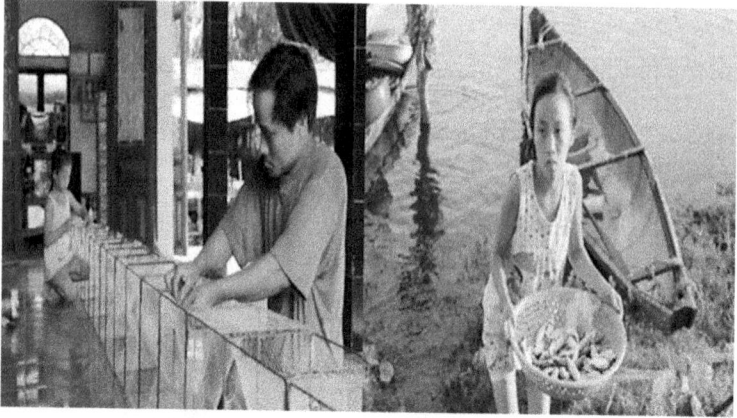

Figure 6.8 Children have to do extra jobs to help their families outside school hours.

support or assistance from relatives, friends, and peers) appear not to be considered by fishers as a way to cope with or respond to climate shocks: this group of strategies was the least frequently adopted. Instead, in addition to resilience strategies, fishery households relied on a combination of fishery-related strategies and livelihood diversification strategies. There was a balance in these two categories that varied between the communes. Among short-term coping strategies, response solutions such as cutting spending, selling assets, borrowing money, and reserving money and food were commonly implemented by fishers at the two study sites. The combined results from the two communes show that financial solutions were used most frequently by fishers (61.13%), followed by reducing general expenditure (60.05%). The poverty rate was relatively high in these fishery communities, so their financial capital was often insufficient and unavailable to deal with the negative impacts of climate risks and shocks. Therefore, the majority of households selected the two financially related responses to restore their livelihoods aftershocks. In particular, reducing spending on leisure activities and daily habits such as tobacco, wine and beer helped fishery households overcome a difficult period because of loss of revenue. In addition, the credit solution helped households reinvest in their livelihood activities after being affected by shocks.

An interesting finding of this study is that in the two communes, the majority of the surveyed households had to reduce their daily living expenses, but they did not reduce education expenditures such as school fees or school supplies for children. Fishers recognized that investing in education for their children was worthwhile and that they must strive to sustain their children's education. Their well-educated children would have more opportunities to obtain better jobs and not have to do strenuous and low-income fisheries activities in the TG–CH Lagoon in the future.

Box 6.1   Investment in children's education

In Van Quat Dong village, a fisher, La Tiem, Huong Phong Commune, stated, 'Fishers like us have been impoverished through generations. Although we have worked very hard, the income is not enough to cover our daily living costs. In addition, the recent negative impacts of climate change and risks make fisheries more difficult. We do not want our children to continue these hard fisheries activities in the future, so at all costs, we have to invest in our children's education so that they can find better jobs.'

There was a large difference between the two communes in terms of priorities in implementing resilience and response solutions following climate shocks. In Loc Binh, choosing to borrow money and reduce spending was the most common response, 75.74% and 73.77% of households. However, in Huong Phong, the local fishers prioritized changing fisheries activities (78.28%) and reserving food (32.58%). The priority given to fisheries changes for long-term adaptation to climate shocks in Huong Phong was very high (78.28%), whereas, in Loc Binh, this was only 40.16%.

Besides fishery-related solutions, a significant proportion of households had adopted non-fishery solutions to cope with and respond to climate shocks, the most popular of which was livelihood conversion. More than 54% of the surveyed households had changed their livelihood activities to avoid relying entirely on fisheries income sources at the two study sites. The rates in Loc Binh and Huong Phong were very different: 65.08% and 25.79%, respectively. The most diversified livelihood activities were temporary or permanent migration to other places to find new jobs. Many fishers and their family members had migrated for seasonal labor to industrial zones and big cities searching for labor hire or construction jobs. Some fishery households had carried out agricultural activities such as rice cultivation, cattle, and poultry raising, but this was not common. The diversification of livelihoods provided significant additional income for local fishers to cope with shocks effectively; however, finding alternative livelihoods was not always easy (Hanh & Boonstra, 2019).

One of the other essential response solutions chosen by fishers was social network-based strategies. Although the percentage of fishery households applying such solutions was not high (19.37% across communes), fishers considered this a crucial solution. They stayed connected by engaging with local government, interacting more often with other fishers in and out of their community, helping local fishers stay informed, and receiving support from stakeholders when dealing with shocks. Local authorities, NGOs, and local organizations provide much timely support so that fishers can cope and recover from aftershocks, including by offering financial and physical assistance.

Box 6.2   The traditional custom of helping each other

According to a leader of the Agriculture and Rural Development Department of Phu Loc District:

Fishers in the TG–CH Lagoon area in particular and fishers in general always have a tradition of helping each other. This is an excellent traditional culture that has been maintained for a long time in the fishery communities. This tradition allows people to receive the support and assistance of other fishers and people in the community to overcome difficulties and disadvantages, including the negative impacts of climate change and shocks.

The survey results also show that despite dealing with difficulties caused by shocks, the percentage of fishers who had to sell their family properties was low, only around 5% across the two communes. This is because most households were poor and thus did not have many valuable assets to sell. Saleable assets such as boats, fishing gear, fish corrals, and aquaculture ponds were closely linked to the livelihood of fishery households, so they could not be sold. Finally, to mobilize financial resources to respond to and recover from shocks, most fishers chose to borrow money from local banks, microcredit institutions, or relatives and friends.

The support and assistance of local authorities were also important in helping fishers to respond promptly and effectively to climate shocks. Access to support resources was an important factor for fishers in restoring their livelihoods and wellbeing; this was especially true for mobile fishing households, most of which were poor and lacked livelihood capital to adapt and recover aftershocks.

Figure 6.5 shows that in response to shocks, solutions involving a reduction in expenditure and borrowing money were applied by most of the fishery households. These groups of solutions helped recover financial capital for households. People used these solutions to reserve money and food over a long period. This was one of the experiences of the fishery community in dealing with natural disasters and shocks. Accordingly, fishery households had to stock up on rice, dry food such as dried fish and meat, and cash just before the annual disaster season (usually July–November each year). Therefore, if natural disasters like storms and floods occurred, fishers would have sufficient food and money to last a few days.

Household strategies for responding to climate shocks (unit %)

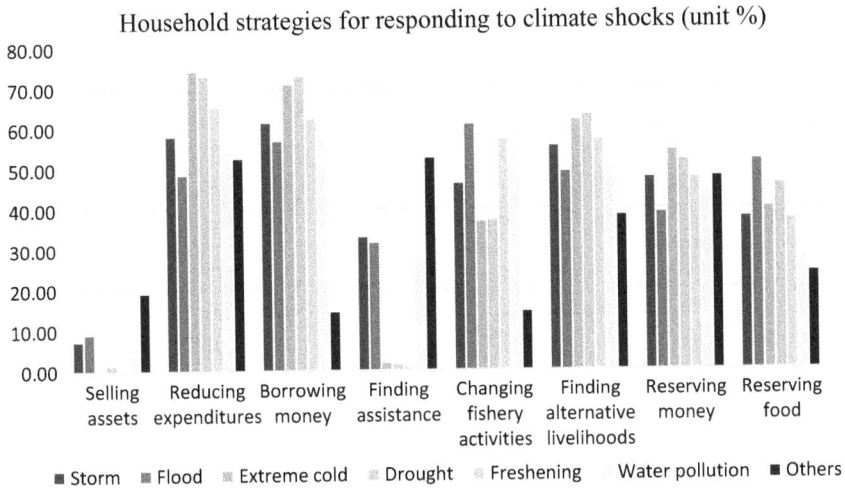

Figure 6.9 Household strategies for responding to climate shocks (grouped by solutions).

Source: Household survey (2018).

Discussion

In this study, the theoretical framework for resilience was approached and considered to analyze the resilience of fishery households. However, this study did not go as far as undertaking a quantitative analysis of resilience indicators such as those reported by (Béné et al., 2016; Smith & Frankenberger, 2017). Thus, in this research, resilience capacity is defined as a group of response solutions, perceptions, or skills that enable fishery households to achieve resilience in the face of climate shocks. This approach is consistent with the IPCC (2012) definition of resilience as 'the ability of a system or individual to anticipate, absorb, accommodate or recover from the effects of threatening shocks in a timely and efficient manner.' Therefore, instead of discussing each performance indicator individually, this discussion addresses them in groups in the following sections.

Social Connection

Social capital can be described as the quantity and quality of social resources (social networks, social connections, and access to wider local organizations) that people can access and use as a platform to deal with shocks. Education and training can improve social capital for resilience (de Bruijn, Buurman, Mens, Dahm, & Klijn, 2017).

Social capital relates to the network and connections of members within and outside the fishery communities at the study sites. In the context of natural disasters and shocks, fishers work closely together and help each other to overcome difficult situations (Binh et al., 2018). Regardless of gender, clan, or age, members of fisher communities are closely linked together. As analyzed in the previous section, there was a tradition of community support among fishers in neighborhoods.[1] This connection ensured timely financial, physical, and information support such as loans of money and food from relatives, friends, and neighbors.

Social cognitive ability, experience and confidence to adapt, and the perceptions and behaviors of fishers when facing difficulties are thought to be essential characteristics of households to promote resilience in the face of climate shocks (Ha & Thang, 2017; Paton & Johnston, 2001). Fishers have lived for a long time in the TG–CH Lagoon study area and have much experience in adapting to and coping with natural disasters (Hoang, Momtaz, & Schreider, 2020). In many cases, the fishers in this study had used their expertise and indigenous knowledge to implement solutions to deal with shocks.

Economic Conditions to Deal with Shocks

The economic sources of resilience capacity for the fishers in this study were property ownership, livelihood diversity, and access to formal and informal loan channels locally. The research results show that most of the fishery households owned assets such as houses, land, boats, stake traps, and motorbikes. These were considered an essential source of collateral and could be used to mobilize financial resources for fishing households to cope with and recover from shocks. However, the majority of households damaged by shocks did not sell their properties (Figure 6.5); rather, they raised money by borrowing from friends, relatives, neighbors, or local credit institutions, consistent with the findings of Béné et al. (2016).

In the past, borrowing money was very difficult because organizations and individuals lending money were not present in the communities. However, with the support of local authorities and organizations, there are now more and more loan channels available to fishers. Notably, in Mai Gia Phuong village, Loc Binh Commune, the local FA can mobilize money and lend it to members facing economic difficulties because of natural disasters and shocks.

Local Governance and Alternative Livelihoods

Flexible and responsible local governance was crucial for fishery households; it ensured their rights and allowed local authorities to provide timely support when natural disasters occurred. Also, it enabled households to access resources, services, markets, and information, thus helping them prepare for and recover from shocks.

Local authorities also play a significant role in promoting the diversification of livelihoods for fishers (Huong, 2010; Tuyen, Armitage, & Marschke, 2010). Fishers had applied a group of solutions to gradually decrease their dependence on lagoon resource exploitation. These solutions were also

Figure 6.10 Mangrove restoration at the study sites.

widely reviewed and implemented by authorities in Huong Phong and Loc Binh communes. In addition, NGOs and local organizations had played a particularly important role in helping fishers find new livelihoods such as community-based ecotourism (CBE) in Loc Binh, polyculture aquaculture, and rehabilitation of mangrove ecosystems to prevent sea-level rise, storms, and floods in Huong Phong (Figure 6.10).

Finally, the research results confirm that the three capacity groups of fishers' resilience were found to be necessary. They played a vital role in helping fishers immediately deal with the adverse effects of shocks and rapidly recover part of their livelihoods and lives.

Conclusion

In Loc Binh and Huong Phong, fishers' proactive solutions and coping strategies were very active in general. In addition, the timely assistance and support of local authorities and other organizations helped fishers at the two research sites to recover quite effectively. This explains why the fishery households' adaptive capacity was relatively high. Also, solutions to diversify livelihoods have brought stability to people's livelihoods and lives. Such solutions provided positive initial results and promise to help people to adapt to long-term climate change and the increased disaster risk situation in the TG–CH Lagoon.

In terms of response strategies, this research has identified several groups of adaptive strategies used by local fishers. Solutions to respond to shocks were quite diverse, ranging from fisheries-related changes to non-fishing strategies. The most important was finding alternative livelihoods. New livelihoods provided fishers with financial security to help them to cope with short-term shocks. However, non-fishery livelihoods reduced resource exploitation pressure in the TG–CH Lagoon, enhancing the lagoon ecosystem's ability to recover sustainably and thereby minimizing the impacts of climate change and shocks to its environment and natural resources.

Acknowledgments

The authors would like to acknowledge The Ministry of Education and Training, Vietnam, and The University of Newcastle, Australia, for their financial and academic support for this research. The authors also acknowledge the partial support of the University of Agriculture and Forestry, Hue University, under the Strategic Research Group Program, Grant No. NCM. DHNL.2021.05.

Notes

1 In Vietnamese, *Tình làng, nghĩa xóm*.
2 Based on *Decision 59/2015/QD-TTG of Vietnam Government on Multidimensional Approach to Poverty Applies to the Period 2016–2020*.

References

Armitage, D., & Marschke, M. (2013). Assessing the future of small-scale fishery systems in coastal Vietnam and the implications for policy. *Environmental Science & Policy, 27*, 184–194. http://dx.doi.org/10.1016/j.envsci.2012.12.015

Béné, C. (2013). Towards a quantifiable measure of resilience. *IDS Working Papers, 2013*(434), 1–27.

Béné, C., Al-Hassan, R. M., Amarasinghe, O., Fong, P., Ocran, J., Onumah, E., ... & Mills, D. J. (2016). Is resilience socially constructed? Empirical evidence from Fiji, Ghana, Sri Lanka, and Vietnam. *Global Environmental Change, 38*, 153–170. http://dx.doi.org/10.1016/j.gloenvcha.2016.03.005

Binh, T. X., Huong, D. T. V., Thien, P. V., & Chau, D. L. M. (2018). *Đánh giá tính dễ bị tổn thương trên hệ thống đầm phá Tam Giang - Cầu Hai, tỉnh Thừa Thiên Huế - Assessment of climate change vulnerability on the Tam Giang - Cau Hai lagoon system in Thua Thien Hue province*. Retrieved from The Green Annamites: https://pdf.usaid.gov/pdf_docs/PA00TQXG.pdf

Chambers, R. (2005). *Rural appraisal: Rapid, relaxed, participatory*. Vikas Publishing House.

de Bruijn, K., Buurman, J., Mens, M., Dahm, R., & Klijn, F. (2017). Resilience in practice: Five principles to enable societies to cope with extreme weather events. *Environmental Science & Policy, 70*, 21–30. http://dx.doi.org/10.1016/j.envsci.2017.02.001

Douxchamps, S., Debevec, L., Giordano, M., & Barron, J. (2017). Monitoring and evaluation of climate resilience for agricultural development – A review of currently available tools. *World Development Perspectives, 5*, 10–23. https://doi.org/10.1016/j.wdp.2017.02.001

GADM (Cartographer). (2018). Global administrative areas: Maps and data. Retrieved from https://gadm.org/maps/VNM.html

Ha, H. D., & Thang, T. N. (2017). Fishery communities' perception of climate change effects on local livelihoods in Tam Giang Lagoon, Vietnam. In Thang, T.N., Dung, N.T., David,. H., Sharma. S., Shivakoti, P.G., Ngo Tri Dung (Eds). In *Redefining diversity and dynamics of natural resources management in Asia, volume 3* (pp. 111–124). Elsevier.

Hanh, T. T. H., & Boonstra, W. J. (2019). What prevents small-scale fishing and aquaculture households from engaging in alternative livelihoods? A case study in the Tam Giang lagoon, Viet Nam. *Ocean & Coastal Management, 182*, 104943. https://doi.org/10.1016/j.ocecoaman.2019.104943

Hoang, H. D., Momtaz, S., & Schreider, M. (2020). Assessing the vulnerability of small-scale fishery communities in the estuarine areas of Central Vietnam in the context of increasing climate risks. *Ocean & Coastal Management, 196*, 105302. https://doi.org/10.1016/j.ocecoaman.2020.105302

Huong, T. T. T. (2010). *Resource access and livelihood resilience in Tam Giang Lagoon, Vietnam*. University of Manitoba.

IPCC. (2012). *Managing the risks of extreme events and disasters to advance climate change adaptation: Special report of the intergovernmental panel on climate change*. Cambridge University Press.

Momtaz, S., & Kabir, S. M. Z. (2018). Chapter 6 - evaluating community participation in environmental impact assessment. In S. Momtaz & S. M. Z. Kabir (Eds.), *Evaluating environmental and social impact assessment in developing countries* (Second Edition) (pp. 123–140). Elsevier.

Paton, D., & Johnston, D. (2001). Disasters and communities: Vulnerability, resilience and preparedness. *Disaster Prevention and Management: An International Journal, 10*(4), 270–277.

Smith, L. C., & Frankenberger, T. R. (2017). Does resilience capacity reduce the negative impact of shocks on household food security? Evidence from the 2014 floods in Northern Bangladesh. *World Development*.

Thanh, N. V., Vu, M. T., Nguyen, A. D., & Zheng, J.-H. (2013). Overview of estuary research and waterway engineering in Vietnam. *Proceeding of the 35th IAHR World Congress*.

Tuyen, T. V., Armitage, D., & Marschke, M. (2010). Livelihoods and co-management in the Tam Giang lagoon, Vietnam. *Ocean & Coastal Management, 53*(7), 327–335.

Yamane, T. (1967). *Statistics, an introductory analysis*. Harper & Row.

7 Relocation as a climate change risk reduction strategy

Socio-political insights from Sri Lanka

Cynthia M. Caron

Introduction

Sri Lanka, formerly Ceylon, is an island lying in the Indian Ocean 40 miles off India's southeastern coast. The country has 25 districts organized into nine provinces. As a lower-middle-income country, plantation-based export agriculture, manufacturing and industry, and tourism and related services contribute to the country's national economic growth. The island's topography includes a central mountainous region, where considerable agricultural, forestry, and tea cultivation takes place. The island has two monsoon seasons. Monsoonal rains with annual rainfall levels often over 3,000 mm in a season fall over these mountainous regions and the country's low-lying coastal metropolitan area, which includes its capital city, Colombo (Kawamura, 2015). Heavy monsoonal rains and subsequent flash floods and landslides not only threaten the country's natural resource base but also temporarily displace thousands of urban and rural residents, compromising their well-being and destroying their accumulated wealth, assets, and livelihoods.

In 2019, the Internal Displacement Monitoring Center (IDMC) calculated 23.9 million *new* global displacements due to weather-related disasters, with 87,000 of these displacements taking place in Sri Lanka (IDMC, 2020). In 2021, Germanwatch ranked Sri Lanka 30th out of 180 countries on the Global Climate Risk Index, based on its "exposure and vulnerability to extreme events" (Eckstein et al., 2021, p. 3). Economists estimate that flooding from heavy rainfall or precipitation across Sri Lanka has caused household-level annual asset losses of US$78 million (Walsh & Hallegatte, 2019). With climate change and accompanying temperature rise, Sri Lanka's gross domestic product (GDP) might decline by 7.7%, representing a loss of $50 billion by 2050 (United Nations Office for Disaster Risk Reduction [UNDRR], 2019, p. 13).

Terms such as "weather-related disasters" and "exposure to extreme events" are relevant to understanding how the Sri Lankan state and local communities contend with current and prepare for future climate-induced change. This chapter considers that climate change effects include both slow-onset processes and rapid-onset events. Slow-onset processes include the incremental changes "that have no clear start or endpoint" (Schafer et al.,

DOI: 10.4324/9781003216476-9

2021, p. 4) and manifest themselves in sea-level rise, increases in mean temperature, and land degradation. Rapid-onset events, often referred to as extreme weather events, are discrete events with a clear beginning and end that may "occur or reoccur in a matter of days or even hours at a local .... scale" (Schafer et al., 2021, p. 4), and have highly visible impacts that result in major damages in a short period. Heavy rains and subsequent flash flooding and landslides are examples.

Relying on these climatic characterizations, I construct three case studies to illustrate state and local community responses to weather-related events and the use of relocation to reduce future damages. Each relocation case study results from a different type of event, two of which scholars expect to become more common with global climate change (Caron et al., 2014; Schafer et al., 2021). Relocation following the 2004 Indian Ocean tsunami, which was a significant state-coordinated relocation event, is the first case. The second case examines the aftermath of heavy monsoonal rains in the Sri Lankan central highlands in May 2016. The third case presents an atypical landslide, the collapse of the Meethotamulla landfill, following heavy rains in May 2017, and prompts planners and policymakers to rethink the contours of natural disasters. These three cases fulfill three aims. The first aim is to explore how the Sri Lankan state used relocation as a governance strategy that also structures its relationship with society. The second aim is to situate the Government of Sri Lanka's (GoSL's) current use of the relocation as a risk reduction strategy historically. The final aim is to highlight how the effects of climate change might emerge in unexpected places.

Ethnicity and Sri Lankan socio-political dynamics

To set the socio-political context for the state's approach to and engagements with local communities vis-à-vis relocation, I briefly lay out important events in the country's ethnic politics. The Portuguese, the Dutch, and the British colonized the island for over 400 years before it won independence in February 1948. Sri Lanka's 2020 population of approximately 21.9 million represents an ethnically and culturally diverse society (World Bank, 2020). Three languages are spoken (Sinhala, Tamil, and English), and many religions are practiced (Buddhism, Hinduism, Islam, Roman Catholicism, and many forms of Christianity). According to the most recent government census, the Sinhalese form the largest ethnic group on the island, constituting 75% of the population, followed by Sri Lankan Tamils 11%, Muslims and Moors 9%, and Indian Tamils (Tamils of Indian descent) 4% (GoSL, 2012). Following Independence, Sri Lanka followed a Socialist-Democratic form of government making literacy and healthcare gains (Silva, 2010). Over time, tensions between ethnic groups emerged from divergent understandings of which group benefited more from government interventions. These ethnic tensions partially grew out of the colonial experience but were also rooted in political, economic, and social insecurities that were class and regionally based. Sinhalese politicians, for example, feared that the Sri Lankan Tamil population would join with the Indian Tamil population (that worked on tea estates) to form a voting bloc and threaten their majority. Despite this ethnically charged climate, the government implemented two multi-purpose agricultural development projects that relocated landless Sinhalese peasants from Southern provinces to the north and east, displacing their original Sri Lankan Tamil and Muslim farmer-occupants that the state deemed "encroachers" (Muggah, 2008, pp. 82–83). These projects significantly altered the demographic composition of the north and east and continue to shape how citizens perceive the intentions of state-led development.

Ethnic tensions led to the emergence of political organizations in the Tamil-dominated North seeking to lead a separatist struggle against the Sinhala government in the South. The country suffered more than 30 years of protracted ethnic conflict. Sri Lankan Armed Forces militarily defeated the Liberation Tigers of Tamil Eelam (LTTE) in May 2009. While many military operations took place in the North and East, citizens across the island contended with militarization in everyday life. The ethnic conflict was not the only source of political violence. From 1971 through the late 1980s, leaders from the revolutionary group, the Janatha Vimukthi Peramua (JVP), harnessed feelings of political alienation and mobilized youth, particularly Sinhalese university students from rural peasant backgrounds, to participate in a series of insurrections aimed at capturing state power (Moore, 1993). Many of the same rural grievances animate political relations today and shape how rural residents perceive state actions. In sum, ethnic and class politics and urban–rural dynamics shape the Sri Lankan government's development agenda, including relocating families to pursue them (Muggah, 2008). Finally, Sri Lanka, like most country governments, has incorporated climate change effects into its economic development planning and its obligation to maintain the social contract (Loewe et al., 2021). Climate change effects threaten the lives and livelihoods of citizens around the world, leading many country governments to implement state-led relocation as a strategy to protect them. In Sri Lanka, socio-political dynamics of ethnicity, class, and urban–rural location certainly influence state-led relocation efforts.

The role of the state: relocation and climate change effects

The Sri Lankan state has a history of using relocation to fulfil its development agenda (Muggah, 2008). While government officials might suggest that families living in areas threatened by climate-change effects such as floods or landslides relocate to protect their lives and livelihoods, this recommendation is often suggested or ordered after the fact, when families have lost members and/or all or some of its wealth and asset base. Unfortunately, financial and/or logistical assistance to relocate is not always provided before disaster strikes. Ferris (2015) argued that just as the government has the responsibility to evict people from unsafe buildings or structures, it too has the responsibility "to move people from areas where it is unsafe for them to live" (p. 110). In such instances, governments might pursue planned relocation (United

Nations High Commission for Refugees [UNHCR], 2014) or preventative relocation (Claudianos, 2014). The Government of Fiji, for example, recognized the threats that climate change presented to its small island nation and created a planned relocation framework in 2018. Drawing on UNCHR work, the Fijian Plan states that planned relocation is

> understood as a solution-oriented measure, involving the State, in which a community (as distinct from an individual/household) is physically moved to another location and resettled permanently there.
> (UNHCR 2014: 10; National Legislative Bodies [NLB] 2018: 7)

Claudianos (2014) coined the term preventative resettlement to propose relocating communities vulnerable to a natural disaster sooner rather than later, thereby preventing disaster altogether. While the Fiji framework recognizes that human mobility is a climate change adaptation (CCA) strategy, one principle of the Fijian framework is that planned relocation will take place "only when all [other] adaptation options have been exhausted (NLB, 2018, p. 7). The term preventative resettlement is not in common use. However, it shares many of the same principles as the more commonly used term planned relocation (i.e., being human-centered, livelihood-based, and preemptive). Therefore, in my analysis below, I will use the term preventative resettlement, when appropriate, to recognize how this conceptualization of relocation with its emphasis on a paternalistic state reveals the socio-political dynamics of relocation in Sri Lanka. In doing so, I contribute to the wider literature on how states employ relocation as both a strategy of governance and a disaster risk reduction (DRR) strategy.

Relocation as disaster risk reduction and climate change adaptation

Around the world, governments have relocated entire neighbors and communities around the world to make areas more attractive for foreign direct investment and reduce vulnerability to recurrent hazards. The emerging scholarship on relocation as either a DRR or CCA strategy is insightful. As scientists and policymakers grapple with the effects of global climate change, scholarship on saving lives, protecting assets, and pursuing livelihoods problematize relocation as both a DRR and CCA measure. King et al. (2014) wrote that "migration is a natural part of life" (p. 284) and considers relocating one's self from an unsafe area as a natural response to feelings of insecurity, fear, or danger. Viewing such migration to reduce the risks of climate change frames relocation as an adaptive strategy rather than as a failure to adapt to risk. Not all potentially affected persons who wish to relocate from a place suspectable to recurrent natural disasters such as floods, drought, or landsides have the economic means and social networks to move. In such cases, they may need to wait for government authorities to recognize risk and engage in consultation measures to plan collaboratively for relocation or wait for disaster to strike and sustain losses in order to be eligible for financial resources to relocate.

Nalau and Handmer's (2018) analysis of DRR and CCA perspectives on relocation shows how DRR practice conceptualizes relocation as a temporary measure such as a short-term evacuation. In contrast, CCA practice views relocation as a permanent mass relocation, including moving communities to other countries. I draw attention to this distinction, as it emphasizes how government integration or formalization of DRR or CCA thinking into national-level policies or action plans might shape how agencies approach and propose relocation measures. The statement made by King et al. (2014) that "relocation is part of a suite of adaptive strategies available to households and communities... formalizing relocation into policy may reduce vulnerability" (p. 89) is an important perspective for the GoSL to consider given the country's socio-economically disadvantaged population that pursues natural resource-based or weather-dependent livelihoods. How the Sri Lankan state approaches local communities with respect to introducing planned relocation as an option raises key questions about how it historically has engaged communities not only in relocation programming in general but in relocation programming following the natural disaster, in particular.

The GoSL has implemented large-scale relocation to fulfill several objectives, including economic development, nation-building, flood risk reduction in the 1990s, and then again for a tsunami risk reduction following the 2004 Indian Ocean tsunami (Muggah, 2008; Silva, 2010). The Mahaweli Development and Irrigation Program of the 1960s is a rich example of relocating entire communities from densely populated areas in southern regions of the island to vast, agricultural settlements in northern, central, and eastern regions for purposes of economic development as well as social-engineered, nation-building (Muggah, 2008). In the 1990s, the government undertook several flood-reduction cum informal settlement removals in and around Colombo, relocating families into some of the city's first high-rise apartment buildings for the urban poor (Fernando, 2018). Since 2010, the Urban Development Authority (UDA) has identified over 68,000 families residing in underserved settlements in Colombo for relocation, partially under the guise of being prone to flooding (Perera, 2019). These settlements also happen to occupy lands with high development or real estate investment potential. To improve their living conditions and to refashion the capital city of Colombo into a "world-class city," a goal of the UDA's Urban Regeneration Project (URP) has been to "eliminate shanties, [and] slums ... from the city of Colombo by relocating dwellers in modern houses" including high-rise complexes (Perera, 2016, 2019).

Relocation programs often fail. Families move back to hazardous areas to pursue livelihoods and reconnect with a strong place attachment to their former homes (Oliver-Smith, 2010). Perera (2016, 2019) questioned the success of Colombo's relocation of families into high-rise complexes, as two years after possession of their new homes, relocated families reported an overall deteriorating quality of life, increased debt, desire to move, and crumbling infrastructure. Perera (2016) also found that even though the URP defined a flood threat, nearly 55% of relocated residents surveyed reported that their prior homes never flooded, and nearly 63% never experienced mosqui-to-borne diseases associated with flooding and stagnating pools of water (2016, p. 14). In a survey of families relocated as part of the 1990 flood relo-cations, Fernando (2018) found that relocated families were selling their new homes illegally to pay off post-relocation accrued debt. The experience of relocated Colombo residents raises questions about and undermines the gov-ernment's argument that relocation is to protect lives, assets, and well-being and illustrates that the government might focus more on the socio-cultural dimensions of relocation, such as preserving social networks and access to jobs rather than just shelter. Officials used flood risk reduction and improve-ments to well-being to justify relocation, thereby clearing the way for real estate investment.

Institutional arrangements and frameworks to address disaster risks and relocation

Many Sri Lankan state institutions and agencies provide emergency services following natural disasters and coordinate with local government authorities to provide relief and any subsequent relocation services. Relevant institutions and agencies to this discussion include the Ministry of National Security and Disaster Management, which oversees both the National Building and Research Organization (NBRO), and National Disaster Relief Services Centre, the UDA, and sub-district level government authorities at the Divisional Secretariat (DS) level. The NBRO, a semi-government research and development unit established in 1984, provides a number of web-based public services to keep citizens up to date on weather concerns, including color-coded landslide warnings at the district level and a live, interactive rain-fall map. The NBRO coordinates with and advises donor institutions, local government officials, and affected communities on the design and implemen-tation of relocation sites, including recent relocation programming in the Kegalle District (Fernando et al., 2020). The UDA, established in 1978, pro-motes and regulates development in urban areas. It handed over some of its powers to the Ministry of Megapolis & Western Development (MoMWD) following its establishment in 2015.

The Government created and certified its first Disaster Management Act, No. 13, in May 2005, as, after the tsunami, it was clear that policies were needed "to protect human life and property of the people and the environ-ment of Sri Lanka from the consequence of … disasters" (GoSL, 2005, p. 2). The Disaster Management Act established the Disaster Management Center and the National Council for Disaster Management. The Act empowers gov-ernment agencies to "evacuate people, property and animals from affected or vulnerable areas" (GoSL, 2005, p. 11) and to draw funds from the Reconstruction and Rehabilitation Act to fund the "reconstruction of prop-erty affected by any disaster" (GoSL, 1993, p. 3).

Since the tsunami disaster and increased damages from extreme weather events, the GoSL has enhanced its capacity to understand natural disasters and has created policy frameworks, including adopting measures promoted in the Sendai Framework for Disaster Risk Reduction (UNDRR, 2019). The government also created a Climate Change Secretariat (2008), a National Climate Change Policy (2012), a Comprehensive Disaster Management Program in 2014, and the National Adaption Plan for Climate Change Impacts (2016–2025). The new National Policy Framework (NPF) Vistas of Prosperity and Splendour (2019) includes ten objectives and the key activities to achieve them. Disaster-related content in the NPF falls under *New Approach in National Spatial System* which emphasizes improving the country's early warning system, protecting lives and property damage, empowering institutions to act swiftly, creating appropriate compensation packages for affected communities, and establishing "care centers" to assist women and other vulnerable groups after a disaster along with management plans to sustain and maintain these centers after disaster response ends (2019). The NPF does not include the planned relocation of vulnerable families or entire communities.

Officials that authored the Adaption Plan for Climate Change Impacts (2016–2025) invoked "gradual relocation" for communities threatened by sea-level rise (Climate Change Secretariat, 2016, p. 94) rather than the concept of planned relocation. The adaption plan noted that landslide frequency will create "psychological trauma and stress due to victimization" (2016, p.128) and still, planned relocation as a preventative, preemptive measure is not mentioned as a strategy to reduce such a threat. As such, Sri Lankan policies and plans neither incorporate nor reflect new advances on climate change and DRR. Below, I discuss my approach to exploring the Sri Lankan state's use of relocation with respect to natural disasters.

Method, data collection, and analysis

To investigate how the Sri Lankan state uses relocation following natural disasters, I employed the case study method (Patton, 2002; Crowe et al., 2011) for two main reasons. First, case studies promote purposeful sampling (Patton, 2002). As I am interested in state implementation of post-disaster relocation as a social phenomenon, I purposefully chose specific disaster events regardless of where they took place across the country. I used the purposeful sampling technique of theoretical sampling, as it allowed me to extend theories on the conceptualization and implementation of relocation following natural disasters. I chose to use the case study approach because case studies have advantages for theory building. Case study creation permits an iterative approach between data and existing literature, allowing for the refining of the construct under investigation, in this case, relocation, as new data becomes available. Part of this iterative process uses a bottom-up approach to creating a theory "about specific phenomena" (Eisenhardt, 1989, p. 547), which, while localized, contributes to broader understandings of how and when states use relocation in natural disaster contexts.

To illuminate the GoSL's use of relocation following natural disasters, I created three *instrumental* case studies "to gain a broad appreciation"

(Crowe et al., 2011, p. 101) of post-disaster relocation in Sri Lanka. The first case briefly outlines the relocation of coastal communities following the 2004 Indian Ocean tsunami followed by two more-recent rainfall and landslide events that occurred in 2016 and 2017 following heavier than normal monsoonal rainfall. In each case, I provide pertinent geophysical characteristics of the disaster and the relocation approach pursued. Examining interactions between the groups that suffered loss and government authorities, such as the Disaster Management Center, the UDA, and the Colombo Municipal Council (CMC), reveals how government officials used relocation to address risks associated with the extreme weather events. The relationship between government officials and affected communities reveals the socio-political dynamics of relocation premised on historically contingent relationships between the state and these societal groups.

Finally, to inform these case studies, I draw on 30 years of work and research experience in Sri Lanka. I have worked on a range of Sri Lankan environmental issues since 1992, including holding an international project manager position for post-tsunami resettlement, relocation, and reconstruction from 2005 to 2009. The tsunami case study draws on secondary materials as well as my practitioner experience rebuilding houses with a community in Sri Lanka's eastern Batticaloa District that the government had relocated inland from their previous seaside location. The second and third cases draw on secondary scholarship, including academic writings, government documents, situational reports, local news media reports, and in-depth investigative journalism that I found online. I traced out citations and bibliographic references to build out these two later cases. After presenting each case, I analyze how relocation emerged and how relocation was carried out.

Case Study 1   The Indian Ocean tsunami (2004)

The 2004 Indian Ocean tsunami struck Sri Lanka's coastline on the morning of 26 December 2004 following a 9.0 magnitude earthquake off the northern Sumatran coast of Indonesia. The tsunami's crushing wave affected five of the country's nine provinces (Silva, 2010). More than 35,000 individuals perished, over 800,000 persons were left homeless, $900 million worth of assets were destroyed, and more than 150,000 persons lost their means of livelihood (GoSL & Development Partners, 2005; Caron, 2009a). Nearly 46% of all tsunami deaths occurred in coastal districts of the eastern province, with a predominance of Tamil and Muslim ethnic minorities (Silva, 2010).

Displaced families along the island's affected coastline languished in camps, as the Government worked with international donors to coordinate post-tsunami resettlement and reconstruction programming. In an age of globalized electronic media, the scale of the destruction generated unprecedented levels of financial assistance to fund humanitarian efforts, which Gamburd (2014) refers to as the second tsunami or "golden" wave. The deluge of foreign actors arriving in the country initially hampered the central government's ability to coordinate recovery and reconstruction at the district level (Caron, 2009b; Silva, 2010). In addition to this coordination challenge, another resettlement and reconstruction obstacle was the central government's indecision on a "no-build" coastal buffer zone (Ingram et al., 2006; Lyons, 2009; Silva, 2010).

Invoked as a safety measure to reduce coastal populations' vulnerability to future tsunamis, central government authorities established an irregular and therefore controversial "no-build buffer zone." Buffer zone widths differed by district, with original widths of 100 meters in the South and 200 meters in the ethnic-minority areas of the north and east (GoSL & Development Partners, 2005). The buffer zone created a lack of clarity among displaced families and donor agencies regarding which among tsunami-displaced families could return "home" and rebuild in their original locations and who was to be relocated elsewhere. Government officials, however, claimed that adherence to the 1981 Coastal Conservation Act, ocean floor topography, and the presence of natural barriers such as sand dunes necessitated regionalized differences in buffer zone widths. However, with wider "no build" buffer zones in districts primarily settled by minority Tamil and Muslim ethnic groups, many analysts suspected that the Sinhala-Buddhist dominated state had set the stage for coastal land grabs and fueling real estate speculation.

Buffer zone creation meant that the government needed to find alternative lands to relocate tsunami-affected households and then funding to construct new homes and associated village infrastructure (toilets, roads, water supply). Finding alternative lands that were large enough to keep original village communities together and surrounded by adjacent host communities that would welcome newcomers was not easy. Eventually, the government's inability to find suitable alternative lands compelled it to reduce the original-buffer zone widths to 55 meters in the south and 50–100 meters in the north and east, which allowed thousands of families to return to their original locations (Lyons, 2009; Silva, 2010). Even with reductions in buffer zone width, finding suitable land for relocation sites continued to be plagued with problems such as seasonal water supply, lack of employment opportunities, lack of social and cultural spaces to promote community building, and conflicts with nearby host communities (Caron, 2009a, 2009b; Dias et al., 2016). With the end of the ethnic conflict's temporary ceasefire, its resumption caused some international organizations to leave the country, causing further delays (United Nations Office for Project Services [UNOPS], 2009). In some districts, relocating communities found and purchased large enough relocation sites for their entire community on their own or visited parcels of available land identified by local government officials and chose which parcel to move to (Caron, 2009a).

Tsunami relocation was partial. In one of the post-tsunami villages that I helped reconstruct, the government relocated families within 3 kilometers of their original homes. They now use that land for fish processing and keeping livestock. In some cases, parents have allowed a child to build small, temporary structures in this "no build" zone, as post-tsunami relocation sites lack space to accommodate growing or extended families (Caron fieldwork notes, 2016). Relocated families use their original homes for livelihood pursuits and feel connected to their lost loved ones, reaffirming the power of place attachment (Caron fieldwork notes, 2016; Oliver-Smith, 2010). During my last visit to former project sites in 2018, relocated families with whom I spoke stated that local government officials either did not know or chose to ignore the fact that former homes that were supposed to be unoccupied as a risk reduction measure were slowly being reoccupied.

Case Study 2 Floods and landslides in Kegalle District, May 2016

With 20% of Sri Lanka's land area vulnerable to landslides (UNDRR, 2019), heavy rainfall will result in flooding and landslides across its mountainous areas. After three to four days of relentless rain in May 2016, a large landslide on Samasarakanda buried parts of several villages in the Aranyaka Divisional Secretariat (DS) Division of Sri Lanka's Kegalle District (GoSL, 2016; Perera, 2016). The Aranyaka Division is approximately 110 kilometers from Colombo in the foothills of Sri Lanka's central highlands. Located in the island's wet zone, the Aranyaka area normally receives an average annual rainfall of 2500–3000 mm, with heavy rains associated with both the southwest and the inter-monsoon periods. In May 2016, however, a low-pressure zone that formed over the island resulted in over 450 mm of rainfall in the Kegalle District between May 15–18; the heaviest recorded rainfall in 18 years (Ministry of National Policies and Economic Affairs & Ministry of Disaster Management [MoNPEA & MoDM], 2016). This unusual amount of heavy rain falling within a short period of time triggered a landslide on Samasarakanda, one to one and one-half days after the rainfall's peak (JICA, 2016).

The National Disaster Relief Services Center reported 89 deaths and 102 missing across the entire district, with 50 deaths and 96 persons missing (over a month later) in the Aranyaka area (Ministry of National Policies, 2016, p. 22). The landslide and associated floodwaters fully damaged 623 houses and partially damaged 4,414 houses (United Nations Office for the Coordination of Humanitarian Affairs [UN OCHA], 2016). The Government of Japan's International Cooperation

Agency followed up with technical assistance ten days after the disaster. In cooperation with the Sri Lankan Air Force and as part of the Building Back Better agenda of the Sendai Framework, JICA specialists conducted an aerial survey to understand the landslide's debris flow, sediment movement, and slope angles (Japanese International Cooperation Agency [JICA], 2016). However, without ground-truthing and site visits, the team could not determine if the NBRO should reclassify Samasarakanda and other hilly areas close by as the most dangerous areas for habitation or, more specifically, areas where "landslides are most likely to occur" (JICA, 2016, p. 6).

My review of press releases, activists' blogs, government statements, and scholarly publications yielded contradictory results about whether or not state agencies notified residents of landslide possibilities to keep them safe or asked them to evacuate the area during those heavy rainfall days in May 2016. The Disaster Management Center (DMC) claims that its field staff had sent messages to the community via local village head persons (Grama Niladharis) about landslide risks for several years. However, displaced persons could only recall warming being given once in the 1980s (Sangasumana, 2018). The Department of Meteorology issued both special weather bulletins about heavy rainfall and severe weather advisories for May 15–18 (JICA, 2016). A geophysical scientist noted that the landslide that occurred in Aranyaka was unique, with an unusual width and multiple secondary failures (multiple downslope slumping events of mass material such as soil and rock), making it a type of landslide that is difficult to predict. Yet, if "exceptionally heavy rainfall" (Petley, 2016) continues to fall in Sri Lanka's mountainous areas, then this same type of landslide might become more common.

As to local risk awareness, Perera et al. (2018) reported that some villagers started seeing new foundation cracks in some homes the morning of the landslide, indicating land shifts. Yet, residents did not evacuate, as they neither had an appropriate evacuation plan nor were they prepared to leave behind the wealth in well-established mixed-species home gardens and plantations cultivated with export crops (tea, rubber, and clove). Clearing the slope's natural vegetation to create these rich agricultural landscapes changed the previous rooting structure that stabilized the hillside's soil (Perera et al., 2018).

After the landslide and flooding across the district, local government officials identified over 2,300 houses to be relocated and estimated that the total relocation costs across the entire Kegalle District would be LKR 5.8 billion inclusive of land acquisition, site preparation, and construction costs (MoNPEA & MoDM, 2016, p. 230). Families directly affected by the Samasarakanda landslide will not return home. Affected families reported that while they may continue to cultivate their lands, they are not allowed to rebuild their homes.

The government will acquire them as a security measure (Fernando et al., 2020, p. 43). Until the government identifies and purchases alternative land for relocation, displaced families remained in one of 42 temporary camps scattered across the district, with World Vision, the Sri Lankan Red Cross, and community-based organizations (CBOs) providing emergency relief.

As with post-tsunami relocation, finding enough suitable land to relocate families was a challenge. The then 2016 Minister for Disaster Management stated that the government might need to rethink its approach to relocation (Perera et al., 2018). Historically, reconstruction policy provides a displaced family with a single-family house and a small garden area. Each landslide-affected family is expected to receive a house and a plot of land between 250 and 500 sq meters depending on the land cost (MoNPEA & MoDM, 2016). Within a few months of the landslide, the government had acquired 374 acres of land, and 48 new houses were under construction.

Case Study 3   The Meethotamulla landfill landslide (2017)

The Meethotamulla landslide event is a unique landslide event in Sri Lanka. On 14 April 2017, following three days of intense rainfall, Meethotamulla, a garbage dump on the outskirts of Sri Lanka's capital, Colombo, collapsed, triggering a landslide of mud, decaying refuse, and debris (GoSL, 2017). The landfill's blocked drainage system, and a sudden release of methane amplified the dump's slope instability (Ministry of Megapolis & Western Development [MoMWD], 2017b). The landfill covered over 21 acres, due to an estimated 900 tons of waste dumped every day (MoMWD, 2017a, 2017b). Saturated soil gave way under the garbage's weight.

Even before the landfill collapsed, residents living nearby learned to live with a range of health and environmental hazards: the smell of rotting refuse, respiratory illness, and high rates of dengue fever, hepatitis, leptospirosis (rat fever), and other mosquito-transmitted diseases. Their homes had foundation cracks as a result of the garbage's weight combined with the force of machine motorized garbage compaction (MacDonald, 2019). However, losing lives and property under the weight of a torrent of crushing garbage was unforeseen. Thirty-two persons lost their lives that day (Gunawardena et al., 2019). Some bodies were never recovered. The debris damaged 82 houses, directly affecting 315 families (MoMWD, 2017b).

The Meethotamulla open landfill served as a temporary site following methane explosions and subsequent closure in 2009 of the island's nearby Bloemendhal landfill. The Urban Development Authority (UDA) authorized the Colombo Municipal Council (CMC) to use two acres as a dumping site for waste collected within Colombo. No agency conducted an environmental impact assessment of the site to assess its suitability to serve as a landfill. The site was marshy, abandoned paddy land. The UDA was aware that the CMC overstepped its permitted dumping area (the two acres). In 2016, the UDA requested the CMC remove the solid waste dumped beyond the two acres granted, pay rent for and rehabilitate that land, and return it to the UDA, which needed to repurpose it for other "development purposes" (Wickrematunge, 2017). The CMC ignored the UDA. Other government bodies ignored Meethotamulla residents.

Residents publicly protested the site and its effects on their health in 2010. They later hired a lawyer to file a Supreme Court case to stop dumping in 2014. Protesters encountered police-propelled tear gas, batons, water cannons, and later, a Court Order banning any further protests against the landfill (Decent Lanka, 2015, 2017; Perera & Nagaraj, 2017; Correspondents, 2018). According to the CMC's Information Officer, one week before the 2017 landslide, a study team comprised of UDA and CMC officials identified several houses that should immediately evacuate the area. However, neither UDA nor CMC officials issued an evacuation order (Wickrematunge, 2017). The Sri Lanka Electricity Board did warn some families to evacuate the area due to damages to a local high voltage line (Samarasinghe & Perera, 2017).

Sri Lankan veteran investigative journalists and policy analysts argued that the residents of Meethotamulla are like other disaster-affected poor communities around the island in so far as the state blames them for their residential choices and that the state claims that they ignore warnings to relocate (Perera & Nagaraj, 2017) or the state breaks its promises (Perera & Nagaraj, 2017; Decent Lanka, 2015). The central government made numerous proposals to close the dump and relocate it outside the district. Provincial council members promised alternative housing and schooling. In 2015, Sri Lanka's Prime Minister claimed that dumping refuse in Meethotamulla would end – a call echoed again by another politician in November 2016 (Decent Lanka, 2015). The CMC claimed that during a 2016 meeting held by the Colombo Municipal Commissioner, the government decided that Rs. 1.5 million in compensation would be given for certain "identified residents" to relocate (Wickrematunge, 2017). However, relocation did not occur.

Following the 2017 collapse, President Maithripala Sirisena announced closing the site. The MoMWD submitted addendums to the Environmental and Social Management Planning Framework (ESMPF) of the World Bank-funded 2012 Metro Colombo Urban Development Plan to facilitate its closure. Within six weeks, the UDA had evacuated and permanently relocated 98 affected families (MoMWD, 2017a, 2017b) and provided another 100 families with temporary shelter. Samarasinghe (2017) noted that three months after the collapse, the government had provided some affected families with compensation ranging between Rs. 50,000 and Rs. 150,000 (US$325 and US$975, respectively) to relocate to apartments in a multi-story "low-income" complex about a kilometer away. As the landslide spilled refuse onto private lands adjacent to its 1,150-meter perimeter, the UDA planned to acquire these lands as part of the remediation process (MoMWD, 2017b). The UDA stated that it would provide rental assistance to keep safe families that might be affected temporarily by the landfill's stabilization, closure, and reclamation processes. The SMF indicated that evacuees would "be consulted adequately and meaningfully on relocation options ... and their preferences for relocation sites and options" (MoMWD, 2017b, p. 8). As of April 2019, the dump had not been removed (MacDonald, 2019), and studies to transform the site into an urban park were underway (Nalin, Premasiri & Karunarathana, 2020).

Discussion: relocation aftermaths

Relocation processes operated slightly differently in each case discussed above. The tsunami case illustrates large-scale involuntary planned relocation, a strategy that scholars have argued can be difficult to implement as it takes place following an emergency when there is not very much time "for proactive planning and stakeholder consultation" (Badri et al., 2006, p. 453). In Kegalle floods and landslides, hundreds of families were given several relocation options again after the emergency. Relocation again was involuntary, but local government officials made allowances for families to continue with their cultivation. In the Meethotamulla case, only a few families were relocated, even though many more asked to be. I argue that one reason why families in Meethotamulla were not relocated as was their preference is that they are members of the working class, urban poor. The state views and upholds the livelihoods of the rural farmer-cultivator affected in Kegalle in higher esteem (Moore, 1989; Ruwanpura & Hollenback, 2014).

Similar to the relocation schemes that the GoSL implemented following the 2004 Indian Ocean tsunami, Kegalle families could choose to participate in one of three housing reconstruction modalities: 1) owner-driven housing reconstruction, 2) the self-settlement model (cash grants for land and housing), and 3) a donor-driven housing reconstruction (Caron, 2009b). Under the owner-driven scheme, families that already owned or had access to an alternative piece of land would receive a cash grant of slightly over Rs. 1 million from the government for the construction of a new house (MoNPEA & MoDM,

2016; Fernando et al., 2021). Affected families that did not like the GoSL's alternative relocation sites could opt for the self-settlement model, a scheme implemented in the Colombo District after the tsunami. Under self-settlement, Kegalle families would receive a cash grant of Rs. 400,000 to purchase a plot of land of their choosing and then, upon proof of purchase (a survey plan and a deed), could request funds for housing reconstruction. However, some families that lost homes in the Samasarakanda landslide were too tired to think about building a home to replace the destroyed one (Fernando et al., 2020). These families opted to receive a house built by a donor that, upon completion, would be handed over to them. Sixty families received houses built with Chinese Government funds on land that the government had acquired by purchasing a cemetery (Fernando et al., 2020).

Relocation and governing climate change dynamics

The GoSL has used its power, policies, and institutions to relocate entire communities to fulfill its economic development goals and proclamations of saving lives. However, many relocations following natural disasters, particularly since the 2004 tsunami, are reactionary. Relocation is not planned even when risks have been identified by residents or relevant disaster-oriented government agencies. The ability of state officials and affected residents to have open and honest discussions about relocation is undermined when the dialogue starts in the immediate aftermath of a disaster. Affected community members are suffering, grieving, and maybe too tired to engage in the work of relocation and reconstruction.

As slow-onset disasters and extreme weather events are expected to increase with global climate change, I argue that the GoSL should change the way it conceptualizes the use of relocation. First, using relocation to reduce harm to lives, assets, and livelihoods entails more than just providing a piece of land and funds to construct a house. Relocation needs a holistic focus to include restoration of livelihoods and physical, social, and emotional well-being. In post-relocation studies among Kegalle-affected families, both Fernando et al. (2020) and Sangasumana (2018) found that a lack of cooperation between affected families and local government authorities and a lack of community participation hampered relocation following the landslide. Families that received the cash grant to purchase a plot of land often could not find an affordable plot of land near their original locations. Families that moved into a government-identified relocation site claimed that they could not pursue their previous agricultural livelihoods and that they were too far away from their children's previous schools. They also reported conflicts over the use of common spaces such as water sources and roads with their new neighbors and deeply missing their friends and neighbors (Sangasumana, 2018). Sangasumana's findings reflect my own (Caron 2009a, 2009b), demonstrating that many of the challenges of the post-tsunami relocation have not been addressed. In recent visits to post-tsunami relocation villages in 2016 and 2017, I found that families continue to utilize their lands in "no build" zones to support the next generation.

Running out of land/identifying area for future relocations

The lack of available land to relocate families from areas exposed to dangerous climate change effects will continue to complicate relocation efforts, especially if the government continues to pursue ad-hoc relocation implemented after disaster strikes and continue with the house-for-house (with garden) policy pursued since the 2004 tsunami. With population growth and urbanization, unoccupied land suitable for construction will be more difficult to identify. After the mass displacements and destructions of the home following the May 2016 floods and landslides, a government used a formula to assess reconstruction and relocation needs that used a conservative estimate of "LKR 400,000 in other areas" [such as Kegalle].... and a land preparation cost of LKR 120,000 per "housing site in landslide affected areas" (MoNPEA & MoDM, 2016, p. 57).

Given supply and demand dynamics, land prices most likely will continue to rise, complicating either temporary (evacuation) or permanent relocation efforts. Whether or not the government decides to formalize planned relocation as part of its approach to protecting lives and assets, land administration officials should proactively inventory state-owned lands for temporary or permanent relocation purposes and document land rights, uses, and claims associated with specific parcels to avoid conflict should the need to relocate arise (Caron et al., 2014). Documenting land tenure relations and customary use rights is important as temporary relocation (evacuation) might become permanent (Nalau & Handmer, 2018).

The search for land and the high "footprint" that the government's house-for-house policy requires indicates that the GoSL still faces the same post-tsunami land acquisition challenges. In densely populated urban areas such as Colombo, relocated families are moved to high-rise apartment buildings. Such housing structures are not fit within the cultural frame of reference in agricultural, rural areas. If the GoSL considers building vertically outside of major urban centers due to land shortages, it most likely will need to prepare a culturally sensitive and ambitious awareness-raising campaign about living in flats.

Seeing climate-change effects in unexpected places

One lesson about the cumulative effects of climate change is that planners and policymakers need to examine what might be considered unexpected places. Meethotamulla is a reminder that climate change outcomes are not based on a "single causal factor ... but ... increases the magnitude of all other contextual pressures" (Nalau & Handmer, 2018, p. 4). This landfill collapse should remind planners and risk reduction specialists to think more broadly about climate change effects and their influence on physical infra-structure. Culverts, storm drains, bridges, and canals are other examples of physical infrastructure that might be damaged, destroyed, or overloaded by heavy rain and flash flooding. The Meethotamulla landslide shows how cur-rent approaches to solid waste management and its associated physical infra-structure are vulnerable to extreme weather. The collapse also forces a rethinking of political proclamations that a garbage collapse cannot be "a natural disaster" (Correspondents, 2018).

The National Adaptation Plan indicates that industry and the energy and transportation sectors must improve their climate resilience and disaster risk preparedness (Climate Change Secretariat, 2016). The effects of climate change on solid waste management must be included too. With extreme weather events expected to become more common, planners and policymakers need to 1) re-assess the physical integrity of infrastructure, 2) explore the fea-sibility of possible alternatives (in this case, for example, ways to reduce solid waste generation), 3) model the multiple threats that climate change effects such as heavy rain, warming temperatures, drought conditions may have on physical infrastructure, and 4) improve community relations with the urban poor as it pursues its goal of being a "world-class" city, which includes its use and conceptualization of relocation. Even if preventative or planned reloca-tion was incorporated into the country's adaptation policy, it might not have prevented the Meethotamulla disaster, as this garbage collapse occurred in an unexpected place, and even after the fact, was not conceptualized by people with decision-making power as a natural disaster (Correspondents, 2018).

Conclusion

Policymakers around the world recognize that moving communities away from harm is one solution to protecting lives, assets, and livelihoods. Given that several of Sri Lanka's government institutions systematically collect and analyze weather data and have remote-sensing capabilities, officials responsi-ble for land use management might consider pursuing preventative resettle-ment or planned relocation according to the risk profiles it created under the Sendai Framework (UNDRR, 2019). Preventative resettlement proactively addresses disaster-related threats and provides a longer lead time to plan and organize financing for possible resettlement rather than spontaneously rec-ommended relocation after disaster-related losses of lives and livelihoods. Claudianos (2014) indicates that preventative resettlement has an ethical basis as the state is responsible for its citizens and holds "life as paramount above" other considerations (Claudianos, 2014, p. 316). Given the ways in which the Sri Lankan state has used relocation to protect citizens, specifically to pave the way for investment, the perceived paternalistic nature of preventa-tive resettlement might not be well received. Silva (2010) argues that while the tsunami buffer zone policy might symbolize good intentions, it also allowed the Sri Lankan state to project the desired image of "paternalistic benevo-lence" (Korf, 2005, p. i) at a time of eroding legitimacy and allowed some members of Sri Lanka's ruling elite to benefit from the post-tsunami investment. Even though Klein's disaster capitalism thesis (2007) did not necessarily play itself out to the fullest extent possible after the tsunami, four and five-star resort hotels mark some former residential areas along Eastern Province's coast (Caron field notes, 2016). Trust in a benevolent, paternalistic state would be highly uncertain, as today's current "government looks increasingly like a family firm" (Mushal, 2021).

National planning systems contend with globalizing land markets, pressures for foreign direct investment, and civil society demands for equity. The Sri Lankan state proactively relocated poor urban residents that occupied valuable real estate, "liberating" such land from its current occupants and placing it into the real estate market. While the government's post-flood and landslide assessment recognizes that such disaster presents an opportunity to "relocate all families from landside risk areas and flood plains" (MoNPEA & MoDM, 2016, p. 226), the state has yet to proactively liberate poor residents from these environmental and health risks even when residents request assistance to move (e.g., Meethotamulla). Examining the ways that the Sri Lankan state and communities encounter relocation as a disaster-mitigation measure illuminates the politics and possibilities surrounding relocation and how the Sri Lankan state approaches managing and responding to the disasters that climate change might exacerbate. A large activist and social justice-oriented community exists in Sri Lanka. This community of activists, advocates, and journalists is essential to amplifying the voices and documenting the experiences of the poor that persons in power would like to hide or forget. Holding government officials and institutions accountable for harms, damages, or neglect that they directly or indirectly create underscores the socio-political struggles that citizens pursue to protect lives, livelihood, and well-being in an era of global climate change.

References

Badri, S.A., A. Asgary, A.R. Eftekhari, and J. Levy (2006) 'Post-disaster resettlement, development and change: a case study of the 1990 Manjil earthquake in Iran'. *Disasters* 30(4). pp. 451–468.

Caron, C., Menon, G., & Kuritz, L. (2014). *Land Tenure & Disasters: Strengthening and Clarifying Land Rights in Disaster Risk Reduction and Post-Disaster Programming*. USAID Issue Brief. Washington, DC: USAID. https://www.landlinks.org/issue-brief/land-tenure-and-disasters/

Caron, C.M. (2009a). Left behind: Post-tsunami resettlement Experiences for women and the urban poor in Colombo. In Fernando, P., K. Fernando, & M. Kumarasiri (Eds.), *Forced to Move: Involuntary Displacement and Resettlement – Policy and Practice* (pp. 177–206). Colombo, Sri Lanka: Centre for Poverty Analysis.

——— (2009b). A most difficult transition: Negotiating post-tsunami compensation and resettlement from positions of vulnerability. In de Mel, N., K.N. Ruwanpura & G. Samarasignhe (Eds.), *After the Waves: The Impact of the Tsunami on Women in Sri Lanka* (pp. 112–152). Colombo, Sri Lanka: Social Scientists Association.

Claudianos, P. (2014). Out of harm's way; preventative resettlement of at risk informal settlers in highly disaster prone areas. *Procedia Economics and Finance*, 18, 312–319.

Climate Change Secretariat. (2016). *National Adaptation Plan for Climate Change Impacts in Sri Lanka*. Colombo: Ministry of Mahaweli Development and Environment. pp. 178.

Correspondents. (2018, 12 April). Sri Lanka workers' inquiry presents findings on Meethotamulla garbage disaster. *World Socialist Website*. https://www.wsws.org/en/articles/2018/04/12/iwic-a12.html

Crowe, S., Cresswell, K., Robertson, A., Huby, G., Avery, A., & Sheikh, A. (2011). The case study approach. *BMC Research Methodology*, 11, 100–109.

Decent Lanka 2015. (2017). Dump, dumber and dumbest. *Daily Mirror Online* 19 April. Retrieved April 2, 2021, from https://www.dailymirror.lk/article/Dump-dumber-dumbest-127342.html

Dias, N.T., Kerminiyage, K., & deSilva, K.K. (2016). Long-term satisfaction of post disaster resettled communities: The case of post tsunami Sri Lanka. *Disaster Prevention and Management*, 25(5), 581–594.

Eckstein, D., Kunzel, V., & Schafer, L. (2021) *Global Climate Risk Index 2021*. Berlin: Germanwatch e.V. Retrieved August 22, 2021, from https://germanwatch.org/sites/default/files/Global%20Climate%20Risk%20Index%202021_2.pdf

Eisenhardt, K.M. (1989). Building theories form case study research. *Academy of Management Review*, 14(4), 532–550. https://doi.org/10.2307/258557

Fernando, N. (2018). Voluntary or involuntary relocation of underserved settlers in the city of Colombo as a flood risk reduction strategy: A case study of three relocation projects. *Procedia Engineering*, 212, 1026–1033.

Fernando, N., Amaratunga, D., Haigh, R., Wise, B., Jayasinghe, & Prasanna, N. (2021). Life Two Years and Relocation: Status Quo of Natural Hazard Indcued Displacement and Relocation in Kegalle, Sri Lanka. In Amaratunga, D., Haigh, R., & Dias, N. (Eds). Multi-Hazard Early Warming and Disaster Risks. Springer Nature Switzerland AG pp. 3–25. https://doi.org/10.1007/978-3-030-73003-1

Fernando, N., Senenayake, A., Amaratunga, D., Haigh, R., Malalgoda, C., & Jayakody, R.R.J.C. (2020). Impact of disaster-induced relocation process on the displaced communities in Kegalle district, Sri Lanka. *European Scientific Journal*, 16(39), 33–53. https://doi.org/10.19044/esj.2020.v16n39p33

Ferris, E. (2015). Climate-induced resettlement. *The SAIS Review of International Affairs Winter-Spring*, 35(1), 109–117.

Gamburd, M.R. (2014). *The Golden Wave: Culture and Politics After Sri Lanka's Tsunami Disaster*. Bloomington and Indianapolis: Indiana University Press.

Government of Sri Lanka. (1993). *Reconstruction and Rehabilitation Fund Act, No. 58 of 1993*. Retrieved September 22, 2021, from https://www.lawnet.gov.lk/reconstruction-and-rehabilitation-fund-3/

———. (2005). *Sri Lanka Disaster Management Act, No. 13 of 2005*. Colombo: Government Publications Bureau. http://www.unlocked.lk/wp-content/uploads/2019/06/Disaster-Management-Act_E.pdf

———. (2016). Disaster management centre - Aranayaka landslide. Retrieved May 1, 2021, from https://disaster.lk/aranayaka/

———. (2019). National policy framework vistas of prosperity and splendour (2020–2025). http://www.doc.gov.lk/images/pdf/NationalPolicyframeworkEN/FinalDov Ver02-English.pdf

———. (2020). Statistical abstract 2020. Chapter 2 population. Retrieved September 1, 2021, from http://www.statistics.gov.lk/abstract2020/CHAP2/2.11

Government of Sri Lanka & Development Partners. (2005). *Tsunami Building Back Better – Achievements, Challenges and Way Forward*. Colombo, Sri Lanka.

Government of Sri Lanka, Disaster Management Centre. (2017). Meethotamulla Municipal Solid Waste Dump Disaster. Retrieved May 1, 2021, from https://disaster.lk/meethotamulla/

Government of Sri Lanka, Ministry of Disaster Management (2016). Flood and Landslide Situation Report on May 2016. Retrieved May 1, 2021, from http://www.ndrsc.gov.lk/web/images/situationreport/2016%20flood%20detail.pdf

Gunawardena, S.A., Samaranayake, R., Dias, V. et al. Challenges in implementing best practice DVI guidelines in low resource settings: lessons learnt from the Meethotamulla garbage dump mass disaster. Forensic Sci Med Pathol 15, 125–130 (2019). https://doi.org/10.1007/s12024-018-0033-4

IDMC (2020). *Global Report on Internal Displacement*. Internal Displacement Monitoring Center. Retrieved September 10, 2021 from https://www.internal-displacement.org/global-report/grid2020/

Ingram, J.C., Franco, G., Rumbaitis-del Rio, C., & Khazai, B. (2006). Post-disaster recovery dilemmas: Challenges in balancing short-term and long-term needs for vulnerability reduction. *Environmental Science and Policy*, 9, 607–613. https://doi.org/10.1016/j.envsci.2006.07.006

JICA (2016). Aerial Survey report on inundation damages and sediment disasters. Japanese International Cooperation Agency (JICA) Survey Team. https://www.jica.go.jp/srilanka/english/office/topics/c8h0vm00006ufwhl-att/160615.pdf

Kawamura, Y. (2015). Landslides along national roads in central highland in Sri Lanka: Review on current situation and suggestion for further development of landslide mitigation along highways in Sri Lanka. https://www.jica.go.jp/srilanka/english/office/topics/c8h0vm00009qnvwr-att/151216_04_01.pdf

King, D., Bird, D., Haynes, K., Boon, H., Cottrell, A., Millar, J., Okada, T., Box, P., Keogh, D., & Thomas, M. (2014). Voluntary relocation as an adaptation strategy to extreme weather events. *International Journal of Disaster Risk Reduction*, 8, 83–90. https://doi.org/10.1016/j.ijdrr.2014.02.006

Klein, N. (2007). *The Shock Doctrine: The Rise of Disaster Capitalism*. London: Penguin Books.

Korf, B. (2005). Sri Lanka: The tsunami after the tsunami. *International Development Planning Review*, 27(3), i–vii. https://doi.org/10.3828/idpr.27.3.1

Loewe, M., Zintl, T., & Houdret, A. (2021). The social contract as a tool of analysis: Introduction to the special issue on "Framing the evolution of new social contracts in Middle Eastern and North African countries" *World Development*, 145, 1–16. https://doi.org/10.1016/j.worlddev.2020.104982

Lyons, M. (2009). Building back better: The large-scale impact of small-scale approaches to reconstruction. *World Development*, 37(2), 385–398. https://doi.org/10.1016/j.worlddev.2008.01.006

MacDonald, C. (2019). A preventable tragedy: The Meethotamulla garbage landslide. Retrieved April 28, 2021, from https://medium.com/@craig.david.macdonald/a-preventable-tragedy-the-meethotamulla-garbage-landslide-6635a9b5dd38

Ministry of Megapolis & Western Development (MoMWD). (2017a). *Addendum to Environmental Management Framework: Metro Colombo Urban Development Project*. Colombo, Sri Lanka. 69 pp. https://documents1.worldbank.org/curated/ar/527471502033353436/pdf/Addendum-to-MCUDP-Environmental-Management-Framework-June-2017.pdf

———. (2017b). *Addendum to Social Management Framework: Metro Colombo Urban Development Project*. Colombo, Sri Lanka. 14 pp. https://documents1.worldbank.org/curated/pt/338231502204204437/pdf/SFG3543-EA-P122735-Box405293B-PUBLIC-Disclosed-8-8-2017.pdf

Ministry of National Policies and Economic Affairs & Ministry of Disaster Management (MoNPEA & MoDM). (2016). *Sri Lanka Post-Disaster Needs Assessment: Floods and Landslides-May 2016.* Colombo: Ministry of Disaster Management. 300 pp.

Moore, M. (1989). The ideological history of the Sri Lankan "peasantry". *Modern Asian Studies,* 23(1), 179–207. https://doi.org/10.1017/S0026749X00011458

———. (1993). Thoroughly modern revolutionaries: The JVP in Sri Lanka. *Modern Asian Studies,* 27(3), 593–642. https://doi.org/10.1017/S0026749X00010908

Muggah, R. (2008). *Relocation Failures in Sri Lanka: A Short History of Internal Displacement and Resettlement.* London and New York: Zed Books.

Mushal, M. (2021, 10 July). In Sri Lanka, the Government looks Increasingly like a Family Firm. *The New York Times.* https://www.nytimes.com/2021/07/10/world/asia/sri-lanka-basil-rajapaksa.html

Nalau, J. & Handmer, J. (2018). Improving development outcomes and reducing disaster risk reduction through panned community relocation. *Sustainability,* 10(10), 1–14. https://doi.org/10.3390/su10103545

Nalin, K., Premasiri, H.D.S., & Karunarathana, A.K. (2020). *Meethotamulla solid waste dump rehabilitation and design challenges.* Paper presented at the 10th Annual Research Symposium -2019. Colombo: National Building Research Organisation.

National Legislative Bodies [NLB]. (2018). *Fiji: Planned Relocation Guidelines – A framework to undertake climate change related relocation.* Government of Fiji. Retrieved September 22, 2021, https://www.refworld.org/docid/5c3c92204.html

Oliver-Smith, A. (2010). *Defying Displacement: Grassroots Resistance and the Critique of Development.* Austin: University of Texas Press.

Patton, M.Q. (2002). *Qualitative Evaluation and Research Methods.* 3rd edition. Thousand Oaks, CA: Sage Publications

Perera, A. (2016a). Shortage of safe land blocks Sri Lanka disaster relocation efforts. Retrieved June 11, 2021, from https://floodlist.com/asia/sri-lanka-flood-landslide-disaster-relocation-efforts

Perera, E.N.C., Jayawardana, D.T., Jayasinghe, P., Bandara, R.M.S., & Alahakoon, N. (2018). Direct impacts of landslides on socio-economic systems: A case study from Aranyaka, Sri Lanka. *Geoenvironmental Disasters,* 5(11), 1–12. https://doi.org/10.1186/s40677-018-0104-6

Perera, I. (2016b). *Living it Down: Life after relocation in Colombo's high rises.* Colombo, Sri Lanka: Centre for Policy Alternatives.

———. (2019). *The Urban Regeneration Project.* The Bank Information Center. 15 pages. Retrieved October 1, 2021, from https://bankinformationcenter.org/en-us/project/aiib-colombo-urban-regeneration-project/

Perera, I. & Nagaraj, V. (2017, April 18). The Meethotamulla tragedy: The face not failure of development. *Groundviews.* https://groundviews.org/2017/04/18/the-meethotamulla-tragedy-the-face-not-failure-of-development/

Petley, D. (2016, 19 July). The Aranyaka landslide disaster in Sri Lanka – a JICA report. *The Landslide Blog.* https://blogs.agu.org/landslideblog/2016/07/19/aranayake-landslide-1/

Ruwanpura, K. & Hollenback, P. (2014). From compassion to the *will to improve*: Elistion of scripts? Philanthropy in post-tsunami Sri Lanka. *Geoforum,* 51, 243–251. https://doi.org/10.1016/j.geoforum.2013.04.005

Samarasinghe, V. (2017, 19 July). Three months after the Meethotamulla disaster in Sri Lanka. *World Socialist Website.* https://www.wsws.org/en/articles/2017/07/19/garb-j19.html

Samarasinghe, V. & Perera, W. (2017, April 17). At least 26 dead in garbage dump collapse in Sri Lanka. World Socialist Website. https://www.wsws.org/en/articles/2017/04/17/garb-a17.html

Sangasumana, P. (2018). Post disaster relocation issues: A case study of Samasarakanda landslide in Sri Lanka. *European Scientific Journal*, 14(32), 1–17. 10.19044/esj.2018.v14n32p1

Schafer, L., Jorks, P., Seck, E., Koulibaly, O., & Diouf, A. (2021). *Slow-Onset Processes and Resulting Loss and Damage – Introduction*. Berlin: Germanwatch e.V. Retrieved January 9, 2022, from https://germanwatch.org/sites/default/files/FINAL_Slow-onset%20paper%20Teil%201_20.01.pdf

Silva, K.T. (2010). Tsunami Third Wave and the politics of disaster management in Sri Lanka. Blaikie, P.M. & R. Lund. (Eds.), *The Tsunami of 2004 in Sri Lanka: Impacts and Policy in the Shadow of Civil War* (pp. 60–71). London and New York: Routledge.

United Nations High Commission for Refugees [UNHCR]. (2014). *Planned relocation, disasters, and climate Change: Consolidating good practices and preparing for the future*. https://www.unhcr.org/54082cc69.pdf

United Nations Office for Disaster Risk Reduction (UNDRR). (2019). *Disaster Risk Reduction in Sri Lanka: Status Report 2019*. https://www.unisdr.org/files/68230_10srilankadrmstatusreport.pdf

United Nations Office for Project Services (UNOPS). (2009). Quarterly progress report. ASB/Solidar housing programme construction of 90 housing units at Vaharai in the Batticaloa district. Colombo, Sri Lanka. 17 pp.

United Nations Office for the Coordination of Humanitarian Affairs [UN OCHA]. (2016, 26 May). Sri Lanka: Floods and landslides situation report No. 2. https://reliefweb.int/report/sri-lanka/sri-lanka-floods-and-landslides-situation-report-no-2-26-may-2016

Walsh, B. & Hallegatte, S. (2019). *Socioeconomic Resilience in Sri Lanka: Natural Disaster Poverty and Wellbeing Impact Assessment*. Policy Research Working Paper 9015. Washington, DC: The World Bank Group.

Wickrematunge, R. (2017, June 20). A tale of incompetence – RTI reveals CMC inaction leading up to the Meethotamulla tragedy. *Groundviews*. https://groundviews.org/2017/06/20/a-tale-of-incompetence-rti-reveals-cmc-inaction-leading-up-to-meethotamulla-tragedy/

World Bank. (2020). Data for Sri Lanka. Retrieved January 5, 2022, from https://data.worldbank.org/country/sri-lanka?view=chart

8 Does local politics have relevance to the local climate action programs in India and Bangladesh?

Review and discussion

Devendraraj Madhanagopal, Mohammed Moniruzzaman Khan and Farhana Zaman

Introduction

In South Asia, the discussions on climate change are becoming more important and widespread as the effects of climate change and risks are more intense and brunt. In recent decades, India and Bangladesh are the two most densely populated nations of South Asia that have largely witnessed climate change impacts at multiple dimensions (ADB, 2017; Bangladesh Climate Strategy and Action Plan (BCCSAP), 2009; Goyal & Surampalli, 2018; Picciariello et al., 2021; Roxy et al., 2017; Roy et al., 2016). Research also warns that both nations are more likely to face the adverse effects of climate change in the future (Almazroui et al., 2020; Dastagir, 2015; Khan et al., 2020; Krishnan et al., 2020; NIC, 2009; UNFCCC, 2012; USAID, 2015). With the brunt effects of climate change and risks in India and Bangladesh, the policies, actions, and strategies have continuously emerged amongst the international, state, and non-state institutions and actors to respond to climate change. Hence, by recognizing the imminent threats of climate change, since the 2000s, both nations have started setting up various institutions, agencies, and action plans to cope and adapt to climate change, respectively.

Climate action has two pillars throughout the world: mitigation and adaptation; both require certain policy changes, and politics intersect in mitigation and adaptation (Dolšak & Prakash, 2018). Both mitigation and adaptation address climate change: mitigation addresses climate change source, whereas adaptation addresses its consequences. The earliest reports of the IPCC, the internationally influential authority on climate change, provided more attention to mitigation. There was limited attention to adaptation, vulnerability, and equity discussions. Emphasizing "adaptation" in international climate change discussions happened only after the late 1990s (Schipper, 2006). Though there have been substantial differences in approach, administrative structure, and actions, notably, the action plans of both the nations have emphasized "Adaptation," and it interlinks the action strategies with "Sustainable development."

DOI: 10.4324/9781003216476-10

Over the last two decades, discussions on the politics of climate change and adaptation have received substantial attention in climate change literature, and several conceptual, theoretical, empirical, and policy-oriented literature emerged on these themes across the world, particularly from Western academia (Adger et al., 2006; Giddens, 2009). Notably, during the last years, there has been growing interest regarding the power and politics in climate change adaptation; there has been substantial conceptual, empirical, and policy-oriented literature on this theme (Eriksen et al., 2015; Funder et al., 2018; Glover & Granberg, 2020; Javeline, 2014; Nagoda, 2017; Nightingale, 2017; Ohja et al., 2016). In tandem with this, there have also been theoretical debates of the politics of climate change during the last years, focusing on political ecology (Birkenholtz, 2012; Taylor, 2014; Tschakert, 2012; Watts, 2015). Since the 2010s, empirical works have been noted focusing on the power and local politics of climate change. It should be noted that the scholars of the West proposed a large proportion of those works, and those concepts/theories/approaches were also applied to the research on climate change and risks in the South Asian context (Afroz et al., 2018; Sanyal, 2006; Sovacool, 2018; Taylor, 2014). In addition, though there has been growing literature focusing on the power and politics of climate change adaptation, Javeline (2014) highlighted several unanswered "political" and "political-related questions" in adaptation to climate change, and only political scientists can address that.

This chapter is one such attempt to address the "political" question of adaptation to climate change. In contrast to the majority of literature focusing on the micro-politics of adaptation, this chapter focuses on the local politics of adaptation. By taking the case of two different regions of India and Bangladesh, it posits to address a central question: How do "politics" play a central role in implementing adaptation programs to climate change at the regional/local levels? In the context of climate change action programs in India and Bangladesh, this chapter attempts to rethink our understanding of the classic relationship between "politics" and "administration." The following sections of this chapter are structured: first, conceptualizing adaptation and the "politics" of adaptation to climate change. Second, to provide an understanding of the adaptation programs/action plans to climate change in India and Bangladesh. Third, to introduce the cases from the selected regions of India and Bangladesh and critically discuss how "local politics" play critical roles in the implementation process of adaptation programs to climate change at the regional/local levels by unraveling the major dimensions of local politics. Finally, a comparative discussion on the scenarios of India and Bangladesh about the relevance of local politics is followed by the conclusion.

Conceptualizing adaptation and the politics of adaptation to climate change

In recent decades, the term "adaptation" has largely been used in climate change literature, notably after the 2000s. In contrast, the roots of the term "adaptation" lie in natural sciences, particularly evolutionary biology. Over the period, this term has widely been used in different streams, including famine, drought, ecology, disaster management (Kelman & Gaillard, 2010; Mortimore & Mortimore, 1989; Sukkary-Stolba, 1989), and climate change. Over the past few decades, several researchers, policy organizations, and climate think tanks have conceptualized adaptation to climate change. Nalaua & Verrall's (2021) review has found that, since the 2000s, five papers were found to be highly cited in adaptation concepts and are known as foundational papers. Some of them are the following: Adger (2003); Smit et al. (1999); Smit et al. (2000); Smit & Wandel (2006). The common element of all these noted definitions of adaptation is that it has emphasized "adjustments" of the system to respond to stressors. Also, they are of the consensus that adaptation processes would vary according to the context, purposefulness, and system in which it occurs. Adaptation to climate can be of various types: incremental, transformational, proactive, reactive, private, and public (Noble et al., 2015). Adaptation to climate change can occur by public and private actors. A person or private firm can take private adaptation for private and public benefits. At the same time, public adaptations are the ones that are undertaken by the public actors (Noble et al., 2015; CoastAdapt, 2017).

The recent IPCC Fifth Assessment Report (AR5) has defined adaptation to climate change as

> In human systems, the process of adjustment to actual or expected climate and its effects, in order to moderate harm or exploit beneficial opportunities. In natural systems, the process of adjustment to actual climate and its effects; human intervention may facilitate adjustment to expected climate and its effects.

Community-based adaptation (CBA) is another category of adaptation that is a relatively recent one (Huq & Reid, 2007; Reid et al., 2009). Noble et al. (2015), in the fourth IPCC assessment report, defined community-based adaptation as the "generation and implementation of locally driven adaptation strategies, operating on a learning-by-doing, bottom-up, empowerment paradigm that cuts across sectors and technological, social, and institutional processes." CBA focuses on local development needs and the contextual nature of climate change vulnerability. In this chapter, our conceptualization of "adaptation" is broad and is largely based on the definition by Noble et al. (2015).

The discussions of politics in climate change and adaptation are not recent. In one of the influential works on adaptation, Smit et al. (1999) asked a few fundamental questions to understand and analyze adaptation, which are as follows: i) Adapt to what? ii) Who or what adapts? iii) How does adaptation occur? The third and second questions of Smit et al. underpin the questions of analyzing the actors, governance, approaches, and decision-making systems in adaptation to climate change and offer the scope of evaluating adaptation in policy development. Thereby, it also paves the way to discuss the politics in adaptation.

By showing cases from the developing world, Adger (2003) has articulated that adaptation to climate change is a "dynamic social process," and it involves agents, institutions, and the resource base. By stressing the collective action and social capital of the community in adapting to climate change and risks, Adger has argued that adaptation decisions are usually taken by the state, private actors, individuals, and the community. However, these decisions prioritize and privilege one over the sets of interests (by other actors), resulting in winners and losers to climate change. Hence, it is essential to note that the questions on "Winners" and "Losers" of climate change were made even before the 2000s. Glantz (1995) made profound questions on such topics in the global climate change regime. Extending this, O'Brien & Leichenko (2003) have discussed by providing examples from climate change and globalization and demonstrating that the Winners and Losers of global change fall into two categories: i) Natural, Inevitable, and Evolutionary (NIE): It perceives the emergence of Winners and Losers through Social Darwinist perspectives and provides explanations along the lines of genetic and evolutionary factors. The groups that look at the Winners and Losers through Social Darwinist perspectives may think compensation to the "losers" of global change is unnecessary. ii) Socially and Politically Generated (SPG): It suggests that the Winners and Losers are socially and politically generated. It analyzes their emergence and contexts through the lens of political ecology and Marxian political economy perspectives. Needless to say, these groups are the critics of Social Darwinism and the proponents of NIE. SPG argues that the winners and losers of global change are the outcomes of unequal and deliberate global processes. They further argue that since the Winners are made up at the expense of Losers, the losers should be compensated.

Case 1 from India: action plan for climate change – national and sub-national levels in India

Since 2007, India has formulated policies, strategies, and institutions to counter climate change. The Union Government of India plays the central in crafting the nation's climate policy, but not state governments. Even though the state governments possess separate action plans to act against climate change, those action plans are essentially the strategies to achieve the targets set by the climate policy of the Union Government. The states were asked to prepare separate state-level climate action plans for them at a conference of state ministers held on 19 August 2009 by the then Prime Minister of India, Dr. Manmohan Singh (Atteridge, 2013; Jörgensen et al., 2015). The state-level climate action plans act as a potentially important policy document at the sub-national level to address climate change vulnerability and increase climate resilience. The water and agriculture sectors are the ones that are highly sensitive to climate change – in India, both the sectors fall under the jurisdiction of Indian states. Hence, the active roles of the state governments are indispensable in implementing climate policies of the union government at the local levels. Rattani et al. (2018) have criticized the lack of

commitments of the union government in implementing the objectives ratified in the National Action Plan on Climate Change (NAPCC). By pointing
out the various missions of NAPCC, including National Water Mission,
National Mission for Sustaining Himalayan Ecosystem (NMSHE), and
National Mission for Strategic Knowledge of Climate Change (NMSKCC),
and criticized how these missions have been crippled due to inadequate funds,
unpreparedness, flawed strategies, and the overlapping of the aims of two
missions (NMSHE and NMSKCC). Similarly, Chatruvedi et al. (2019) have
raised particular concerns about implementing state-level climate action
plans. A few of them are as follows: i) There have been challenges in implementing state-level climate actions due to the complications of decentralized
governance systems, and there has been a mismatch at the local levels. Due to
this, the districts and cities have largely been ignored in the climate change
action plans of the states. ii) The states do not possess adequate in-built
design to upscale the pilot projects and demonstrations that are envisaged
and implemented by the state climate action plans.

A very recent analysis of 28 state-level climate action plans shows that
"gender" discussions in the sub-national adaptation policy of India have
inadequately been incorporated, and there has been an uneven treatment of
women and gender in the sub-national climate action plans of many states.
Also, the action plans of many states have homogenized the roles, activities,
and experiences of women in many instances, leading to the overlook of the
changing roles of men and women and changing masculinities and
intra-household heterogeneity (Singh et al., 2021).

The implementation of climate action plans at the sub-national levels
largely falls under the responsibility of the state governments. For example,
the government of Tamil Nadu implements climate action plans through several sectoral projects under adaptation and mitigation. Along with the fund
support from the Union Government, the Ministry of Finance of the state
government of Tamil Nadu has regularly allocated fund support with loan
assistance from national and international organizations for the implementation of various adaptation and mitigation projects to fight against climate
change (for example, Government of Tamil Nadu, 2019; Government of
Tamil Nadu, 2021). The Department of Environment of the Government of
Tamil Nadu is the nodal agency for implementing all climate change-related
programs and actions of the state (Government of Tamil Nadu, 2020). The
seven vulnerable sectors identified by this department are sustainable agriculture, water resources, coastal area management, and others.

Climate change vulnerability in coastal Tamil Nadu – local politics and the state

Tamil Nadu, a state located on the Southeast coast of India, has a coastline
of around 1,076 km, and it is endowed with bountiful marine resources and
ecosystems. The coastal regions of Tamil Nadu are also home to millions of
coastal populations, including fisherfolk, whose livelihoods are directly

dependent on those resources (SoE Report I, 2017). These coastal regions have faced various threats from natural and anthropogenic factors over the last few decades. A few factors that pressure the coastal areas are commercial fishing, domestic and industrial pollution, industrial activities (including ports, harbors, and factories) across the coasts, coastal tourism, sea-level rise, climate change, and coastal hazards. In particular, the rising trends of heavy rains, cyclones, and the resulting floods push the coastal fisherfolk to live in an ecologically and economically vulnerable position. Some of the policy responses of the state government of Tamil Nadu to respond to such pressures along with state-level climate action plans are Integrated Coastal Zone Management (ICZM) Plan, mangrove restoration, annual fishing ban, in-situ conservation, and the implementation of Tsunami rehabilitation activities (SoE Report I, 2017). The State of Environment reports (Part I& II) of the government of Tamil Nadu provide a detailed outline of climate change vulnerability of coastal Tamil Nadu and the responses of the state to climate change (SoE Report I, 2017; SoE Report II, 2017). However, the discussions of "politics" in implementing sub-national action plans to climate change have largely been under-discussed in all such policy reports. Though the Department of Environment is the nodal department to implement the state's sub-national climate action plan[1], other ministries, departments, agencies, and organizations are included in the participation and implementation process. Some of them are as follows: Department of Fisheries, Tamil Nadu Forest Department, Tamil Nadu Maritime Board, Tamil Nadu State Disaster Management Agency (TNSDMA), and Gulf of Mannar Biosphere Reserve Trust (GoMBRT).

Coastal erosion is one of the direct impacts of climate change. The coasts of Tamil Nadu have been witnessing the brunt effects of erosion over the last few decades due to both natural and anthropogenic factors. The recent nationwide report on shoreline changes shows that Tamil Nadu lost 41% of its coast due to coastal erosion over the past two decades. Nagapattinam is one of the coastal districts of Tamil Nadu that have witnessed high to low erosion. The major responsibility of the state to coastal erosion is the construction of seawalls and groins along the coasts (Kankara et al., 2018).

Along with these state institutions, some fishers' non-formal institutions pressurize the state institutions to construct protection walls to prevent sea erosion. Like this, at the local levels, power and politics play an essential role in fastening the construction of seawalls along with the fishing villages of the Tamil Nadu coast, which are largely undocumented and under-discussed in scientific literature. "Politics" in this context means not only the memberships or assuming the key posts in the political parties by the fishers of the study villages but also mean how the power relations determine the implementation of coastal protection projects of the state at the village and block levels across the coastal regions.

To discuss the intricacies of power and politics in adaptation to climate change at the local levels, we show one case from our field research based on the Nagapattinam district of Tamil Nadu, India (Madhanagopal, 2019). This region is part of the Coromandel Coast of Tamil Nadu, and marine fisheries are one of the major sources of income and livelihood for around one million fishers of this district. Several recent studies demonstrate the high vulnerability of Nagapattinam district to sea-level rise as most of the areas of Nagapattinam district are either below the sea level or between 0–5 meters below the sea level. This district is estimated to be at risk of a one-meter sea-level rise, and it is more prone to sea-water intrusion (ArivudaiNambi & Sekhar Bahinipati, 2013; Byravan et al., 2010). From November 1952 to January 2021, this district was hit by extreme climate events and coastal hazards around 17 times. Among them, one of the worst effects was the deadliest tsunami disaster in 2004, which caused massive damage to the coastal ecosystems and fishing populations of this region. Some other recent cyclones that hit this district and caused massive damages are as follows: 2011 Thane Cyclone, 2012 Nilam Cyclone, and 2018 Gaja Cyclone (TNSDMA, 2021). All these extreme events and coastal hazards, and increased sea levels have accelerated erosion along the Nagapattinam coast.

The responses of the state to protect the coast from erosion have been multiple methods, including hard measures, soft measures, and a combination of both over the last few decades. Rubble mound seawalls, groin fields, or a combination of both were constructed, and plantation activities were carried out to protect the Tamil Nadu coast (Government of India, 2020; Sundar & Sundaravadivelu, 2005). World Bank and Asian Development Bank provided substantial assistance, including fund support to the Tamil Nadu government to construct the seawalls (Rodriguez et al., 2008). The critiques against hard engineering measures are not something new. The side effects of seawalls and bulk waters were also noted in the report submitted by the M.S. Swaminathan committee to review the coastal regulation zone notification 2011 (Ministry of Environment and Forests, 2005).

We[1] conducted fieldwork in the selected coastal villages of the Nagapattinam district. In this coastal stretch, marine fishing is primarily a caste-based occupation; almost all the fishing households of this coastal stretch belong to "Pattinavar," a traditional ocean fishing caste. In this chapter, we selectively take the case of two study villages of the Nagapattinam district, which are as follows: Chinnamedu and Kodiyampalayam. As per the census of CMFRI (2012), the population of both the fishing villages are less than 1,000. Over the past two decades, the fishing villages have continuously been hit by various extreme climate events and the 2004 Indian Ocean Tsunami disaster. Open interviews and focus group discussions[2] were conducted.

According to the local fishers, the overall infrastructure and public transportation facilities of both the villages have been grossly inadequate, making them more vulnerable during extreme climatic conditions. Besides that, they noted that they have been losing their fishing and living spaces over the past two to three decades due to coastal erosion. According to them, their villages' beaches have been eroded for two reasons: i) Their beaches are not protected by seawalls. ii) Seawalls protect the beaches of the nearby fishing villages. Due to this, the velocity of the sea waves is largely getting transferred to their coasts, and the intensity of the erosion has been high in recent years. Similar voices were echoed by the fishers of other regions of the Coromandel Coast, and they asked for long-term comprehensive solutions to protect their coast (Yamunan, 2010). In this context, it is also to be noted that the marine fishers of Tamil Nadu possess contrasting opinions about the efficiency of rubble mound sea walls. For example, the fishers of the Kanyakumari district of Tamil Nadu favored sea groins instead of the construction of rubble mound seawalls by pointing out the inefficiency of rubble mound sea walls in protecting their coast from intense erosion (Arockiaraj, 2013).

The fieldwork showed that though the social capital of these fishing villages is invariably similar, the political capital of the fishing villages vastly differs among others. This is because the political capital of the fishing villages largely depends on the overall economic capacity of the fishing households and the political influence of the local leaders of the particular fishing villages. It plays a major role in expediting the construction of coastal seawalls on the beaches of their villages. The discussions with the local leaders of the study villages showed that the influence of the local leaders plays a prominent role in reaching out to the fisheries department for the construction of seawalls on their village coasts. In general, the fishermen leaders based on the fishing villages primarily engaged in mechanized fishing possess strong political networks and connections with the state departments. Three political parties have been active across the coastal stretch of the Coromandel Coast of Tamil Nadu. However, these political parties are only mostly visible during elections. Hence, the internal governance systems of the fishing villages are largely under the control of non-state fishermen councils, which have a rich historical background.

The discussions with the fishers of the study villages revealed that these two villages are surrounded by fishing villages primarily involved in purse-seine fishing. According to them, these villages receive more support from the state and other non-state actors to construct and renovate the coastal bio-shield to protect their coasts, which in turn helps them to save their fishing and living spaces. Like this, the public infrastructure and road transportation facilities of these villages are relatively better than the study villages, which increases their adaptive capacity to respond to climate change and risks.

According to the respondents, the political networks of the local leaders provide support to these fishing villages to demand the state to establish seawalls on their coasts (Madhanagopal, 2023). Noted national news outlets have reported the demands of the near-shore artisanal fishing villages across Tamil Nadu to the state and their vulnerability to climate change impacts, including sea erosion (Muralidharan, 2017; Ragunathan, 2014a; Ragunathan, 2014b; *The New Indian Express*, 2014).

In Kodiyampalayam, the influence of political parties over other local institutions is significant. However, such local influence has not transferred to influence the district- and state-level institutions to protect the seawalls on their village coasts. It shows that mere political memberships and control over the other local institutions in the same fishing villages are not enough to establish the demands of the artisanal fishing villages to the state. The local political capital of the fishing villages needs to be coupled with economic and human capital to make their demands to protect their coasts possible (Madhanagopal, 2023).

Case 2 from Bangladesh: climate change vulnerability in the southwest coastal zone of Bangladesh

Bangladesh, a country with 147,570 square km, is the largest delta globally (Rabbani et al., 2018). Having one-third of the total landmass along the coastal shoreline, the country is one of the hotspots of climate change-induced hazards and the associated disasters for its low-lying condition along the sea, particular geophysical characteristics, unstable landforms coupled with rapid population growth, unplanned urban growth proliferation, deforestation, and non-engineered constructions. Among the various zones of Bangladesh, the southwest coastal zone is increasingly exposed to a wide range of climate change-driven hazards and consequent disasters. Every year, the zone faces multiple stressors, many of which are very frequent and severe in nature, such as cyclones, saline water intrusion, river erosion, heavy rainfall, storm surges, tidal surges, and coastal flooding. These disasters cost many physical, social, and economic losses and adversely affect huge areas and populations. Khulna City of the southwest zone, being one of the most vulnerable coastal cities in the world, is a quintessential example in this context (Esraz-Ul-Zannat et al., 2020). With a population of 1.4 million, estimated to rise to 2.9 million by 2030, the city is more prone to coastal flooding and waterlogging due to increasing sea-level rise, frequent and heavy rainfall, cyclone, and storm surges (Esraz-Ul-Zannat et al., 2020). Dacope Upazila under this division, located between 22°12' and 23°59' North latitude and between 89°14' and 89°45' East longitudes, is not exceptional in this regard (BBS, 2011; Esraz-Ul-Zannat et al., 2020). Several devastating cyclones named *Sidr, Aila, Nargis, Mahasen,* and *Amphan* lashed this Upazila during the last few decades, causing significant damage to agricultural land, property, livestock, and fish farms (Debnath, 2020; Rabbani et al., 2018). Salinity due to the episodic surges caused by these cyclones is a pervasive problem imposing significant challenges on coastal agriculture and overall food insecurity (Zaman, 2018). Also, the geographic position of agricultural lands within or near the river and the exposure to the frequent occurrences of riverbank erosion cause irreparable loss to agricultural lands (Hoque et al., 2019; Khan et al., 2021).

Moreover, frequent inundation of grazing lands by seawater results in an extreme shortage of fodder for livestock (Rabbani et al., 2018; Zaman, 2018). Besides, the long-term saline water stagnation keeps detrimental effects on the health and hygiene of the coastal people, particularly women and children (Rabbani et al., 2018; Zaman, 2017). The government of Bangladesh has taken essential adaptation policies to boost the adaptation capacity and minimize the vulnerability of the coastal communities. Besides, development organizations are playing an active role in this regard. The coastal communities have benefitted mainly from various training provided by many development organizations, particularly agriculture-based training. However, the adaptation and resilience efforts of GOs, NGOs (Non-Government Organizations), and INGOs face unique challenges due to the undesirable interferences of local politics. Therefore, it is vital to highlight the national climate change action plans and policies of Bangladesh. The subsequent sections attempt to exhibit various dimensions of local politics manifested in the southwest coastal zone, particularly in Dacope, and show how local politics trap strategies to achieve national goals.

Climate change policies and action plans in Bangladesh: a brief overview

Bangladesh, as a newly independent (1971) country, actively considered strong measures regarding climate change that include the Bangladesh Climate Change Strategy and Action Plan (BCCSAP, 2009a), National Adaptation Plan of Action (NAPA, 2009), ccGAP: Bangladesh (2013), Climate Change and Disaster Management (2015), Vision-2021 (a strategic plan/road map for Bangladesh till 2021), and the sixth (2011–2015) and seventh (2016–2021) Five Year Plans (FYPs) (Khan, 2022). The BCCSAP is committed to achieving its goals through the implementation of six pillars, including "food security, health care, and social protection, all-inclusive disaster management strategies, infrastructural development and better communication, in-depth research for new knowledge production and generation, mitigation and low carbon development, capacity building, and institutional development" (MoEFCC, 2018: 5). The NAPA is designed to promote adaptation to coastal crop agriculture, introduce salt-tolerant fish and diversified fish culture practices, promote forest plantation, and ensure safe drinking water to coastal communities (GoB, 2010).

Focusing mainly on building adaptation capacity, the Bangladesh government established a National Climate Change Fund (from national and international sources including the UN) with an initial capitalization of $45 million, which was later raised to $100 million (Ahmed et al., 2015; Khan, 2019). The Bangladesh government prioritized developing a pro-poor climate change strategy to achieve the people's economic and social well-being, where the poorest and most vulnerable groups received special attention. In the Climate Change Action Plan (2009–2018), the needs of poor and vulnerable people have been prioritized in all sectors to meet the challenges of climate change (GoB, 2009a). The sixth FYP (2011–2015) indicates that the Bangladesh government supports communities and rural people all the way to strengthen resilience and adaptive capacity to face climatic challenges (GoB, 2013; Khan, 2019). The seventh FYP (2016–2020) adopted a pro-poor Climate Change Management strategy which prioritizes adaptation and disaster risk reduction through achieving environmental sustainability, quality of life for people of all regions, preserving agricultural land to ensure production growth for food security, reducing the rate of salinity intrusion and mitigating impacts of salinity on human health, and minimizing potential economic losses due to climate change-induced hazards (GoB, 2015). Regarding sustainable hazard and disaster management, the GoB (2009b), GoB (2009a), and GoB (2016) emphasized strengthening the capacity of the state, civil society, volunteer organizations, and communities to manage climatic disasters.

Climate change affects women and children adversely in terms of health issues and food insecurity. Efforts toward including women into the process of development in Bangladesh are grounded on a wide range of international commitments too that include CEDAW (Convention on the Elimination of all forms of Discrimination Against Women) (1979), Beijing Conference (1995), UN Decade of Women (1976–1985), MDGs (2000–2015), etc. In addition, by signing, and ratifying the agreements and conventions of UNCBD (United Nations Convention on Biological Diversity) and UNFCCC (UN Framework Convention on Climate Change), the Bangladesh government has reshaped its policies (such as FYPs) focusing on gender equality and women's empowerment at every stage, including disaster leadership, disaster management, capacity building, resilience, etc. At the microlevel under the Community Disaster Management (CDM) pillar programs, the development of a gender-responsive disaster management policy, ensuring women's participation at all levels, increasing financial allocation for women, capacity-building activities for women, and ensuring healthcare for women is considered as a prerequisite for sustainable development. The nexus between climate change and women empowerment policies matters because these policies reflect the country's commitment to addressing the needs and challenges of the often invisible section (i.e., women) of the population (GoB, 2013; Khan, 2019; Khan, 2022).

Local politics and climate change adaptation challenges in coastal Bangladesh

The vulnerability of the southwest coastal zone is characterized by the combination of some root causes, dynamic forces, and exposure to unsafe natural conditions. Among these components, root causes are the initial most profound causes, many of which seem latent in nature, rooted in the social structure, and badly affect the adaptation capacity. In Dacope of Khulna district, local politics is such a deep and underlying root cause that contributes to intensifying the effects of natural hazards manifolds and is consequently keeping disastrous effects on the lives and livelihoods of the coastal people. Thereby, priority should be given to unmasking the nature of local politics to minimize the potential consequences of climate change-induced risks and support the analyses of possible responses. In this section, we[3] explore the nature of local politics of the southwest coastal region of Bangladesh, particularly Dacope Upazila of Khulna city, and its unfavorable impact on the adaptation capacity of the local people based on both primary and secondary sources of information. For quantitative data of this study, 263 coastal women as respondents were interviewed face to face using a survey method with a structured questionnaire where few questions were unstructured/open-ended. For the qualitative data, a total of nine focus group discussions and nine in-depth individual interviews were conducted among disaster-affected young women, mothers, older women, female household heads, NGO activists, political leaders, community leaders, elected representatives, and with different committees which manage the local response to disasters. The sample size for each technique was appropriate for the study because the population of the study area is homogeneous in terms of their occupation, religion, education, lifestyle, disaster experience, etc. Data were conducted in a natural setting so that respondents could feel comfortable providing authentic data without disrupting their daily life. In each FGD, the number of participants ranged between 5 to 15 (Khan, 2019; Khan, 2022).

Social supremacy

Most of the coastal people of Dacope Upazila are landless who are subjugated in many ways by the comparatively richer population groups. The elite groups locally known as *sardars, matabbars, mahajan,* etc. By their higher socioeconomic status, they often become the elected local political representatives who have the monopoly of land ownership and ultimately hold the authority to suppress the poor class. This societal power is defined as "social supremacy" (Mallick and Vogt, 2011). This supreme class is the major decision-makers and most often manipulates the participation of community people in various governmental and non-governmental development projects.

Both the discussants of the FGDs and survey respondents (70%) of the study area argue that local elites do not consider the coastal societal system, practices, experiences, and norms at the time of projects' planning and implementation. As a result, projects cannot achieve their goals and become unsuccessful. While conducting focus group discussion with the members of the Project Implementation Committee (PIC) who are supposed to be involved actively with the project planning and programs, one male member talked as follows:

> We don't know the details about the programs taken for our village. They (GO officials) don't come to our villages before implementing any projects. Suppose now you are here to talk to us, and we are sharing our

views with you. If the decisions are taken in this way projects and programs will be more effective for us.

(Male, 48, (Education: Master's degree) FGD with PIC members)

(Khan, 2019)

Development workers and external project developers also consider the
"social supreme" as the local resource persons who play the gatekeeping role
of the community (Mallick & Vogt, 2011). The existing feudal power relationships of Bangladesh's coastal region are so complex that it sidetracks the
all-inclusive disaster management strategies adopted by BCCSAP and
thereby affecting the process of disaster preparedness and climate adaptation
plans in multifarious ways.

Illegal land grabbing

Coastal sediment regions, where small islands known as char lands are continuously shifting, are an important source of power-play regarding land
conflicts. Local elites or politically affiliated people often employ gangs to
forcibly uproot the landless inhabitants from the rich alluvial char lands
where they were living without any document or official records. A joint
report from the Human Development Research Centre and the Association
for Land Reform and Development estimated that 95% of fishing areas
belong to "waterbody-grabbers." In contrast, only 5% have been leased to
poor fishers (Barkat et al., 2007). Rashid (2014) shows a connection between
the Coastal Development Strategy ([CDS] 2006) under the National
Development Plan of Action (NDPA) and the illegal grabbing of lands without compensation and the consequent population displacement. This is also
related to the process of "enclosure" that transfers a public asset (common
lands/khas lands or state-owned lands/char lands) into private hands (a few
land grabbers) or extends the role of a few private actors into the public
sphere. As a part of the wealth accumulation strategy, approximately 53% of
the respondents stated that coastal embankments are illegally fractured for
the direct interests of the political leaders and local elites who allow the saline
water to inundate the non-saline lands for the expansion of shrimp farms.
Gradually, they become the owner of these salt-contaminated lands. A male
participant described,

> We are getting used to accepting the process of marginalization and
> landlessness as our irony. No one is taking initiative to stop the process
> of shrimp cultivation and salinity in the locality and there is no policy for
> land management. Sometimes the shrimp farm owners or influential of
> the local area grab the land by force. I have lost some of my land and still,
> I am having problem with them. As a result, poverty is increasing among
> poor people in the coastal area.
>
> (Male, 47, Honors, in-depth individual interview)

Thus, it further widens the economic gap between the rich and the poor and creates a bureaucratic administrative structure that enables them to act with increased autonomy or sovereignty (Sovacool et al., 2017). Though the sixth FYP (2011–2015) of Bangladesh has acknowledged the issue and set some objectives, such as rehabilitation of climate victims, char development, and settlement (Rashid, 2013), the problem is still acute due to the interference of local politics and corruption.

Top-down versus bottom-up approach

The drivers or agents of change can emerge or be initiated in society either from "top-down" or "bottom-up." But the bottom-up approach entails shifting accountability directly to the citizens and service receivers through exit mechanisms, i.e., the scope of service users to exit for unsatisfactory service and voice mechanisms, i.e., articulation of preferences and demands by the service users (Haque and Ali, 2011; Khan, 2019; Khan, 2022). Thereby, pressure and expectations from the bottom in implementing the programs make NGO programs effective (Kabeer, 2011; Panday, 2016). However, the reality is different to a large extent. Many development plans and adaptation projects are directly or indirectly connected to the "encroachment" of char lands, forests, farms, or other public commons (World Bank, 2013). This situation indicates a gross distraction from the policies adopted by the Climate Change Action Plan (2009–2018) and the pro-poor climate change strategy where most vulnerable groups are supposed to receive special attention (Khan, 2019; Khan, 2022).

Social exclusion

Social exclusion is observed at every level of the climate change adaptation process ranging from planning at the national level to the implementation at the community level (Afroz et al., 2016; Huq and Khan, 2006; Mallick and Vogt, 2011; Sovacool et al., 2018). A study conducted in Dacope Upazila found that "participation of all" was not ensured in selecting the location of a cyclone shelter. It was observed that poor people lacked the rights of having access the cyclone shelters as these shelters were constructed on the donated lands by powerful people who contributed lands close to their houses to ensure their first access to these shelters (Mallick and Vogt, 2011). A male during the focus group discussion with Cyclone Preparedness Programme (CPP) members describes:

> "Imagine the situation of the shelter centre in disastrous time. In the shelter centre if women members are deployed, they can manage better since women have easy access to women and women can solve their problems best."
>
> (Male, 25, Grade VIII, FGD with CPP) (Khan, 2022)

Apart from cyclone shelter, local people's opinion was not considered during the reconstruction process of their houses after the super cyclone *Aila*. They claim that the structure of the houses was faulty, and the materials used were of very low quality. A female member of the PIC committee though raised her voice against this but was ignored. She replied in an interview as follows:

> We could not say anything and if we wanted to say anything we were told that we are getting things free of cost. So, we don't have the right to say anything, and we accept low quality, and we offer a bribe sometimes.
> (Woman, 45, Grade V, FGD with PIC members)
> (Khan, 2019; Khan, 2022)

Besides, women's participation in disaster management goes unrecognized as political interference refrains women from attending disaster management committee meetings. Out of the total 263 respondents, a significant number of the respondents (32%) didn't know about women's involvement in the meeting. Forty percent (40%) of the respondents opined that the degree of attendance of women in the meetings was affected by different social discriminatory norms including social norms, religious norms/*purdah* systems, patriarchal systems, transport and communication, political corruption, and politicization of everything. The problem of politicization is described by a woman as follows:

> I want to attend the meeting, but the problem is my political ideology is different from most of the members of the committee including the president of the committee. Whatever I say is ignored just because of political ideology and that's why I don't feel like attending meetings anymore and they don't want me to attend as well.
> (Woman, 33, FGD with vulnerable women) (Khan, 2022)

Climate change action strategies are entangled with the process of entrenchment that traps many of the poor and powerless into a patrimonial system of authority promoting insecurity and violence by disseminating the benefits of adaptation unevenly (Sovacool et al., 2017). Such entrenchment may worsen gender inequality. The recently National Plan for Disaster Management (2016–2020) has considered inclusion as an underlying strategy and given adequate importance on the gender inclusion policy, the crucial role of women in disaster management has been still neglected (Hasan et al., 2019; Sovacool et al., 2017; Zaman, 2021).

The political economy of shrimp cultivation

The suitable climatic condition has made the coastal areas of Bangladesh one of the hot spots of industrial shrimp farming in the world. But this farming has raised serious concerns about the impact of saline water intrusion on agricultural lands leading to significant yield loss (Zaman, 2018). In the southwest coastal zone, shrimp farming is an essential source of capital accumulation for the wealthy people, resulting in the exploitation of the poor. For the expansion of shrimp farms, some politically powerful elites dig artificial canals to channel seawater into the agricultural lands that were not naturally affected by seawater before (35% of respondents mentioned), resulting in increased salinity in the coastal areas (84.5% of respondents stated). While passing through this channel, seawater often overflows the banks and affects the lands of the poor farmers along the banks. The small homesteads, which serve as major support for household food and fodder for the poor farmers, become unproductive because of seawater. At some point, the helpless poor sometimes become bound or sometimes forced to sell these salt-contaminated lands to the powerful elites and become landless. The artificial inundation of saline water ultimately results in a process that transfers poor people's land into a few private hands. Besides, climate-resilient agricultural activities failed to bring remarkable outcomes in many villages of Dacope Upazila due to the artificial inundation of saline water.

Corruption and nepotism

Local politics plays a significant role in disaster risk management, especially in identifying victims to receive relief (Uddin et al., 2020). The presence of significant corruption has been observed in selecting the victims and relief distribution. Survey respondents report that nepotism, politicization, personal relation, and corruption by local elected representatives and government officials are manifested in different projects such as cash financial support (83%), food support (69%), *kabikha* (19%), distribution of construction materials (10%), medical facilities (11%), clothes (11%), agricultural accessories (4%), etc. According to a vulnerable male participant (aged 60) of FGD, the list of vulnerable people is unreal, where physically challenged people are shown fit and physically fit people are shown disabled. Discussants described that the government allocated taka 20,000 for each family in the locality. This cash was distributed by elected local union parishad members who kept some money from the allocated taka 20,000. Still, none raised their voice lest they should not get any aid by way of retaliation. Moreover, there was a house-building loan project by the government to construct and rebuild houses. But only people with cultivated lands were eligible to get the loan, which was also discriminatory. Though the Bangladesh government increased National Climate Change Fund, developing adaptation capacity and building resilience may not be possible unless free and fair local governance can be formed for the effective and even distribution of aid and resources (Khan, 2019; Khan, 2022).

The gap in the implementation of national policies at the local level

In Bangladesh, many national-level policies are implemented at the local level. Thereby, the success of any national climate change policy depends on how successfully it has been implemented at the local level. While national

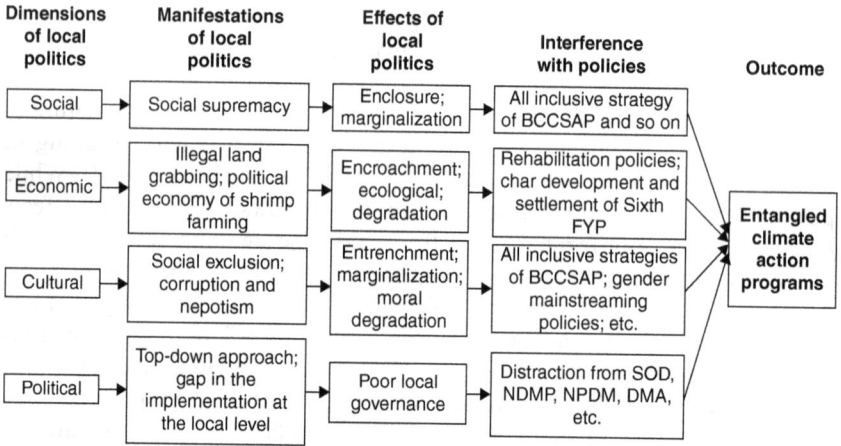

Figure 8.1 Effects of local politics on climate action programs.

Source: Developed by the authors based on the research findings.

policies are expected to keep great strides toward disaster mitigation, most of the shortcomings in disaster management plans result from poor implementation at the local level (Ahmed, 2019).

The UDMC, according to SOD (Standing Orders on Disaster) (2010) is headed by the chairman of the Union Parishad and must consist of 36 total members with the additional three members that the UP chairman can co-opt. A UDMC is supposed to include members from all sectors of society, including UP members, teachers, government officials, vulnerable women, representative of CPP, representative of the Red Crescent, representatives of NGOs, representative of peasant and fishermen, representative of civil society, religious leader, and representative of Village Defence Party (VDP). Discussants in the focus group with vulnerable women, female-headed households, and community people stress that the number and representation of female members in UDMC are poor. As regards under-representation and inequitable access to the meeting, a woman discussant describes as follows:

> In the meeting sometimes no female members attend and sometimes one and sometimes two. I tried to put my suggestions several times but could not and no attention was given to me by the male members. Nowadays I normally do not attend meetings or even if I attend the meeting I don't speak and the same for other female members as well.
>
> (Woman, 38, class IX, FGD with UDMC)

The problems and obstructions to the inclusion of women in the committees and programs are fabricated or exaggerated by the male members of the family/ community/committee and local elected representatives and therefore women cannot use their maximum potential (Kabeer, 2011). The under-representation of women can lead to an unintended reinforcement of gender inequalities by dooming all gender mainstreaming policies of the government into failure.

Conclusion

The government of Bangladesh developed the National Adaptation Programme of Action in 2005, consulting with stakeholders, professional groups, and civil society extensively. Consequent to this, the country prepared the comprehensive action strategy, BCCSAP in 2008, revised in 2009, along the lines of the vision and priorities of the newly formed democratic government of Bangladesh. This revised plan BCCSAP, 2009 has acted as a guideline and mission document for all the action plans and strategies of the government to combat climate change impacts over the past decade. It emphasizes climate change adaptation, and its strategy lies in sustainable development, poverty reduction, and the promotion of the well-being of vulnerable groups with gender sensitivity (BCCSAP, 2009). Like this, by recognizing the global challenge of climate change, the Union Government of India announced the NAPCC in June 2008. The action plan was devised by the then Union Government of India by adopting a top-down approach, and there was no wide consultation that happened in the developing process of this plan. It laid down eight national missions to promote solar energy, sustainable habitat, sustainable agriculture, and others. The subjects of some missions laid down here come under the purview of the states. Also, to decentralize the actions of all the missions, the Union Government of India directed the state to build their action plans on climate change in alignment with the goals and aims of the National Action Plan. By 2014, all states had developed their action plans. Some industrial states in India provided robust contributions to renewable energy plans in these state action plans. Hence, in recent years, the trends have been changing. The states of India have started playing crucial roles in planning and implementing the climate policies, and actions of the Union Government through State Action Plans on Climate Change (Atteridge et al. 2012; Dubash & Jogesh, 2014; Government of India, 2008; Jörgensen et al. 2015; Swenden & Saxena, 2020).

Nevertheless, in India and Bangladesh, adaptation strategies to climate change are still evolving, and the politics of adaptation are limitedly understood and discussed. Drawing on primary and secondary sources of information, the chapter shows three key insights. In both cases of India and Bangladesh, the tradition of social exclusion is found as a powerful barrier to climate action programs. Despite taking adequate initiatives at the policy level, there has been a gross under-representation of women in implementing climate change adaptation projects in coastal India and Bangladesh. Such gross under-representation of women and limited understanding of the real vulnerabilities of women is also reflected in the action plan and policy documents on climate change both in Bangladesh and India.

Our second insights come from our discussion on social and political capital, closely associated with a community's overall well-being. In the case study from Tamil Nadu in India, we found that it is not the social capital but rather the political capital of the coastal villages that plays an influential role in bringing coastal adaptation projects to their villages from the state. Although the fishers across most villages along the coastal stretch of India belong to a single caste, their adaptive capacity to respond to climate change varies along the lines of economic capital and political capital as power relations flow through these two capitals and it determines the politics of adaptation to climate change. From this case study, we also found that the CBA strategies to respond to climate change vary irrespective of the socioeconomic backgrounds of the fishing villages. In the case of Bangladesh, it is found that the traditional feudal landholding social system is so closely attached to the sociopolitical power structure that it results in the landlessness of many poor in the coastal belt for various commercial purposes. Illegal land grabbing, forceful land selling, artificial inundation of saline water, and the political economy of shrimp farming are a few processes of landlessness of which coastal poor are the direct or indirect victims. In fact, the existing feudal power relationships of Bangladesh's coastal region are so complex that it sidetracks the all-inclusive disaster management strategies adopted by BCCSAP and thereby are affecting the process of disaster preparedness and climate adaptation plans in multifarious ways.

Our third insight come from our analysis of the stakeholders' backgrounds that influence adaptation strategies/projects to climate change at the local levels. In the case of coastal Tamil Nadu in India, there are no significant social differences between the "powerful" elites and "powerless" local people – whereas, from the case study in Bangladesh, we found that there have been significant socioeconomic differences lie between the local elites and the powerless victims of disasters and climate change. The "powerless" victims of disasters and climate risks in Bangladesh are further marginalized by the existing social inequalities and the resulting complexities in distributing relief aids, having access to cyclone shelters and state-owned lands, and getting loans, grants, and other benefits during disasters and post-disaster phases. We also found that the multiple manifestations of elite-centric local politics have substantial implications for sub-national climate action plans and national climate change policy. As Stock et al. (2021) noted, the Bangladesh government took the top-down sectoral approach in climate change adaptation policy-making and disregarded the multiple local drivers of vulnerability. Also, the power symmetry among the influential factors blocks the participation of socially excluded and economically marginalized local people. In the southwest coastal zone of Bangladesh, social elites play a vital role in systematically excluding the vulnerable poor from the decision-making process and thereby contribute to fabricating the inclusionary approach to policymaking. Similarly, India's climate change action plans at the sub-national levels are unwilling to locate the vulnerabilities within sociopolitical causes and provide limited or no emphasis on the vulnerabilities that arise from sociopolitical differences.

Since India and Bangladesh share common historical backgrounds, their economic, social, cultural, and political nature is closer. The cases have revealed almost similar characteristics of the coastal rural power structure, the nature of social exclusion, the role of social and political capital, and their influences on climate change adaptation policies. However, the degree of influence may vary based on the current social context, changing social structure, and the impact of globalization. In recent years, there has been emerging research to understand the adaptation policies of South Asia, particularly India and Bangladesh, focusing on power, politics, and decision-making (Ayers, 2011; Ishtiaque et al., 2021; Stock et al., 2020; Vij et al., 2017; Vij et al., 2018). We hope that the case studies of this chapter and the resulting comparative insights will bolster the growing but limited knowledge of local politics of climate change in India and Bangladesh. In this context, we also note that the case studies that this chapter shows were conducted at micro-levels. Hence, generalizing the insights of these case studies at the national level seems like a difficult task. However, the scope of this chapter lies only in providing insights into the local politics of climate change adaptation by taking the cases from selected regions in India and Bangladesh. Both cases reflect that there have been loopholes in the policy and administration in addressing the climate change issues at the local levels. The state and policymakers have given limited understanding and emphasis to examining the influence of local politics and power in climate change adaptation at the local levels.

Notes and Acknowledgements

1 The first author of this chapter conducted fieldwork in seven fishing villages of Nagapattinam district of Tamil Nadu as a part of his Ph.D. research from 2014 to 2018 in different phases.
2 To discuss the case from Tamil Nadu, this chapter selectively uses a few insights from the doctoral research of the first author. The respondents are local leaders, senior fishermen, and representatives of local institutions of the fishing villages. Detailed discussions on these themes and topics are available in his forthcoming book. The details of the book are as follows (Local Adaptation to Climate Change in South India: Challenges and the Future in the Tsunami-hit Coastal Regions, Routledge, UK. 2023. Forthcoming, In Press).
3 The discussions on Bangladesh in this chapter largely come from the Ph.D. research of the second author. To ensure consistency and the flow of reading, we prefer to use the term "we." The details of the doctoral dissertation of the second author of this chapter are as follows: (Disaster and Women: The Wave of Change in the Position and Status of Women in the Coastal Bangladesh. Submitted to the University of Canterbury, 2019. Available at: https://hdl.handle.net/10092/100703). This thesis has recently been published as a book monograph in Springer Singapore. The details are as follows: Disaster and Gender in Coastal Bangladesh: Women's Changing Roles, Risk and Vulnerability, 2022. https://doi.org/10.1007/978-981-19-3284-7. A few selected contents and field narratives of the present chapter focusing on Bangladesh were published in this book. The second author gratefully acknowledges and thanks Springer Singapore for providing permission to reuse the contents of the fifth chapter (Women's Empowerment, Role of GO and NGO in Disaster Management) of this book in the present chapter.

References

ADB. (2017). *A Region at Risk: The Human Dimensions of Climate Change in Asia and the Pacific.* Asian Development Bank. Retrieved May 24, 2021, from https://www.adb.org/sites/default/files/publication/325251/region-risk-climate-change.pdf

Adger, W. N. (2003). Social capital, collective action, and adaptation to climate change. *Economic Geography,* 79(4), 387–404. https://doi.org/10.1111/j.1944-8287.2003.tb00220.x

Adger, W. N., Paavola, J., Huq, S., & Mace, M. J. (Eds.). (2006). *Fairness in Adaptation to Climate Change.* Massachusetts: MIT Press.

Afroz, S., Cramb, R., & Grünbühel, C. (2016). Ideals and institutions: Systemic reasons for the failure of a social forestry program in southwest Bangladesh. *Geoforum,* 77, 161–173. https://doi.org/10.1016/j.geoforum.2016.11.001

Afroz, S., Cramb, R., & Grünbühel, C. (2018). Vulnerability and response to cyclones in coastal Bangladesh: A political ecology perspective. *Asian Journal of Social Science,* 46(6), 601–637. https://doi.org/10.1163/15685314-04606002

Ahmed, A. U., Haq, S., Nasreen, M., & Hassan, A. W. R. (2015). *Climate Change and Disaster Management: Sectoral Inputs towards the Formulation of Seventh Five Year Plan (2016-2021).* Dhaka: Planning Commission. Retrieved December 27, 2021 from http://nimc.portal.gov.bd/sites/default/files/files/nimc.portal.gov.bd/page/6c53bd88_ad69_4ccf_bbae_d45b70dbc0bf/017%207th%20FYP%20and%20 2021%20Climate-Change-and-Disaster-Management.pdf

Ahmed, I. (2019). The national plan for disaster management of Bangladesh: Gap between production and promulgation. *International Journal of Disaster Risk Reduction,* 37(101179), 1–8. https://doi.org/10.1016/j.ijdrr.2019.101179

Almazroui, M., Saeed, S., Saeed, F., Islam, M. N., & Ismail, M. (2020). Projections of precipitation and temperature over the South Asian countries in CMIP6. *Earth Systems and Environment,* 4(2), 297–320. https://doi.org/10.1007/s41748-020-00157-7

Arockiaraj, J. (2013, July 04). Rubble-mound sea walls ineffective against sea erosion: Coastal villagers. *The Times of India.* https://timesofindia.indiatimes.com/city/madurai/rubble-mound-sea-walls-ineffective-against-sea-erosion-coastal-villagers/articleshow/20902929.cms

Atteridge, A. (2013). The evolution of climate policy in India: Poverty and global ambition in tension. In D. Held, C. Roger, & E.-M. Nag (Eds.), *Climate Governance in the Developing World.* Cambridge, UK & Malden, USA: Polity Press/Wiley.

Atteridge, A., Shrivastava, M. K., Pahuja, N., & Upadhyay, H. (2012). Climate policy in India: What shapes international, national and state policy?. *Ambio,* 41(1), 68–77. https://doi.org/10.1007/s13280-011-0242-5

Ayers, J. (2011). Resolving the adaptation paradox: Exploring the potential for deliberative adaptation policy-making in Bangladesh. *Global Environmental Politics,* 11(1), 62–88.

Barkat, A., Ara, R., Taheruddin, M., Hoque, S., Hoque, S., & Islam, N. (2007). *Towards a Feasible Land Use Policy of Bangladesh.* A report prepared for the Association for Land Reform and Development (ALRD). Retrieved January, 8, 2022, from https://www.hdrc-bd.com/wp-content/uploads/2018/12/05.-Towards-a-Feasible-Land-Use-Policy-of-Bangladesh.pdf

BBS. (2011). *District Statistics, Khulna District.* Bangladesh Bureau of Statistics, Statistics and Informatics Division, Ministry of Planning, Government of the People's Republic of Bangladesh. Retrieved January, 4, 2021, from http://203.112.218.65:8008/WebTestApplication/userfiles/Image/District%20Statistics/K.hulna.pdf

BCCSAP. (2009). *Bangladesh Climate Change Strategy and Action Plan 2009*. Ministry of Environment and Forests, Government of the People's Republic of Bangladesh. Retrieved April 24, 2021, from https://www.iucn.org/downloads/bangladesh_climate_change_strategy_and_action_plan_2009.pdf

Birkenholtz, T. (2012). Network political ecology: Method and theory in climate change vulnerability and adaptation research. *Progress in Human Geography*, 36(3), 295–315. https://doi.org/10.1177%2F0309132511421532

Byravan, S., Rajan, S. C., & Rangarajan, R. (2010). *Sea Level Rise: Impact on major infrastructure, land and ecosystems along the Tamil Nadu Coast. Indian Institute of Technology (Madras) & Institute of Financial Management and Research (Madras)*. https://www.indiawaterportal.org/sites/default/files/iwp2/Sea_level_rise__Impact_on_major_infrastructure_ecosystems_and_land_along_the_Tamil_Nadu_coast_IFMR_IIT_Madras_2010.pdf

Chaturvedi, A., Rattani, V., & Awasthi, K. (2019). State action plans on climate change need upscaling and capacity enhancement. *Down To Earth*. Retrieved August 21, 2021, from https://www.downtoearth.org.in/blog/climate-change/state-action-plans-on-climate-change-need-upscaling-and-capacity-enhancement-66796

CMFRI. (2012). *Marine Fisheries Census 2010 Part II. 4 Tamil Nadu*. Central Marine Fisheries Research Institute. Retrieved August 21, 2021, from http://eprints.cmfri.org.in/9002/1/TN_report_full.pdf

CoastAdapt. (2017). *What is Adaptation to Climate Change?* CoastAdapt. Retrieved August 21, 2021, from https://coastadapt.com.au/overview-of-adaptation

Dastagir, M. R. (2015). Modeling recent climate change induced extreme events in Bangladesh: A review. *Weather and Climate Extremes*, 7, 49–60. https://doi.org/10.1016/j.wace.2014.10.003

Debnath, P. K. (2020). Demonstrate the capacity, vulnerability and community based disaster risk management scenario of the vulnerable groups in the coastal belt area Dacope and Chalna in Khulna District of Bangladesh. *Research Square*, 1–15. http://dx.doi.org/10.21203/rs.3.rs-30111/v1

Dolšak, N., & Prakash, A. (2018). The politics of climate change adaptation. *Annual Review of Environment and Resources*, 43, 317–341. https://doi.org/10.1146/annurev-environ-102017-025739

Dubash, N. K., & Jogesh, A. (2014). From margins to mainstream? State climate change planning in India. *Economic and Political Weekly*, 86–95.

Eriksen, S. H., Nightingale, A. J., & Eakin, H. (2015). Reframing adaptation: The political nature of climate change adaptation. *Global Environmental Change*, 35, 523–533. https://doi.org/10.1016/j.gloenvcha.2015.09.014

Esraz-Ul-Zannat, M., Abedin, M. A., Pal, I., & Zaman, M. M. (2020). Building resilience fighting back vulnerability in the coastal city of Khulna, Bangladesh: A perspective of climate-resilient city approach. *International Energy Journal*, 20(3A), 549–566.

Funder, M., Mweemba, C., & Nyambe, I. (2018). The politics of climate change adaptation in development: Authority, resource control and state intervention in rural Zambia. *The Journal of Development Studies*, 54(1), 30–46. https://doi.org/10.1080/00220388.2016.1277021

Giddens, A. (2009). *Politics of Climate Change*. Polity.

Glantz, M. H. (1995). Assessing the impacts of climate: The issue of winners and losers in a global climate change context. *Studies in Environmental Science*, 65, 41–54. https://doi.org/10.1016/S0166-1116%2806%2980193-7

Glover, L., & Granberg, M. (2020). The politics of adapting to climate change. In L. Glover & M. Granberg (Eds.), *The Politics of Adapting to Climate Change* (pp. 3–22). London, UK: Palgrave Pivot.

GoB [Government of the People's Republic of Bangladesh]. (2009a). *Bangladesh Climate Change Strategy and Action Plan (BCCSAP)*. Ministry of Environment and Forests. Retrieved January 4, 2022 from https://www.iucn.org/downloads/bangladesh_climate_change_strategy_and_action_plan_2009.pdf

GoB [Government of the People's Republic of Bangladesh]. (2009b). National Adaptation Programme of Action (NAPA). Dhaka: Ministry of Environment and Forests. Retrieved January 5, 2022 from https://unfccc.int/resource/docs/napa/ban02.pdf

GoB [Government of the People's Republic of Bangladesh]. (2010). *National Plan for Disaster Management 2010–2015*. Dhaka: Disaster Management Bureau. Retrieved January 11, 2022 from http://nda.erd.gov.bd/en/c/publication/national-plan-for-disaster-management-2010-2015

GoB [Government of the People's Republic of Bangladesh]. (2013). *Bangladesh Climate Change and Gender Action Plan (ccGAP: Bangladesh)*. Dhaka: Ministry of Environment and Forests. Retrieved January 11, 2022 from http://nda.erd.gov.bd/en/c/publication/climate-change-and-gender-action-plan-ccgap-2013

GoB [Government of the People's Republic of Bangladesh]. (2015). *The 6th Five Year Plan 2016–2020. Accelerating Growth and Reducing Poverty*. Dhaka: Planning Commission, Ministry of Planning. Retrieved January 8, 2022 from https://policy.asiapacificenergy.org/node/285

GoB [Government of the People's Republic of Bangladesh]. (2016). *The 7th Five Year Plan FY2016–FY2020: Accelerating Growth, Empowering Citizens*. Dhaka: General Economic Division (GED), Planning Commission. Retrieved January 8, 2022 from https://policy.asiapacificenergy.org/node/2443

Government of India. (2008). *National Action Plan On Climate Change*. Prime Minister's Council on Climate Change. Retrieved August 21, 2021, from http://www.nicra-icar.in/nicrarevised/images/Mission%20Documents/National-Action-Plan-on-Climate-Change.pdf

Government of India. (2020). *Guidelines for "Protection and Control of Coastal Erosion in India." CSIR-National Institute of Oceanography (Council of Scientific & Industrial Research) Dona Paula, Goa*. Central Water Commission, Government of India. Retrieved August 21, 2021, from http://cwc.gov.in/sites/default/files/guidelines-%E2%80%9Cprotection-and-control-coastal-erosion-india%E2%80%9D.pdf

Government of Tamil Nadu. (2019). *Speech of Thiru O. Panneerselvam, Hon'ble Deputy Chief Minister, Government of Tamil Nadu, Presenting the Budget for the year 2019–2020 to the Legislative Assembly on 8th February, 2019*. Tamil Nadu Government Portal. Retrieved August 21, 2021, from https://cms.tn.gov.in/sites/default/files/documents/fin_budget_speech_e_2019_20.pdf

Government of Tamil Nadu. (2020). *Policy Note Demand No. 15. Policy Note 2021–21*. Environment and Forests Department, Government of Tamil Nadu. Retrieved August 21, 2021, from https://cms.tn.gov.in/sites/default/files/documents/env_e_pn_2020-2021.pdf

Government of Tamil Nadu. (2021). *Speech of Thiru O. Panneerselvam, Hon'ble Deputy Chief Minister, Government of Tamil Nadu, presenting the Interim Budget for the year 2021-2022 to the Legislative Assembly on 23rd February, 2021*. Tamil Nadu Government Portal. Retrieved December 21, 2021, from https://cms.tn.gov.in/sites/default/files/documents/fin_budget_speech_e_2019_20.pdf

Goyal, M. K., & Surampalli, R. Y. (2018). Impact of climate change on water resources in India. *Journal of Environmental Engineering*, 144(7), 04018054. http://dx.doi.org/10.1061/(asce)ee.1943-7870.0001394

Haque, S. K. T. M., & Ali, K. S. (2011). NGO governance: Accountability and transparency issues for the development sector. *Public Management Review*, 5(2), 267–278.

Hasan, M. R., Nasreen, M., & Chowdhury, M. A. (2019). Gender-inclusive disaster management policy in Bangladesh: A content analysis of national and international regulatory frameworks. *International Journal of Disaster Risk Reduction*, 41(101324). https://doi.org/10.1016/j.ijdrr.2019.101324

Hoque, M. Z., Cui, S., Xu, L., Islam, I., Tang, J., & Ding, S. (2019). Assessing agricultural livelihood vulnerability to climate change in coastal Bangladesh. *International Journal of Environmental Research and Public Health*, 16(22), 4552. https://doi.org/10.3390/ijerph16224552

Huq, S., & Khan, M. R. (2006). Equity in national adaptation programs of action (NAPAs): The case of Bangladesh. In W. N. Adger, J. Paavola, S. Huq, & M. J. Mace (Eds.). *Fairness in Adaptation to Climate Change* (pp. 181–200). Cambridge, MA: The MIT Press.

Huq, S., & Reid, H. (2007). *Community-Based Adaptation: A Vital Approach to the Threat Climate Change Poses to the Poor*. London: IIED Briefing. Retrieved August 21, 2021, from https://www.osti.gov/etdeweb/servlets/purl/22059781

Ishtiaque, A., Eakin, H., Vij, S., Chhetri, N., Rahman, F., & Huq, S. (2021). Multilevel governance in climate change adaptation in Bangladesh: Structure, processes, and power dynamics. *Regional Environmental Change*, 21(3), 1–15. https://doi.org/10.1007/s10113-021-01802-1

Javeline, D. (2014). The most important topic political scientists are not studying: Adapting to climate change. *Perspectives on Politics*, 12(2), 420–434. https://doi.org/10.1017/S1537592714000784

Jörgensen, K., Mishra, A., & Sarangi, G. K. (2015). Multi-level climate governance in India: The role of the states in climate action planning and renewable energies. *Journal of Integrative Environmental Sciences*, 12(4), 267–283. https://doi.org/10.1080/1943815X.2015.1093507

Kabeer, N., (2011). Between Affiliation and autonomy: Navigating pathways of women's empowerment and gender justice in rural Bangladesh. *Development and Change*, 42(2), 499–528. https://doi.org/10.1111/j.1467-7660.2011.01703.x

Kankara, R. S., Ramana Murthy, M. V., Rajeevan, M. (2018). *National Assessment of Shoreline Changes Along Indian Coast. Status Report for 26 Years. 1990 – 2016*. Ministry of Earth Sciences, National Centre for Coastal Research, Chennai. Retrieved August 21, 2021, from https://www.nccr.gov.in/sites/default/files/schangenew.pdf

Kelman, I., & Gaillard, J. C. (2010). Embedding climate change adaptation within disaster risk reduction. In R. Shaw, J. M. Pulhin & J. J. Pereira (Eds.), *Climate Change Adaptation and Disaster Risk Reduction: Issues and Challenges* (pp. 23–46). Bingley, WA: Emerald Group Publishing Limited. https://doi.org/10.1108/S2040-7262(2010)0000004008

Khan, M. A., Kabir, K. H., Hasan, K., Sultana, R., Al Imran, S., & Karmokar, S. (2021). Household's agricultural vulnerability to climate-induced disasters: A case on south-west coastal Bangladesh. *Research Square*. https://doi.org/10.21203/rs.3.rs-399084/v3

Khan, M. J. U., Islam, A. K. M., Bala, S. K., & Islam, G. M. (2020). Changes in climate extremes over Bangladesh at 1.5° C, 2° C, and 4° C of global warming with high-resolution regional climate modeling. *Theoretical & Applied Climatology*, 140. https://doi.org/10.1007/s00704-020-03164-w

Khan, M. M. (2019). Disaster and Women: The Wave of Change in the Position and Status of Women in the Coastal Bangladesh (Doctoral dissertation). Department of Sociology and Anthropology, School of Language, Social & Political Sciences, University of Canterbury. https://hdl.handle.net/10092/100703

Khan, M. M. (2022). *Disaster and Gender in Coastal Bangladesh: Women's Changing Roles, Risk and Vulnerability*. Singapore: Springer Nature.

Krishnan, R., Sanjay, J., Gnanaseelan, C., Mujumdar, M., Kulkarni, A., & Chakraborty, S. (2020). *Assessment of Climate Change Over the Indian Region: A Report of the Ministry of Earth Sciences (MOES), Government of India* (p. 226). Springer Nature.

Madhanagopal, D. (2019). Vulnerability and adaptation of marine fishers to climate change: A study of local institutions in Nagapattinam district, Tamil Nadu (Unpublished Doctoral Dissertation). Department of Humanities and Social Sciences, Indian Institute of Technology Bombay.

Madhanagopal, D. (2023). *Local Adaptation to Climate Change in South India: Challenges and the Future in the Tsunami-Hit Coastal Regions*. London: Routledge.

Mallick, B., & Vogt, J. (2011). Social supremacy and its role in local level disaster mitigation planning in Bangladesh. *Disaster Prevention and Management: An International Journal*, 20(5), 543–556. https://doi.org/10.1108/09653561111178970

Ministry of Environment and Forests. (2005). *Report of the Committee Chaired by M.S. Swaminathan to Review the Coastal Regulation Zone Notification 1991*. New Delhi: Ministry of Environment and Forests. Retrieved August 21, 2021, from http://iomenvis.in/pdf_documents/MSS_Report.pdf

MoEFCC. (2018). *Roadmap and Action Plan for Implementing Bangladesh NDC: Transport, Power and Industry Sectors*. Areport prepared by the Ministry of Environment, Forest and Climate Change. Dhaka: Ministry of Environment, Forest and Climate Change Building # 6, Level # 13, Bangladesh Secretariat, Dhaka 1000. Retrieved January 8, 2022 from https://moef.portal.gov.bd/sites/default/files/files/moef.portal.gov.bd/page/ac0ce881_4b1d_4844_a426_1b6ee36d2453/NDC%20Roadmap%20and%20Sectoral%20Action%20%20Plan.pdf

Mortimore, M., & Mortimore, M. M. (1989). *Adapting to Drought: Farmers, Famines and Desertification in West Africa*. Cambridge: Cambridge University Press.

Muralidharan, R. (2017). Open letter: Why is Tamil Nadu neglecting its artisanal fishers? *The Wire*. https://thewire.in/environment/artisanal-fishers-tamil-nadu-sri-lanka-trawling

Nagoda, S. (2017). *Reproducing vulnerability through climate change adaptation? policy processes, local power relations and food insecurity in North-Western Nepal*. [Doctoral dissertation: Norwegian University of Life Sciences]. https://nmbu.brage.unit.no/nmbu-xmlui/handle/11250/2447872

Nalau, J., & Verrall, B. (2021). Mapping the evolution and current trends in climate change adaptation science. *Climate Risk Management*, 32, 100290. https://doi.org/10.1016/j.crm.2021.100290

Nambi, A., & Bahinipati, C. S. (2012). Adaptation to climate change and livelihoods: An integrated case study to assess the vulnerability and adaptation options of the fishing and farming communities of selected east coast stretch of Tamil Nadu, India. *Asian Journal of Environment and Disaster Management*, 4(3), 297–321. doi:10.3850/S1793924012001691.

NAPA. (2009). *National Adaptation Plan of Action.* A report published by Ministry of Environment and Forests Government of the People's Republic of Bangladesh. https://unfccc.int/resource/docs/napa/ban02.pdf

NIC. (2009). *India: The Impact of Climate Change to 2030: A Commissioned Research Report.* Joint Global Change Research Institute and Battelle Memorial Institute, Pacific Northwest Division. Retrieved April 24, 2021, from https://www.dni.gov/files/documents/climate2030_india.pdf

Nightingale, A. J. (2017). Power and politics in climate change adaptation efforts: Struggles over authority and recognition in the context of political instability. *Geoforum*, 84, 11–20. https://doi.org/10.1016/j.geoforum.2017.05.011

Noble, I.R., S. Huq, Y.A. Anokhin, J. Carmin, D. Goudou, F.P. Lansigan, B. Osman-Elasha, and A. Villamizar, 2014: Adaptation needs and options. In: Climate Change 2014: Impacts, Adaptation, and Vulnerability. Part A: Global and Sectoral Aspects. Contribution of Working Group II to the Fifth Assessment Report of the Intergovernmental Panel on Climate Change [Field, C.B., V.R. Barros, D.J. Dokken, K.J. Mach, M.D. Mastrandrea, T.E. Bilir, M. Chatterjee, K.L. Ebi, Y.O. Estrada, R.C. Genova, B. Girma, E.S. Kissel, A.N. Levy, S. MacCracken, P.R. Mastrandrea, and L.L. White (eds.)]. Cambridge University Press, Cambridge, United Kingdom and New York, NY, USA, pp. 833–868.

Noble, I. R., Huq, S., Anokhin, Y. A., Carmin, J. A., Goudou, D., Lansigan, F. P., ... & Chu, E. (2015). Adaptation needs and options. In C. B. Field, V. R. Barros, D. J. Dokken, K. J. Mach, and M. D. Mastrandrea (Eds.), *Climate Change 2014 Impacts, Adaptation and Vulnerability: Part A: Global and Sectoral Aspects* (pp. 833–868). Cambridge, UK and New York, USA: Cambridge University Press.

O'Brien, K. L., & Leichenko, R. M. (2003). Winners and losers in the context of global change. *Annals of the Association of American Geographers*, 93(1), 89–103. https://doi.org/10.1111/1467-8306.93107

Ojha, H. R., Ghimire, S., Pain, A., Nightingale, A., Khatri, D. B., & Dhungana, H. (2016). Policy without politics: Technocratic control of climate change adaptation policy making in Nepal. *Climate Policy*, 16(4), 415–433. https://doi.org/10.1080/14693062.2014.1003775

Panday, P. (2016). *Women's Empowerment in South Asia: NGO Interventions and Agency Building in Bangladesh.* London & New York: Routledge.

Picciariello, A., Colenbrander, S., Bazaz, A., & Roy, R. (2021). *The Costs of Climate Change in India.* Overseas Development Institute. Retrieved April 14, 2021, from http://www.indiaenvironmentportal.org.in/files/file/cost%20of%20climate%20change%20in%20india.pdf

Rabbani, G., Munira, S., & Saif, S. (2018). Coastal community adaptation to climate change - Induced salinity intrusion in Bangladesh. In N. K. Surendra (Ed.), *Agricultural Economics-Current Issues* (Chap 6, pp. 87–100). IntechOpen. https://doi.org/10.5772/intechopen.78437

Ragunathan, A. V. (2014a, December 24).a Ten years of Tsunami: A sea wall with yawning gaps. *The Hindu.* https://www.thehindu.com/news/national/tamil-nadu/ten-years-of-tsunami-a-sea-wall-with-yawning-gaps/article6720608.ece

Ragunathan, A. V. (2014b, July 8).b Sea erosion poses threat to coastal villages. *The Hindu.* https://www.thehindu.com/news/national/tamil-nadu/sea-erosion-poses-threat-to-coastal-villages/article6187468.ece

Rashid, M. M. (2013). Lives and livelihoods of riverbank erosion displaces in Bangladesh: Need for protection framework. *Journal of Internal Displacement*, 3(1), 19–35.

Rashid, M. M. (2014). Political commitments and aspirations of grassroots coastal communities: A micro-level study in Bangladesh. *American Journal of Rural Development*, 2(2), 24–33. https://doi.org/10.12691/ajrd-2-2-2

Rattani, V., Venkatesh, S., & Pandey, K. (2018). India's national action plan on climate change needs desperate repair. *Down to Earth*. Retrieved August 21, 2021, from https://www.downtoearth.org.in/news/climate-change/india-s-national-action-plan-on-climate-change-needs-desperate-repair-61884

Reid, H., Alam, M., Berger, R., Cannon, T., Huq, S., & Milligan, A. (2009). Community-based adaptation to climate change: An overview. *Participatory Learning and Action*, 60(1), 11–33.

Rodriguez, S., Subramanian, D., Sridhar, A., Menon, M., & Shanker, K. (2008). *Policy Brief: Seawalls*. UNDP/UNTRS, Chennai and ATREE, Bangalore, India, 8. Retrieved August 21, 2021, from https://www.seaturtlesofindia.org/wp-content/uploads/2016/10/Sea-Walls_Policy-Brief.pdf

Roxy, M. K., Ghosh, S., Pathak, A., Athulya, R., Mujumdar, M., Murtugudde, R. ... Rajeevan, M. (2017). A threefold rise in widespread extreme rain events over central India. *Nature Communications*, 8(1), 1–11. https://doi.org/10.1038/s41467-017-00744-9

Roy, M., Hanlon, J., & Hulme, D. (2016). *Bangladesh Confronts Climate Change: Keeping our Heads Above Water*. London: Anthem Press. https://anthempress.com/bangladesh-confronts-climate-change-pdf

Sanyal, N. (2006). *Political ecology of environmental crises in Bangladesh* (Doctoral dissertation, Durham University).

Schipper, E. L. F. (2006). Conceptual history of adaptation in the UNFCCC process. *Review of European Community & International Environmental Law*, 15(1), 82–92. https://doi.org/10.1111/j.1467-9388.2006.00501.x

Singh, A. P., Murty, T. S., Rastogi, B. K., & Yadav, R. B. S. (2012). Earthquake generated tsunami in the Indian Ocean and probable vulnerability assessment for the east coast of India. *Marine Geodesy*, 35(1), 49–65. https://doi.org/10.1080/0149041 9.2011.637849

Smit, B., Burton, I., Klein, R. J., & Street, R. (1999). The science of adaptation: A framework for assessment. *Mitigation and Adaptation Strategies for Global Change*, 4(3), 199–213. https://doi.org/10.1023/A:1009652531101

Smit, B., Burton, I., Klein, R. J., & Wandel, J. (2000). An anatomy of adaptation to climate change and variability. In *Societal Adaptation to Climate Variability and Change* (pp. 223–251). Dordrecht: Springer. https://doi.org/10.1023/A: 1005661622966

Smit, B., & Wandel, J. (2006). Adaptation, adaptive capacity and vulnerability. *Global Environmental Change*, 16(3), 282–292. https://doi.org/10.1016/j.gloenvcha. 2006.03.008

SoE Report I (2017). *State of Environment Report for Tamil Nadu*. ENVIS Centre, Department of Environment, Government of Tamil Nadu. Retrieved August 21, 2021, from https://www.environment.tn.gov.in/template/ngc-reports/soer1.pdf

SoE Report II. (2017). *State of Environment Report for Tamil Nadu*. ENVIS Centre, Department of Environment, Government of Tamil Nadu. Retrieved August 21, 2021, from https://www.environment.tn.gov.in/template/ngc-reports/soer2.pdf

Sovacool, B. K. (2018). Bamboo beating bandits: Conflict, inequality, and vulnerability in the political ecology of climate change adaptation in Bangladesh. *World Development*, 102, 183–194. https://doi.org/10.1016/j.worlddev.2017.10.014

Sovacool, B. K., Tan-Mullins, M., & Abrahamse, W. (2018). Bloated bodies and broken bricks: Power, ecology, and inequality in the political economy of natural disaster recovery. *World Development*, 110, 243–255. https://doi.org/10.1016/j.worlddev.2018.05.028

Sovacool, B. K., Tan-Mullins, M., Ockwell, D., & Newell, P. (2017). Political economy, poverty, and polycentrism in the global environment facility's least developed countries fund (LDCF) for climate change adaptation. *Third World Quarterly*, 38(6), 1249–1271. https://doi.org/10.1080/01436597.2017.1282816

Stock, R., Vij, S., & Ishtiaque, A. (2021). Powering and puzzling: Climate change adaptation policies in Bangladesh and India. *Environment, Development and Sustainability*, 23(2), 2314–2336. https://doi.org/10.1007/s10668-020-00676-3

Sukkary-Stolba, S. (1989). Indigenous Institutions and Adaptation to Famine. In R. Huss-Ashmore., & S.H. Katz. (Eds.), *African Food Systems in Crisis* Part One: Microperspectives. (pp. 281–294). London: Routledge.

Sundar, V., & Sundaravadivelu, R (2005). *Protection Measures for Tamil Nadu Coast, Final Report*. Submitted to Public Works Department, GoTN. Department of Ocean Engineering, Indian Institute of Technology Madras. Retrieved August 20, 2020, from https://tnsdma.tn.gov.in/app/webroot/img/document/library/15-Protection-Measures-For-Tamil-nadu-cost.pdf

Swenden, W., & Saxena, R. (2020). Environmental competencies in India's federal system. In N. Ciecierska-Holmes, K. Jörgensen, L. Laura Ollier, & D. Raghunandan (Eds.), *Environmental Policy in India*. London and New York: Routledge.

Taylor, M. (2014). *The Political Ecology of Climate Change Adaptation: Livelihoods, Agrarian Change and the Conflicts of Development*. London and New York: Routledge.

The New Indian Express. (2014, August 9). Rubble mound wall can protect Dansborgfort. *The New Indian Express*. https://www.newindianexpress.com/states/tamil-nadu/2014/aug/09/Rubble-Mound-Wall-Can-Protect-Dansborg-Fort-645742.html

TNSDMA. (2021). *Nagapattinam District District Disaster Management Plan 2021*. Tamil Nadu State Disaster Management Authority. https://tnsdma.tn.gov.in/pages/view/ddmp-2021

Tschakert, P. (2012). From impacts to embodied experiences: Tracing political ecology in climate change research. *Geografisk Tidsskrift-Danish Journal of Geography*, 112(2), 144–158. https://doi.org/10.1080/00167223.2012.741889

Uddin, M. S., Haque, C. E., & Khan, M. N. (2020). Good governance and local level policy implementation for disaster-risk-reduction: Actual, perceptual and contested perspectives incoastal communities in Bangladesh. *Disaster Prevention and Management: An International Journal*, 30(2), 94–111. https://doi.org/10.1108/DPM-03-2020-0069

UNFCCC. (2012). *Climate Displacement in Bangladesh The Need for Urgent Housing, Land and Property (HLP) Rights Solutions*. Displacement Solutions. Retrieved April 24, 2021, from https://unfccc.int/files/adaptation/groups_committees/loss_and_damage_executive_committee/application/pdf/ds_bangladesh_report.pdf

USAID. (2015). *Climate Change Information Fact Sheet BANGLADESH*. United States Agency for International Development. Retrieved August 04, 2021, from https://www.climatelinks.org/sites/default/files/asset/document/Bangladesh%20Climate%20Info%20Fact%20Sheet_FINAL.pdf

Vij, S., Biesbroek, R., Groot, A., & Termeer, K. (2018). Changing climate policy paradigms in Bangladesh and Nepal. *Environmental Science & Policy*, 81, 77–85. https://doi.org/10.1016/j.envsci.2017.12.010

Vij, S., Moors, E., Ahmad, B., Arfanuzzaman, M., Bhadwal, S., Biesbroek, R., ... & Wester, P. (2017). Climate adaptation approaches and key policy characteristics: Cases from South Asia. *Environmental Science & Policy*, 78, 58–65. https://doi. org/10.1016/j.envsci.2017.09.007

Watts, M. J. (2015). 2Now and then: The origins of political ecology and the rebirth of adaptation as a form of thought. In M. J. Watts (Ed.), *The Routledge Handbook of Political Ecology* (pp. 41–72). London and New York: Routledge.

World Bank. (2013). *Resettlement Action Plan. Bangladesh*. Retrieved January 5, 2022 from   http://documents.worldbank.org/curated/en/2013/02/17364117/bangladesh-first-phase-coastal-embankment-improvement-project-resettlement-plan

Yamunan, S. (2010, June 24). Coastal villages grapple with sea erosion. *The Hindu*. https://www.thehindu.com/news/national/tamil-nadu/Coastal-villages-grapple-with-sea-erosion/article16265675.ece

Zaman, F. (2017). Impact of salinity on poor coastal people's health: Evidence from Bangladesh. *Journal of Asiatic Society of Bangladesh*, 62(1), 1–14. http://www. asiaticsociety.org.bd/journal/1_H_952%20_Farhana.pdf

Zaman, F. (2018). Salinity: A social problem. In S. Ruby (Ed.), *Contemporary Social Problems in India* (Chapter 13, pp. 243–263, Vol II). Kolkata: Readers Service.

Zaman, F. (2021). *Sociology of Disaster and Vulnerability: Concepts, Perspectives, and Theories*. Dhaka, Bangladesh: Public Relations, Information and Publication Department, Jagannath University.

9 Communication tools to tackle cascading effects of climate change

Evidence from Eastern Bihar, India

Alankrita Anand and Eila Romo-Murphy

Defining cascading effects of disasters

Climate disasters are not isolated events. Instead, they have become interdependent with other systems. For example, an earthquake disrupts the infrastructure, causing contamination of water, which causes diseases to spread, worsening the local economy; or landslides block or damage the roads (Pescaroli & Alexander, 2015; Yadav et al., 2021). Instead of using the metaphor of toppling dominoes to describe the cause-and-effect relationship of catastrophic events, Pescaroli and Alexander (2015, p. 58) prefer to suggest that "interdependencies, vulnerability, amplification, secondary disasters, and critical infrastructure are important factors needing to be addressed in risk reduction practices to limit cascading during disasters."

According to Alexander and Pescaroli (2019, p. 1),

> The high level of dependency of modern populations on critical infrastructure and networks allows the impact of disasters to propagate through socio-economic systems. Where vulnerabilities overlap and interact, escalation points are created that can create secondary effects with greater impact than the primary event.

Critical infrastructures include energy supply, transportation, information, communication, and water supply (IRGC, 2007). For example, when the transportation system gets damaged in floods, it prevents relief support from reaching critical remote areas. In societies, there are weak points where cascading effects form. Therefore, it is essential to identify the critical spots and direct support to increase resilience. According to industry studies, in India, the gas and water supply form vulnerable spots for cyberattacks due to possibilities of human error or outdated installations of software and/or third-party data leaks (Deepanjii, 2021).

This chapter aims to show that information and communication can tackle disasters and slow or stop some of the cascading events that can unfold without the intervention of information and communication. In the history of disasters, both negative and positive examples of how information and

DOI: 10.4324/9781003216476-11

communication have affected the cascading effects of disasters. On the negative side, disruption of information relations has been the most common trigger of cascading events in some disasters. The 2011 Tohoku earthquake and the following tsunami disaster and the 2014 Hiroshima landslides in Japan are two examples of disruption of information relations triggering cascading effects in other elements of the socio-technical system, i.e., how information flows from citizens to emergency management and responders (Hagen et al., 2015; Ray-Bennett & Shiroshita, 2014; Reilly & Atanasova 2016; Pescaroli & Alexander, 2015). Both citizen and professional journalists are needed to contribute to the information flow that emerges during each stage of a large-scale emergency (Reilly & Atanasova, 2016).

According to flood prognosis analysis, 12–17% of Bihar was inundated during the 2017 floods, with the Katihar and Darbhanga districts being the worst affected (Tripathi et al., 2019). During these floods, telephone, road, and rail lines were all disrupted for the first two days, and the response network could not reach the flood-affected people. Due to the lack of immediate communication and rehabilitation facilities, affected people had to find elevated areas for themselves, which made many of them, especially women, vulnerable as there were no security arrangements or basic sanitary amenities. As research shows, unsanitary conditions in the aftermath of floods further lead to the widespread incidence of diarrhea and other water-borne diseases (Dhara et al., 2013; Hunter, 2003).

A positive research example of communications media mitigating cascading effects comes from the Philippines, where radio broadcasts on health-related information and psychosocial health constituted a positive component toward the recovery of typhoon survivors. Information and practical advice are delivered by disaster radio, via a temporary radio station, which consists of a suitcase that can be converted into a functional radio studio, to enable the production of content in the crisis environment, which can then be broadcast with a small transmitter; helping in recovery and enabling a sense of life control. Information and music played an important positive effect in not feeling worse in the disaster situation. Survivor interviews showed how music played an important role in increasing feelings of hope and trust, reducing stress, and contributing to mental recovery and feeling of normalcy (Hugelius, 2017).

The research findings in Indonesia, three years after the Great Indian Ocean tsunami, showed that radio broadcasting could contribute to individual and community information-sharing and decision-making communication processes (Bryar, James, and Adams, 2006; Tait, 2000), as well as to strengthen the capacity of communities through the creation of a "culture of resilience" (Tolentino, 2007, p. 147). In addition, it can facilitate mental and physical preparedness for future disasters, particularly when messages are designed with the participation of vulnerable communities (Romo-Murphy et al., 2011).

Cascading disasters in our societies

In a natural calamity, electricity is one of the first facilities to be cut off even in developed countries. From that, follows all else that needs electricity—appliances, lights, internet, and trains.

> No electricity grid is entirely immune to all the threats that it faces: cyber-attacks and coordinated-sabotage terrorism, major storms and flooding, space weather, even sudden excessive demand, for example, during a very hot summer. Rather than taking electricity for granted, we should be developing scenarios of how we would cope without it for extended periods.
>
> (Birkmann et al., 2014)

Both cross-disciplinary and integral approaches have been called for to tackle cascading events (Pescaroli et al., 2018; Vos, 2017). Critical infrastructures (CIs)—including buildings, services, supply chains, and cyber assets—have been recognized as having a determining role when a disaster occurs. These critical infrastructures matter, once given help or support by local or international disaster responders and can stop the escalation of disasters (Pescaroli & Kelman, 2016).

One of the recognized cascading effects of a disaster is its influence on the economy. For example, in the warming of the eastern equatorial Pacific Ocean, Africa, Central America, Southeast Asia, and the Pacific islands were affected.

> The 2015–2016 El Niño event had significant socio-economic impacts such as extending losses in agricultural and fisheries production, revenue, and GDP far beyond its active years. This is particularly the case when looking at examples of mass unemployment and even instances where primary education was halted, showing its generational impacts which will eventually translate to economic losses for both individuals and countries.
>
> (Thomalla & Boyland, 2017, p. 43)

On the one hand, in the case of Bihar, floods cut people off from critical infrastructure like electricity, markets, and health services, which had a cascading effect on existing grave issues like maternal mortality. The state already ranks among the worst in the country.[1] On the other hand, research reveals that floods also lead to cascading effects in Bihar; for example, the 2017 floods led to more school dropouts, leading to early marriage. Dowry, an illegal but prevalent practice, also propels early marriage in the aftermath of floods as it helps families tide over income shocks (Khanna & Kochhar, 2020). Pescaroli and Kelman (2016), in looking at the disaster cases of 2002 floods in Europe, 2005 hurricane Katrina, and the 2011 Tohoku earthquake and tsunami, verify that the disruption of energy and communication can be the main drivers of cascading failures, as effectively demonstrated by some authors (Luiijf et al., 2009; Van Eeeten et al., 2011).

Communications helping the vulnerable

Disaster research has moved from viewing calamities as "fate" to looking at crises overwhelming those vulnerable in terms of either physical or economic qualities or infrastructural aspects of their living conditions. Wisner et al. (2004, p.11) define vulnerability as the characteristics of a person or group in terms of their capacity to anticipate, cope with, resist, and recover from the impact of a natural hazard (an extreme natural event or process). It involves a combination of factors that determine the degree to which someone's life and livelihood are put at risk by a discrete and identifiable event (or series or "cascade" of such events) in nature or society. Alexander and Pescaroli (2019) consider vulnerable people as those with disabilities, illnesses, the aged, or inadequate incomes. They end up suffering most in the face of the disaster due to the weakening or failure of their support mechanisms. Wisner's situational approach views vulnerability from the point of understanding how complex political, infrastructural, ideological, and economic factors form the conditions that make some people of a certain class, race, ethnicity, gender, and age vulnerable to different levels of risk from the same hazard.

Oliver-Smith (2009) concludes that "not only the processes associated with creating vulnerability but also the process of post-disaster reconstruction reveals how deeply disasters are embedded in the political economy of a region." He analyzes Hurricane Mitch of 1998 in Honduras and Katrina of 2005 in the United States to pinpoint how disasters are embedded in the region's political economy. In Honduras' case, the use of nontraditional export crops was part of the reason for vulnerability. In the case of Louisiana, environmental degradation and the building of canals for the petrochemical industry increased the vulnerability of citizens, whose health indicators, and other measures such as educational attainment and the share of the population with special needs, were already among the lowest in the USA. Both disasters show how "there needs to be a greater understanding about the structural components of vulnerability to disasters and how they contribute as much or more to the unfolding of a disaster as the actual hazard agent itself" (Oliver-Smith, 2009, p. 26). He calls for action by local communities, people reclaiming agency, and deciding about the direction of their communities.

Looking into Katrina and Mitch from the communications point of view, research points to the need for media to focus more on preparedness (Baker, 2011; Barnes et al., 2008; Lachlan et al., 2009; Savova, 2004). However, Miles & Morse (2007) found that the media focused on the role of political and economic systems, following the public's concerns in building up the human and social capital component of hurricane vulnerability and risk after Katrina. In Indonesia, after the tsunami, local Radio Djati trained disabled persons and other vulnerable groups in designing and disseminating radio-based disaster mitigation messages. Research reveals that disabled people, more so than the rest of the sample, are confident that they can do something to reduce the negative effects of disasters by themselves (Romo-Murphy,

James & Adams, 2011). Ongoing radio-based messages designed by and for vulnerable communities in Aceh have transformed mitigation concepts into tangible realities, such as recognizing tsunami signs, preparing an emergency bag, knowing where to get information when the emergency siren sounds, minimizing injury by keeping a residence and surrounding areas free of refuse and being aware of how to help others in a crisis. In India, the role of mobile phones has facilitated early warning communication between officials and their departments (Acharya & Prakash, 2019), even though solutions for effective communication systems are called for (Kishorbhai & Vasanthbhai, 2017). At an institutional level, communication can be hampered by bureaucracy and transboundary sensitivities between countries, as with the Kosi floods in 2008 (Shreshtha et al., 2010).

History of floods in Bihar

The Kosi River belongs to India's largest river system called the Ganges–Brahmaputra—Meghna basin in the Northern region of India, covering an area of 1.7 million square kilometers, drawing its water from the Himalayas. This basin is situated in Uttar Pradesh, Madhya Pradesh, Bihar, Rajasthan, West Bengal, Haryana, Himachal Pradesh, Delhi, Arunachal Pradesh, and Assam. Hydro-political decisions made by these states and other relevant stakeholders over water resources in this area are crucial since these states and the neighboring countries of Pakistan and Bangladesh are agrarian economies needing water-fed irrigation facilities. Moreover, it's not just the agrarian economy but also industrialization and urbanization that need mismanaged water resources in the region (Mathur, 2011).

In 2008, the embankment breached upstream of Kosi Barrage (dam) for the first time, changing its natural river course and covering the so-called protected area with water, affecting three million people. Flooding of this magnitude had happened only once before, in 1953–1954, after which the Indian and Nepalese governments signed an agreement to regulate the river flow, and build embankments on either side of the river. Additionally, to control floods, a barrage was built on Kosi 1959–1963, changing the morphology of the river. Previously seven breaches occurred downstream of the barrage, discharging the water along the river: in 1963, 1968, 1971, 1980, 1984, 1987, and 1991. But in 2008, it was the first time that the embankment was broken upstream of the dam, and the water came to the floodplains (Mishra, 2008). The 2017 flood, however, was caused by "a sudden increase in water discharge due to torrential rain in the foothill of the Himalayas near Nepal and adjoining areas" (Tripathi et al., 2019, p.4).

The Indian subcontinent forms a zone of the Asian monsoon's most intensive activity. Historically, monsoon winds blow between the wet and dry seasons, shedding 70% of total South Asian rainfall between June and September. However, due to human intervention such as intensive irrigation and the planting of new crops, the monsoon itself has destabilized. Seeking to model the behavior of the monsoon has become difficult due to climate change in

India (Amrith, 2018). Based on the most recent National Family Health Survey of India (2020), every second household in Bihar is multidimensionally poor (NITI Aayog, 2021). Bihar has historically been one of India's most deprived states and, as a result, sees massive out-migration to other states. Recurring floods add to the vulnerability of the population, particularly those from marginalized communities. A needs assessment study in the aftermath of the 2017 floods found that more than half of the flood-affected respondents in Kishanganj, one of the worst affected districts, were from households living below the poverty line (National Dalit Watch-National Campaign on Dalit Human Rights, All India Dalit Mahila Adhikar Manch & Jan Jagran Shakti Sangathan, 2017).

Study background

The Bihar floods of 2017 severely affected lives and livelihoods across the northern half of the state, especially the districts bordering Nepal. The floods occurred in six rivers—Bagmati, Burhi Gandak, Gandak, Kamla, Kosi, and Mahananda—affecting over 17 million people and killing over 500 (Government of Bihar, 2017). Our study site included three of the affected districts—Bhagalpur, Katihar, and Purnia—which were selected based on First Response Radio's broadcast coverage area, all primarily in the Kosi River basin. The data collection for the study was carried out between January and March 2019 in the districts mentioned above and the state capital of Patna. Health Communication Resources (H-C-R) led the study, which equips communities with training and resources to use community-centered media, especially to strengthen health and disaster response.[2] Nav Jagriti (translation: *New Awakening*), a Bihar-based NGO for whom disaster risk reduction is a key programmatic area, acted as the field collaborator in India.

The floods struck different parts of Bihar between August 12–20, 2017, after which First Response India (FR India), a member of the First Response Radio network, deployed two mobile production teams in the state. The purpose of FR India was to support local communities in recovering from the flood by producing a disaster response radio program The program comprised 31 episodes, each about 15-minute long, which were broadcast from the government-owned All India Radio (AIR) stations in Bhagalpur, Darbhanga, Patna, and Purnia in the evenings between 26 August and 25 September 2017 (Figure 9.1).

Our study aimed to answer the central question: "To what extent did First Response broadcast content help in affected citizens' recovery and influence the response network's decision-making and impact?" But the larger objective of the study was to understand the role of emergency radio communication in the aftermath of disasters and how to strengthen it. We also paid particular attention to changes in the health status of the affected citizens due to the health information they received in the wake of the floods. Our study exclusively focused on radio instead of other forms of media as it was an evaluation of First Response Radio's disaster communication model.

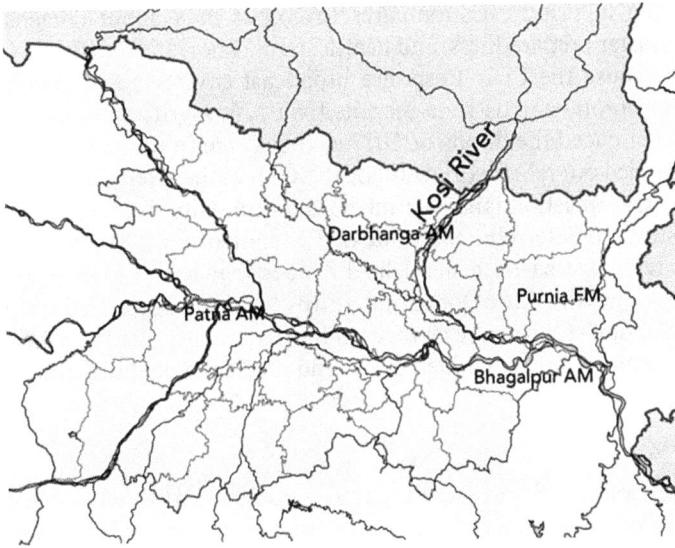

Figure 9.1 A map of Bihar.

Source: NordNordWest, https://creativecommons.org/licenses/by-sa/3.0/de/deed.en, via Wikimedia Commons. Text added by authors.

Methodology

Our study began with content analysis of 14 of the 31 First Response Radio episodes. The content was coded and thematically analyzed under different categories such as speech from officials, psychological encouragement, and media updates. Second, we recruited ten key informants to carry out interviews with. The key informants represented the term "response network," which includes the government's disaster management bodies, non-governmental stakeholders, and radio personnel. We particularly recruited key informants from the radio as opposed to other media as the study centered on the efficacy of radio in disasters.[3] The selection of the key informants was based on recommendations from Nav Jagriti and our research on Bihar's disaster management mechanism to obtain a relevant mix of experts and decision-makers. Our final sample included two radio program executives, two radio station managers, two district-level disaster management officials, two NGO personnel working on disaster preparedness, and two disaster management specialists who requested that their affiliations be kept confidential. The key informants from radio were selected from the AIR stations that aired the First Response content. All key informants had given verbal consent to be interviewed. The interviews were based on a semi-structured topic guide; nine of them were conducted face-to-face and one telephonically. The interviews were then translated and transcribed into English and jointly analyzed by the authors.

Third, we recruited and trained 25 volunteers (graduate students and young professionals) in the flood-affected districts of Bhagalpur, Katihar, and Purnia in using questionnaires to collect data about citizens' media needs, disaster preparedness, and health status. These three districts were also chosen because the First Response broadcast covered them. The students and young professionals were recruited via Nav Jagriti's networks and had either experienced the floods of 2017 as they came from the affected villages or had carried out relief work with local NGOs in the aftermath of the floods. We used two questionnaires—a media questionnaire developed specifically for the study to determine the value of FR content for the citizens. A stand-ardized reference-based general health questionnaire is known as EQ-5D Instrument by the EuroQol group (https://euroqol.org/eq-5d-instruments/eq-5d-3l-about/) to measure changes in the community members' health. The health questionnaire was added to the end of the media questionnaire.

Sampling

Both questionnaires were administered among 449 affected villagers in the three districts. The villages from where respondents would be recruited from were purposely selected using two criteria: the community had been affected by the Bihar flood in 2017, and the area was one where Nav Jagriti had vol-unteered. Volunteers would approach the villages near theirs, but not their own. In that way, the volunteers would be familiar with the surroundings and behave in a culturally acceptable manner. Within the villages, the volunteers were asked to choose every fifth household on the left, and the respondent from within the household was selected through a random sample using the Kish Grid method. After obtaining verbal consent from the respondents and key informants, all data were collected. Out of 449 interviews, 43% were from Katihar, the district most severely affected in the 2017 Kosi flood; 35% were from Bhagalpur; and 22% were from Purnia. Socially, about 13% of the respondents belonged to Scheduled Caste communities, 66% to Other Backward Classes, 2% to Scheduled Tribes, 3% to Economically Backward Class, and 5% to the general category.[4] Of the remaining respondents, 2% described their community affiliation as "other," and 10% did not share their affiliation. In terms of literacy, 64% of the respondents were literate, and 36% were illiterate. Among men, 74% were literate, while 55% were literate among women. Comparison with census data gives confidence that the sample was reasonably representative of the flood-affected population in the state.

Out of the 449 respondents, 346 had not listened to the First Response Radio broadcast. 63% of the non-listeners cited not having a radio set as their reason, as opposed to 30% who cited not knowing about the broadcast. Among the remaining non-listeners, 4% listed their reason as "other," and 3% said they did not want to listen to the broadcast. To determine the impact of First Response broadcasts, the responses of those who listened to First Response content were compared to those who did not listen using statistical crosstabs and t-tests on SPSS (Statistical Package for the Social Sciences) software. The authors primarily carried out the research, except the data col-lected via the two questionnaires, which were collected by the team of 25 volunteers whom the authors trained. The authors coded and analyzed the FR broadcast content and the media questionnaires, developed the key informant interview topic guide and the media questionnaire, and jointly conducted and analyzed the interviews.

Bias considerations

First, some of the volunteers we trained in questionnaire-based data collection had previously been part of flood relief efforts, independently or in association with an aid-providing NGO. So, when they went to collect data, respondents often associated them with flood relief rather than research. This might have affected what the respondents said about the availability of government and NGO services and what services they expected, as they might have been reluctant to be critical of flood relief efforts. For example, when the volunteers in Katihar went to collect data, many of the villagers identified them as Oxfam volunteers because, during the floods, the same group of people had volunteered with Oxfam for relief work. From some of the subjective answers on the completed questionnaires, it was obvious that the respondents thought that the interviewers were Oxfam volunteers. The group of volunteers from Katihar also mentioned that the respondents thought that they had come to distribute aid again or gather data to distribute aid.

Second, in the media questionnaire, where questions were being asked about First Response Radio, respondents found it difficult to recognize the English language phrase "First Response," which was used at the beginning and the end of the 15-minute FR episodes. Therefore, the volunteers were asked to play an audio clip of the First Response program to make the content block more recognizable for the respondent. It is possible that this introduced a slight skew in results based on false positives—i.e., respondents claiming to have listened to the FR Content because the audio ID sounded similar to other radio programs they had heard. On the other hand, respondents who could not understand the "First Response" ID but who heard the FR Content at the time of the flood might have thought it was AIR-produced disaster relief programming.

Role of media and communications in the Bihar floods

Media consumption in Bihar

Bihar has a population of 90.02 million, making it the third-most populous state in India. Of this population, 63.82% are literate, and when the figure is disaggregated by gender, 73.39% of men and 53.33% of women are literate. According to the Registrar of Newspapers in India (RNI), newspaper circulation in the state stood at a little over 9.5 million in 2019–2020. Hindi and Urdu newspapers are most widely circulated, followed by English-language newspapers. Annual data shows that between 2015–2016 and 2019–2020, newspaper circulation declined by 30% (RNI, 2020).

As for electronic media, a Broadcast Audience Research Council of India study, as cited in news reports, found that less than 30% of the population of Bihar owned a television set in 2018. This was the lowest figure in the country alongside the neighboring state of Jharkhand (The Indian Express, 2018). The move to the internet and social media has also been slow in the state. Internet subscription data from the Telecom Regulatory Authority of India shows that Bihar is among the states with the lowest percentage of subscribers (GoI, 2021). Their finding is corroborated by the fifth and latest National Family Health Survey, which found that Bihar has the lowest percentage of women who have ever used the internet (20.6%) and the third-lowest percentage of men who have ever used the internet (43.6%), among the 22 states surveyed (IIPS, 2020).[5]

Radio as a source of information

A background of such low media penetration begets the primary channels of information and communication, more so in the context of emergencies and disasters such as floods. Despite the low media penetration, seven out of our ten key informants from radio and response networks noted that media forms a crucial link between affected people and the response network in the face of disasters. When asked about radio, in particular, eight key informants identified it as a medium that plays a crucial role during disasters:

> Radio is the most commonly used media during disasters. In rural areas, radio is the prime media.
>> (Radio personnel, Bhagalpur, 31 January 2019)

> In rural areas, radio is definitely the best.
>> (Radio personnel, Darbhanga, 14 February 2019)

> Radio can provide important information as points; it is not a platform to fight for a particular opinion. Sharing information is what is most important for the community in times of a disaster.
>> (Response network personnel, Patna, 10 February 2019)

Three key informants were critical of the role of the different kinds of media operating during the floods:

> The media plays a negative role during floods. They publish and broadcast stories about officials taking bribes. While that is their work, they also should focus on the positive things that the government is doing. If the media highlights good things, it incentivizes us to work. If the media does not talk about good things, we get scared and do not want to inform the media. It gets discouraging.
>> (Response network personnel, Purnia, 5 February 2019)

The other key informants who had reservations about the role of the media argued that the media often sensationalized the floods and particularly highlighted aspects such as the death toll. They also mentioned that people in rural areas were sometimes scared of the media and apprehensive about giving too much information. Although radio was identified as a key source of information by most of our key informants, interestingly, our study found that a vast number of rural citizens did not have a radio set. Apart from our interviews with the key informants, our media questionnaire also revealed that listenership is low (346 out of 449 respondents had not listened to FR content). However, as stated before, the role of radio in disasters was acknowledged by eight out of ten key informants and is supported by the data our media questionnaire gathered from citizens about the usefulness of emergency broadcasts and what else they would have liked to hear on air.

> People don't keep radios anymore because they think it is obsolete. They do not understand that radio is the only means of information in times of disaster. Nowadays there are many television and mobile sets, so they do not keep radios. The administration (government authorities) always suggests that people keep radios. If they don't keep radios, how can we communicate the information? There is no electricity in the flood relief camps, but radios will still work.
>
> (Radio personnel, Bhagalpur, 31 January 2019)

In India, there are three types of radio services—government-owned radio, private radio, and community radio—licensed by the Ministry of Information and Broadcasting. AIR, India's national public service broadcaster, has a population reach of 99.18% and an area reach of 91.85% (GoI, n.d.). With such a high level of coverage, coupled with the popularity of private radio in urban areas and the grassroots issues covered by community radio, radio can be a crucial link between affected people and the response network. Compared to smartphones and television sets, radio is also an inexpensive form of media. A basic radio set in India costs anywhere between 300–500 INR (equals 4–7 USD). However, listenership is declining, as many of our key informants pointed out. According to some of them, the reason behind the decline may be radio's lack of inadaptability to new technology. For instance, although FM radio can be accessed on most mobile phones, it requires wired earphones, which many people may not want to buy only to listen to the radio on their mobile phones. According to a key informant, a possible reason behind the reluctance to adapt and upgrade could be that AIR does not have any competitors in rural India.

Information flow during disasters

Before discussing how communications can be strengthened in the context of Bihar's recurring floods, we look at how information flows between different media systems during disasters, between media (particularly radio) and the response network, and between citizens. Tracing the flow of information and analyzing the content of such information sheds light on how radio broadcasts can help in times of disaster and in mitigating cascading effects. At the All-India Radio station in the state capital of Patna and the stations catering to the flood-prone districts of Bihar, flood-related broadcasting begins around the same time as the onset of monsoon. Initially, the broadcasts carry information about the water levels in the rivers. As the situation escalates, their reporters go to the affected areas to record inputs from residents and the district administration to produce the preparedness program. When giving information about preparing flood kits, the AIR broadcast encourages citizens to monitor their radio sets. The flood kit is supposed to have dry rations, torches, candles, matchboxes, and a radio set. Once the flood strikes, the stations broadcast health-related information such as the importance of boiling drinking water, along with information about relief camps. All of the above information is broadcast in Hindi—the predominant language in the state, and regional dialects.

According to radio personnel, they receive information about water levels and relief camps from the Meteorological Department, the Water Resources Department, and the Disaster Management Department of the state government. Our key interviewees from the Disaster Management Department confirmed that they give information to the media in a press briefing at the district headquarters every evening during the floods. Their source of information, too, is the Meteorological Department, the Water Resources Department, and the Health Department. During floods, a control room is set up in the district administration's office and acts as the hub of all information coming in from the affected areas. Typically, this includes callers seeking information about food, shelter, and missing persons or giving information about the situation in their respective areas. AIR broadcasts the telephone numbers of these control rooms, as do community radios.

> We are the connection between the response network and the masses, so we publicize the former's numbers.
>
> (Radio personnel, Patna, 30 January 2019)

One of the radio personnel we interviewed, who liaised with the First Response Radio team during the 2017 floods in Bihar, explained how emergency broadcasts improved citizens' response to the flood, highlighting the importance of a holistic information system:

> The First Response Radio broadcast carried information about where to get relief and what to do in times of flood, and citizens wanted more information like that, based on the feedback letters received at the broadcast station.
>
> (Radio personnel, Darbhanga, 15 February 2019)

Our content analysis found that the broadcast also carried significant portions where the presenters provided mental encouragement and emphasized community resilience. Broadcast segments like these were complemented with songs of encouragement toward the end.

Local knowledge and community-centered media

However, in all the communication plans and broadcast messages, citizens are largely viewed as receivers of information. While information is crucial for affected people during a disaster, studies argue that their knowledge can be equally valuable, especially in the early stages of floods (Acharya & Prakash, 2019; Jha, 2021; Johnson et al., 2018).

According to Acharya & Prakash (2019), who studied local knowledge in flood forecasting in Bihar, local knowledge systems interact with the official information systems, which eventually feed into media broadcasts. While it is true that radio is often the first source of official information for many citizens, communities that have lived in flood-affected areas for years are deeply knowledgeable about phenomenological, meteorological, ecological, celestial, and riverine indicators.

> Disaster mitigation activities have to involve the community and utilize their knowledge. In the villages, people know of the signs of a flood before the official communication about rising water levels from the Meteorological Department.
>
> (Response network personnel, Patna, 10 February 2019)

Therefore, the flood-affected community is central not only in relief operations but can also be central in producing knowledge about prediction and perhaps prevention and mitigation.

In community broadcasts, citizen participation was measured by asking whether First Response listeners had heard content that included citizens' voices in general, in the form of an interview, or as stories. Each of these types of content was heard by at least half of the First Response listeners (Figure 9.2).

First Response listeners were also asked to estimate the value of community content and the value of citizens being heard by the response network. Community content garnered a little higher value than that of citizens being heard (Figure 9.3).

Community radio, therefore, can be a crucial tool to center the voices and the needs of affected communities. Our broadcast content analysis found a continuous emphasis on community action and well-being. Information about health, livelihoods, and relief camps or content on psychosocial support all highlighted the importance of togetherness and community resilience on the road to recovery. For instance, in a broadcast on alternative livelihoods, a response network personnel from an NGO encourages agricultural farmers to switch to dairy farming as growing crops becomes infeasible in the aftermath of floods. Many agricultural farmers in the area own some cattle for personal use but do not know how to turn it into a livelihood. The response

HEARD CITIZEN VOICES/INTERVIEWS/STORIES N=82

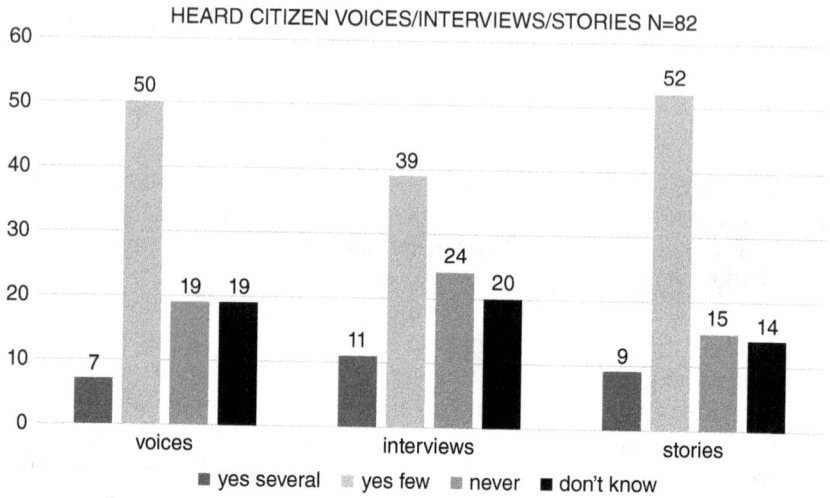

Figure 9.2 Heard citizen voices/interviews/stories.

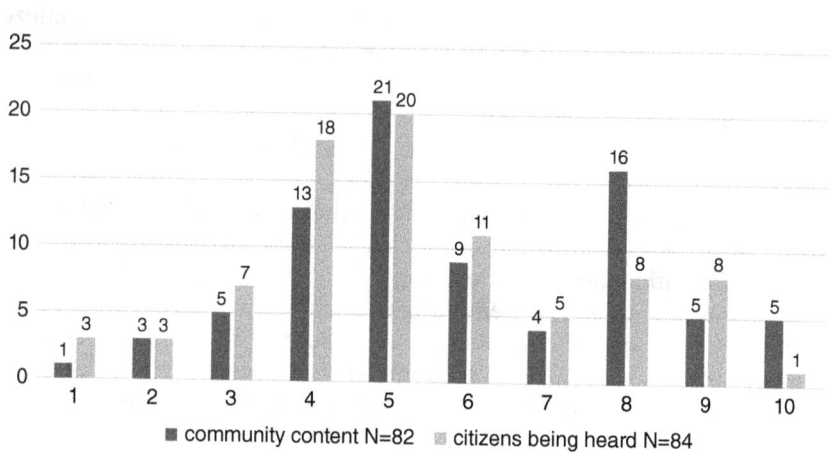

Figure 9.3 Value of community content versus value of citizens being heard by response network.

network personnel emphasizes that the switch to dairy farming can be made only by collectivizing and selling their produce to the state government's dairy agency, which is not possible if done individually, thereby highlighting the importance of community and collaboration in the aftermath of disasters.

Empathetic communication

Broadcast content centered on psychosocial support encouraged listeners to stay resilient as a community. In one of the episodes we analyzed, a response network personnel empathetically explains to listeners how a drastic event like floods can make one feel depressed and alone and that these feelings can only be overcome by sticking together.

> In my understanding, providing psychosocial support is extremely important as it comes with a personal touch, and the listener feels like someone is concerned about them. It is different from providing information, much as that is also important.
>
> (Response network personnel, Patna, 10 February 2019)

Although radio is a one-way channel, empathy was recognized as an important component of disaster programming both by First Response hosts and the response network.

Health information to tackle cascading effects

First Response Radio content was used to provide health-related information and psychosocial support following the typhoon Haiyan in the Philippines in 2013. Subsequently, research with survivors shows that while information helped in recovery and prevented mental health problems, musical content on the broadcast helped return to a semblance of normalcy (Hugelius, 2017). Our study found that the best way to disseminate health-related information to tackle cascading effects was to advise basic sanitary practices, especially home-based practices such as boiling water. It could significantly prevent the further spread of diseases. Simultaneously, we also found that broadcast content like mental encouragement and music aired by First Response Radio brought back the feeling of normalcy. Research shows that communities with poor socio-economic status and poor hygiene and sanitation facilities are the most vulnerable to flood-related diarrhea. In Bihar, the catastrophic 2008 floods led to outbreaks of malaria, diarrhea, measles, and cholera, which were exacerbated due to existing vulnerabilities caused by malnutrition, population density, literacy and awareness, and poor availability of health facilities (Gupta & Kumar, 2011).

Overall, our study found a substantial convergence between broadcast content on health, what citizens found most useful, and what they would like to hear more of. Health advice and information on water and sanitation together constituted 26% of the broadcast content that we analyzed. According to our media questionnaire, 25% of the respondents found information on healthcare most useful, and an additional 11% said that they found information on boiling water most useful, which is a crucial health measure in the aftermath of floods. When respondents were asked what information, they would like to see added to the broadcast, healthcare was again ranked the highest by 46% of the respondents. Although many of the broadcast interviews with response network personnel or experts explored matters of health, respondents wanted to hear fewer interviews and more citizen voices. Citizen voices constituted the second largest chunk of the broadcast, after interviews with experts. Our media questionnaire asked listeners to rate the value of the community content and the value of citizens being heard by the response network. On a scale of 1 to 10, the average rating for community content on-air was rated 5.9, and 5.4 for citizens being heard by the First Response Radio network.

Strengthening emergency communications

In our key informant interviews with personnel from AIR, we asked if they knew of community radios in the state. Two out of three radio informants mentioned a station in Gopalganj, a district in western Bihar. However, all three of them maintained that community media functioned separately from AIR and that they were not in touch with them. Our analysis of the AIR key informant interviews and First Response Radio broadcast suggest that the content on both platforms had many similarities, such as information about health and sanitation, which is a topic listeners reported benefitting from most. Likewise, both broadcasts connected the affected communities to the response network, especially by reporting on missing persons, which is crucial during floods.

Radio has the potential to reconnect people during emergencies, especially in places where social media is not as commonly used or when electricity has been cut off. Our media questionnaire respondents ranked information about missing persons very highly, with over 30% listing it as the "most beneficial broadcast topic" on the First Response Radio broadcast. Our key informants from AIR and the response network confirmed that information about missing persons via radio and text-message services is a key feature of flood programming on radio. Text-message services were first started in the region during the 2008 floods. Hugelius et al. (2019) also note that radio can play an important role in helping people re-establish contact with loved ones but add that such a system of information has to be complemented by other services like the internet, and mobile phones, likewise, would have to be complemented with the necessary services and infrastructure.

A key advantage of community media is that it can air content that critically evaluates the government's management of disasters. For instance, in the First Response Radio broadcast, citizens had the opportunity to air their concerns and talk about the lack of support from the government in certain scenarios. In the broadcasts that we analyzed, this did not amount to media sensationalization or exclusively highlighting negative news about the government but was an example of why citizen voices matter. As stated earlier, our study respondents also reported wanting to hear more local citizens. Our research shows that radio was vital as a source of information and encouragement for listeners during the 2017 floods and was particularly seen as a great source of information in rural areas by members of the relief network. But the lack of radio ownership and use emerged as the key reason behind low listenership.

Our key informant from Katihar explained:

> Flood-affected people carry their valuable items like gold because those are the things that can get stolen when you are not in the house. So people don't carry their radios and televisions, those articles are left to their fate.
>
> (Response network personnel, Katihar, 9 February 2019)

Hugelius et al. (2019), in their review of literature on the role of radio, found that people use the same media during floods that they are accustomed to in their daily life. Therefore, the use of radio must be popularized not only before and during disasters like floods but all year round. This is perhaps possible if radio upgrades its technology to become compatible with and complement other forms of media.

Disasters have also been known to strengthen communication and cooperation with the non-governmental sector. Corbacioglu and Kapucu (2006) found no significant organizational learning within Turkish disaster management due to the earthquakes Ceyhand in 1998, the Dinar in 1995 and Erzincan in 1992. Responses tended to be linear: more resources were applied to the same strategies. The 1999 Marmara earthquake, however, changed this pattern by making the continuation of past failures blatantly (and internationally) apparent. It resulted in new and more effective approaches to communication and cooperation with the non-governmental sector, which showed great improvements in response to the next earthquake that followed in a few months. As a result, Corbacioglu and Kapucu conclude that it is not simply the economic costs of a series of disasters that motivates institutional learning, but the concentration of these costs informs that demands the attention of economic and political elites.

The research based on Odisha, a state in the eastern region of India, found that besides providing information on how to prepare for the floods and heatwaves, emphasis should be on the non-governmental stakeholders as well as on the state disaster mitigation authority to empower communal capability (Mishra & Suar, 2012). Our analysis shows that most of the presenters in the First Response Radio broadcast were from local non-governmental organizations working on community health or local livelihoods, which is a perfect example of coordination with other members of the non-governmental sector, especially those whose work is anchored in the community.

Recent research in flood-affected areas in Pakistan also shows how present early warning systems are gender insensitive in their spread and communication and how it intensifies vulnerability during flood events and argues in favor of "tailoring risk messages more effectively for different gendered contexts" (Mustafa et al., 2015, p. 328). Like this, in India, all media outlets must be cautious that information is not being sourced only from dominant groups such as upper caste men, as demonstrated by Acharya & Prakash (2019).

Knowledge, cascading effects, and vulnerabilities are all affected by social positions and identities, which the media must account for.

There is a need to widen the development and reach of risk communication to communities with less power and privilege. Similarly, there is a need for a gendered consideration of risk communication, gendered production of local knowledge of early warning systems needs further investigation (Schware, 1984). Acharya and Prakash (2018) contend that gender and caste disaggregated data collection brings nuance to local knowledge research in South Asia and allows researchers to engage with knowledge exclusion within communities.

Conclusion

The information and communication channels used by the radio and response networks in the 2017 Bihar floods provided a good platform for multi-stakeholder collaboration. Our broadcast content analysis showed that representatives from local non-governmental organizations were regularly brought in to provide advice on health, sanitation, livelihoods, and resilience. Doctors from the local health centers were also brought in when required. The content, however, did not always reach the intended audience due to a lack of media devices or electricity. At the same time, our research shows that in cases where the broadcast content did reach the intended audience, it greatly helped them, especially in physical and mental health matters.

The Bihar Disaster Risk Reduction Roadmap 2015–2030, commissioned by the state government and developed with local organizations and international bodies like UNICEF, serves as an example of a long-term and holistic policy response. It is the first roadmap of its kind in India, and it highlights strengthening healthcare and preparedness (such as risk assessments of buildings) in non-disaster times, along with a targeted focus on vulnerable sections of the population in post-disaster rehabilitation. It also calls for a multi-stakeholder approach to mitigating disasters, which would involve the state disaster management authority as experts, the state government as the program implementer, and the media, the non-governmental sector, and the private sector as external stakeholders (Government of Bihar, 2015).

The key informants we interviewed from the relief network and AIR confirmed that citizens in flood-prone rural areas have systems of information, monitoring, and preparedness in place, owing to repeated floods, and that information channels like radio have to complement their communication flows and efforts. At the same time, some of our key informants (from outside the government) noted that the collaboration between government and non-government stakeholders, such as government radio and community radio, could be made stronger. During disasters, community media's strength is that affected citizens get a voice on-air and are not afraid of losing legal entitlements if they are critical of the government. Therefore, multi-stakeholder approaches that involve citizens should enable centering their voices and needs without any roadblocks.

In 2017, Bihar suffered from a lack of a comprehensive communications plan. However, the continuous collaboration between media systems, communities, and government response networks, can help mitigate disasters promptly and alleviate their cascading effects. Our findings reveal what value flood-affected citizens, radio listeners, and non-listeners assign to disaster-related radio programming and how the response network perceives radio as a means of communication in today's times. The findings also point to weak points in the information flow; for example, the overlaps between the aims of the government-owned AIR and independent community radios, but the lack of a mechanism whereby they can work together. Likewise, the meteorological department not working collaboratively with communities also leads to a weak information flow which can impact prevention as well as cascading effects.

While this chapter focuses on the role of radio communication in the mitigation of cascading effects, further research can examine the role of radio communication in disaster prevention strategies. Equally important is the involvement of community voices on air, and future research can also explore how citizens can sustainably participate in local communication plans. While we appreciate the importance of timely communication and emergency communication, research on disaster preparedness can make communication plans stronger and more sustainable.

Notes

1 According to the most recent survey by the Government of India, Bihar has a high maternal mortality rate of 149 deaths per 100,000 live births, compared to the countrywide average of 113 (Sample Registration System, 2020).
2 The study was funded by H-C-R and was an independent evaluation of First Response Radio's emergency broadcast.
3 Disaster management in the state is carried out by governmental, quasi-governmental, and non-governmental bodies such as international and regional NGOs. The Disaster Management Department of the state government is the key executive body. In contrast, the State Disaster Management Authority (a statutory body created under the National Disaster Management Authority) is responsible for developing policy and providing technical support to disaster management efforts. Lastly, the State Disaster Relief Force is a police force that carries out operations during and after disasters.
4 Scheduled Caste, Other Backward Class, Scheduled Tribe, Economically Backward Class, and General are government categories based on caste, economic status, and indigeneity. Scheduled castes comprise "lower" castes who have been historically marginalized and relegated to menial labor. Only Hindu, Sikh, and Buddhist castes are given the Scheduled Caste status. Scheduled Tribes come from indigenous groups outside of the caste-Hindu society, and they, too, have been historically marginalized. Other Backward Classes are socially, economically, and educationally backward groups but do not fall under the groups listed under Scheduled Castes and Scheduled Tribes; they can come from any religion. The Constitution of India protects SC, ST, and OBC groups from discrimination and atrocity and provides affirmative action in reservations in education and jobs. General refers to the "upper" castes who may belong to any religious community, and Economically Backward Class is a newer category created to provide affirmative action to economically weak "upper" caste households.

5 The National Family Health Survey is a periodic district-level survey, which is part of USAID's Demographic and Health Surveys Program. The fifth and most recent round of the survey was conducted in 2019–2020, but the findings have been released only for 22 states so far owing to the COVID-19 pandemic.

References

Acharya, A. & Prakash, A. (2019). When the river talks to its people: Local knowledge-based flood forecasting in Gandak River basin, India. *Environmental Development.* 31 (September 2019), 55–67. https://doi.org/10.1016/j.envdev.2018.12.003

Alexander D. & Pescaroli G. (2019). What are cascading disasters? *UCL Open: Environment.* 1 (03). https://dx.doi.org/10.14324/111.444/ucloe.000003

Amrith, S. (2018). *Unruly Waters.* New York: Basic Books.

Baker, E. J. (2011). Household preparedness for the aftermath of hurricanes in Florida. *Applied Geography.* 31 (1), 46–52. https://doi.org/10.1016/j.apgeog.2010.05.002

Barnes, M. D., Hanson, C. L., Novilla, L. M. B., Meacham, A. T., McIntyre, E., & Erickson, B. C. (2008). Analysis of media agenda setting during and after Hurricane Katrina: Implications for emergency preparedness, disaster response, and disaster policy. *American Journal of Public Health.* 98 (4), 604–610. https://doi.org/10.2105/AJPH.2007.112235

Birkmann, J., Kienberger, S., & Alexander, D. E. (Eds.). (2014). *Assessment of Vulnerability to Natural Hazards: A European Perspective.* Elsevier. https://doi.org/10.1016/C2012-0-03330-3

Bryar, T., James, R., & Adams, M. (2006) *Critical Information for Communities Following a Natural Disaster.* Health Communication Resources, Perth, Australia

Corbacioglu, S. & Kapucu, N. (2006). Organisational learning and self-adaptation in dynamic disaster environments. *Disasters.* 30 (2), 212–233. https://doi.org/10.1111/j.0361-3666.2006.00316.x

Deepanjii, P. (ed.) (2021). Abysmal state of global critical infra security: Supply of gas, water and government services at high risk. CloudSEK. https://cloudsek.com/whitepapers_reports/abysmal-state-of-global-critical-infrastructure-security/

Dhara, V. R., Schramm, P., & Luber, G. (2013). Climate change & infectious diseases in India: Implications for health care providers. *Indian Journal of Medical Research.* 136 (6), 847–852.

Government of Bihar. (2015, 2nd October). Roadmap for disaster risk reduction: 2015–2030. *UNICEF.* Retrieved 17th February, 2022, from https://www.unicef.org/india/media/2786/file/DRR-Roadmap-Bihar.pdf

Government of Bihar. (2017, 7th September). 2017 Floods Report. *Disaster Management Department, Bihar.* Retrieved 17th February, 2022, from https://web.archive.org/web/20170910125920/http://disastermgmt.bih.nic.in/press%20notes/PR07092017Evening.pdf

Government of India. (n.d.). *Milestones.* Prasar Bharati. https://prasarbharati.gov.in/milestones/

Government of India (GoI). (2021, 21st May). Highlights of telecom subscription data as on 28th February, 2021. *Telecom Regulatory Authority of India.* Retrieved 17th February, 2022 from https://www.trai.gov.in/sites/default/files/PR_No.27of2021_0.pdf

Gupta, S., & Kumar, A. (2011). (A251) Factors of diarrheal outbreak, Bihar floods-2008. *Prehospital and Disaster Medicine* 26 (S1), S68–S69. https://doi.org/10.1017/S1049023X11002354

Hagen, K., Tzanetakis, M., and Watson, H. (2015). Cascading effects in crises: Categorisation and analysis of triggers, ISCRAM, Short Paper - Planning, Foresight and Risk Analysis Proceedings of the ISCRAM 2015 Conference - Kristiansand, May 24–27.

Hugelius, K. (2017). *Disaster Response for Recovery: Survivors' Experiences, and the Use of Disaster Radio to Promote Health After Natural Disasters.* Örebro Studies of Care Science 69. Örebro University, Sweden. Retrieved 25th January, 2022, from http://urn.kb.se/resolve?urn=urn:nbn:se:oru:diva-52653

Hugelius, K., Adams, M., & Romo-Murphy, E. (2019). The power of radio to promote health and resilience in natural Disasters: A review. *International Journal of Environmental Research and Public Health* 16 (14), 2526. https://doi.org/10.3390/ijerph16142526

Hunter, P. R. (2003). Climate change and water-borne and vector-borne disease. *Journal of Applied Microbiology* 94, 37–46. https://doi.org/10.1046/j.1365-2672.94.s1.5.x

International Institute for Population Sciences (IIPS). (2020). *National Family Health Survey - 5* (Vol. 1). Mumbai, Maharashtra: Ministry of Health and Family Welfare. Retrieved 17th February, 2022, from Ministry of Health and Family Welfare: http://rchiips.org/NFHS/NFHS-5_FCTS/NFHS-5%20State%20Factsheet%20Compendium_Phase-I.pdf

International Risk Governance Council (IGRC). (2007). *Managing and Reducing Social Vulnerabilities from Coupled Critical Infrastructures.* International Risk Governance Council. Retrieved 25th January, 2022, from https://beta.irgc.org/wp-content/uploads/2018/09/IRGCinfra_site06.11.07-2.pdf

Jha, P. K. (2021). State, floods and politics of knowledge: A case of the Mahananda Basin of Bihar. *Studies in Indian Politics* 9 (1), 91–104. https://doi.org/10.1177/2321023021999177

Johnson, R. M., Edwards, E., Gardner, J. S., & Diduck, A. P. (2018). Community vulnerability and resilience in disaster risk reduction: An example from Phojal NALLA, Himachal Pradesh, India. *Regional Environmental Change* 18 (7), 2073–2087. https://doi.org/10.1007/s10113-018-1326-6

Khanna, M. & Kochhar, N. (2020) Do Marriage Markets Respond to a Natural Disaster? The Impact of Flooding of River Kosi in India Available at SSRN: https://ssrn.com/abstract=3644052 or http://dx.doi.org/10.2139/ssrn.3644052

Kishorbhai, V. Y. and Vasanthbhai, N. N. (2017). AON: A survey on emergency communication systems during a catastrophic disaster. *Procedia Computer Science* 115, 838–845. https://doi.org/10.1016/j.procs.2017.09.166

Lachlan, K. A., Burke, J., Spence, P. R., & Griffin, D. (2009). Risk perceptions, race, and Hurricane Katrina. *Howard Journal of Communications* 20 (3), 295–309. https://doi.org/10.1080/10646170903070035

Luiijf, E., Nieuwenhuijs, A., Klaver, M., van Eeten, M., & Cruz, E. (2009). Empirical findings on critical infrastructure dependencies in Europe. *Lecture Notes in Computer Science*, 302–310. https://doi.org/10.1007/978-3-642-03552-4_28

Mathur, A. (2011). Water disputes in South Asia. In Yang Razali Kassim (Ed.) *Issues in Human Security in Asia* (pp. 87–92). ed. by. Singapore: Institute of Southeast Asian studies. S. Rajaratnam School of International Studies.

Miles, B., & Morse, S. (2007). The role of news media in natural disaster risk and recovery. *Ecological Economics* 63 (2), 365–373. https://doi.org/10.1016/j. ecolecon.2006.08.007

Mishra, D. K. (2008). The Kosi and the embankment story. *Economic and Political Weekly* 43 (46), 47–52. https://doi.org/10.2307/40278181

Mishra, S. & Suar, D. (2012). Effects of anxiety, disaster education, and resources on disaster preparedness behavior. *Journal of Applied Social Psychology* 42 (5), 1069–1087. https://doi.org/10.1111/j.1559-1816.2011.00853.x

Mustafa, D., Gioli, G., Qazi, S. Waraich, R., Rehman, A. & Zahoor, R. (2015). Gendered flood early warning systems: The case of Pakistan. *Environmental Hazards* 14 (4), 312–328. https://doi.org/10.1080/17477891.2015.1075859

National Dalit Watch-National Campaign on Dalit Human Rights, All India Dalit Mahila Adhikar Manch & Jan Jagran Shakti Sangathan. (2017, 27th October). *Immediate Needs Assessment and Inclusion Monitoring of Responses towards Affected Dalits, Minorities & Adivasis in Araria and Kishanganj.* Counterview. Retrieved 18th February, 2022, from https://counterview1.files.wordpress.com/2017/10/16th-final.pdf

NITI Aayog. (2021). *National Multidimensional Poverty Index.* New Delhi: NITI Aayog, Government of India. Retrieved 17th February, 2022, from https://www. niti.gov.in/sites/default/files/2021-11/National_MPI_India-11242021.pdf

Oliver-Smith, A. (2009). Anthropology and the political economy of disasters In E. C. Jones and A. D. Murphy (Eds.) *The Political Economy of Hazards and Disasters* (pp. 11–30). Lanham, MD: AltaMira Press.

Pescaroli, G. & Alexander, D. (2015). A definition of cascading disasters and cascading effects: Going beyond the "toppling dominos" metaphor. GFR Davos planet@ risk, Vol. 3, Number 1, Special Issue on the 5th IDRC Davos March 2014.

Pescaroli, G. & Kelman, I. (2016). How critical infrastructure orients international relief in cascading disasters. *Journal of Contingencies and Crisis Management* 25 (2), 56–67. https://doi.org/10.1111/1468-5973.12118

Pescaroli, G., Nones, M., Galbusera, L. & Alexander, D. (2018). Understanding and mitigating cascading crises in the global interconnected system. *Disaster Risk Reduction* 30 (B), 159–163. https://doi.org/10.1016/j.ijdrr.2018.07.004

Ray-Bennet, N & Shiroshita, H. (2019). Disasters, deaths and the Sendai Framework's target one A case of systems failure in Hiroshima Landslide 2014, Japan. *Disaster Prevention and Management* 28 (6). https://doi.org/10.1108/DPM-09-2019-0302

Registrar for Newspapers in India (RNI). (2020). *Press in India - 69th Annual Report* (Vol. 1, pp. 48–62). New Delhi: Ministry of Information and Broadcasting. http:// rni.nic.in/pdf_file/pin2019_20/pin2019_20_eng/Chapter%204.pdf

Reilly, P. & Atanasova, D. (2016). *A Report on the Role of the media in the Information Flows that Emerge During Crises.* CascEff. University of Sheffield. Retrieved 15th January, 2022, from http://casceff.eu/media2/2017/02/D3.4-Media-in-the-information-flows-during-crisis-situation.pdf

Romo-Murphy, E., James, R. & Adams, M. (2011). Facilitating disaster preparedness through local radio broadcasting. *Disasters* 35 (4), 801–815. https://doi. org/10.111/j.1467-7717.2011.01234.x

Rural beats urban India in growth of homes with TVs; Bihar, Jharkhand have least. (2018, 27th July). *The Indian Express.* https://indianexpress.com/article/india/ rural-beats-urban-india-in-growth-of-homes-with-tvs-bihar-jharkhand-have-least-5278531/

Sample Registration System. (2020). *Special Bulletin on Maternal Mortality in India 2016–2018.* New Delhi: Office of the Registrar General of India. Retrieved

17th February, 2022 from https://censusindia.gov.in/vital_statistics/SRS_Bulletins/ MMR%20Bulletin%202016-18.pdf

Savova, N. (2004). Small river valleys step ahead of nature. *Americas* 56 (4), 56–57.

Schware, R. (1984). Flood information systems: Needs and improvements in Eastern India. *Environmental Management* 8 (1), 55–66. https://doi.org/10.1007/BF01867873

Shreshtha, Rashmi Kiran, Rhodante Ahlers, Marloes Bakker, and Joyeeta Gupta (2010). Institutional dysfunction and challenges in flood control: A case study of the Kosi flood 2008. *Economic and Political Weekly* 45 (2), 45–53. http://www.jstor.org/stable/25663989

Tait, A. (2000) The Feasibility of a Rapid Response Radio Unit. Health Communication Resources, Perth.

The human cost of disasters: An overview of the last 20 years (2000–2019). Back page. (n.d.). Retrieved 25th November, 2021, from https://www.undrr.org/publication/ human-cost-disasters-overview-last-20-years-2000-2019

Thomalla, F. & Boyland, M. (2017). *Enhancing Resilience to Extreme Climate events: Lessons from the 2015–16 El Nino event in the Asia Pacific.* Stockholm, Sweden: Stockholm Environment Institute.

Tolentino, A.S. (2007) The challenges of tsunami disaster response planning and management. International Review for Environmental Strategies 7(1), pp. 147–154.

Tripathi, G., Parida, B. R., Pandey, A. C. (2019). Spatio-temporal rainfall variability and flood prognosis analysis using satellite data over North Bihar during the August 2017 Flood Event. *Hydrology* 6 (38). https://doi.org/10.3390/hydrology 6020038

Vahanvati, Mittul (2015). *Owner-driven Reconstruction in India: A case-study of Kosi River Floods in Bihar.* Paper ID: 367. In the *5th International Conference on Building Resilience.*

Van Eeeten, M., Nieuwenhuijs, A. Luiijf, E., Klaver, M., & Cruz, E. (2011). The state and the threat of cascading failure across critical infrastructures: The implications of empirical evidence from media incident reports. *Public Administration, 89*(2), 381–400. https://doi.org/10.1111/j.1467-9299.2011.01926.x

Vos, M. (2017). *Communication in Turbulent Times: Exploring Issue Arenas and Crisis Communication to Enhance Organisational Resilience.* Reports from the School of Business and Economics No. 40. Jyväskylä. http://urn.fi/URN:ISBN:978-951-39-7147-2

Wisner, B., Blaikie, P., Cannon, T. & Davis, I. (2004). *At Risk: Natural Hazards People's Vulnerability and Disasters.* London. Routledge.

Yadav, P., Saville, N., Arjyal, A., Baral, S., Kostkova, P., & Fordham, M. (2021). A feminist vision for transformative change to disaster risk reduction policies and practicecs. *International Journal of Disaster Risk Reduction* 54 (2021) 102026.

10 Climate Change Communication in Kolkata

Applying Communication Theories to Address Climate Change Displacement

Lisha Samuel and Randall S. Abate

Introduction

Climate change communication is an emerging field that addresses how educating and mobilizing the public on climate change can promote awareness of this imminent crisis. According to the Yale Program on Climate Change Communication, the perceived threat that climate change poses is shaped by the different perspectives and underlying values and worldviews of the public. Communication should be considered a two-way process in which the message is sent and the receiver receives the message and transcribes it; however, communication is much more complex than this simple model. Climate change communication aims to develop ways to present relevant and digestible information to the public and address effective strategies to help the public respond to the threat of climate change (Yale Program on Climate Change Communication, n.d.).

Climate change is "a complex, persuasive, and uncertain phenomenon that people find difficult to understand, conceptualize, and relate to" (Ballantyne, 2016, p. 330). Despite growing global awareness that climate change represents a profound threat to the health and well-being of human and nonhuman species (Maibach et al., 2008), the contrast between developed and developing countries on climate change awareness is striking. In North America, Europe, and Japan, more than 90% of the public is aware of climate change, whereas in many developing countries, relatively few are aware of the issue. A developing economy like India might see the repercussions of climate change more readily because of the potential havoc that climate change can wreak on India's poor population and natural resources and the threat it poses to its economic development (Mittal, 2012, p. 227).

India has the highest level of disaster displacements in South Asia and one of the highest globally (Internal displacement monitoring center, n.d.). Over the next 40 years, India will experience one of the most dramatic settlement transitions in history, compounded by the projected growth of its urban population from about 300 million to more than 700 million (Revi, 2007, p. 208). Existing governance structures and the institutional culture of most cities are incapable of addressing the challenge of climate change adaptation and mitigation (Revi, 2007, p. 222).

DOI: 10.4324/9781003216476-12

This chapter examines the importance of climate change communication in Kolkata, India. It addresses how sea-level rise and flooding are increasing in Kolkata and the importance of developing effective strategies to raise awareness of climate-induced migration. The chapter first considers the background of climate change communication and why it is important to create public awareness. It then examines the current physical threats in Kolkata, including sea-level rise and flooding, the vulnerability of the Sundarbans, and the impending challenge of climate migration. Next, it addresses the existing legal framework in India and West Bengal on climate change adaptation and migration, and it considers the gaps in communicating the significance of climate change in India. It concludes by addressing how two communication theories – agenda-setting theory and narrative paradigm – can enable the media to promote more effective climate change communication in Kolkata. It also proposes strategies to strengthen media effects for effective climate change communication to help focus more attention on the risk of forced migration due to sea-level rise and flooding.

Climate Change Communication

Communication is defined as the production and exchange of information to inform, influence, or motivate individual, institutional, and public audiences (Maibach et al., 2008). Climate change communication focuses on educating, informing, warning, persuading, and mobilizing to address this critical problem (Yale Program on Climate Change Communication, n.d.). Moreover, climate change communication should be simple because the more people understand and gain knowledge about this topic, the more awareness of the challenge will increase. The Indian people need to use communication strategies to make informed decisions and stimulate public awareness of climate change and its physical impacts (Leiserowitz & Thaker, 2012).

Many climate scientists and climate change communicators have recently been broadcasting a message that boils down the state of things to just 12 words: "It's warming. It's us. We're sure. It's bad. We can fix it" (Deweerdt, 2020). After surveying 509 German adults, research based on an online study revealed that they know the basics: they know the climate is changing and that human activities are responsible. But they have a harder time distinguishing false statements about climate change from true ones. Researchers suggest that communication strategies can act as a "vaccine" against climate denial (Deweerdt, 2020).

Research into climate change communication has broadened significantly in recent years and now addresses interdisciplinary scientific communication, the scientist–policymaker interface, international diplomacy, media coverage, and public understanding (Russill & Nyssa, 2009, p. 337). Researchers have understood that the news media is an important source of scientific information if put in simple language for public understanding (Russill & Nyssa, 2009).

Climate change communication in the media is more likely to inform the public about climate change than the latest scientific projections from the

Intergovernmental Panel on Climate Change (IPCC). Public polling suggests that print media is the primary vehicle to convey information to the public on climate change (Billett, 2010, p. 4). Media coverage of climate science is context-dependent, influenced by ideological, organizational, and sociopolitical factors, and may evoke different public and political responses.

According to Johnson (2012), climate change communication covers three strategies: (1) persuasion of individuals to change their own energy use and adaptations, (2) mobilization to foster collective action to change individuals' and institutions' behavior and deliberation, and (3) collective efforts to identify problems and solutions (p. 974). Climate change communication prioritizes the relationship of communication to motivation and social change (Russill & Nyssa, 2009). Public attention rises from a series of events that draw the public's attention to the specifics of a situation and harmful consequences (Arnold, 2018, p. 16). When an event affects people at a personal level, however, they are most likely to react to the events.

Climate Change Vulnerabilities in Kolkata

Global climate change impacts are unequally experienced by people in poor and developing countries and threaten their livelihood that is tied to subsistence (McCarthy, 2020). This section addresses how Kolkata is particularly vulnerable to sea-level rise and flooding. Kolkata, the capital city of the state of West Bengal, is situated on the east bank of the Hooghly River, is about 96 miles north of the Bay of Bengal, and consists of a population of more than 14 million. It also considers the Sundarbans, 109 miles from Kolkata, an important and highly vulnerable coastal region experiencing sea-level rise and the threat of forced migration. "Sundarbans was simply described as a 'cyclone shield' for the megacity of Kolkata" (Ghosh & Boykoff, 2019, p. 148). Finally, this section examines the imminent threat of forced climate-induced migration into Kolkata. The Sundarban region in Bangladesh is closest to the Northern part of Kolkata. Due to sea-level rise, an increased number of people from this region will migrate to Kolkata because of its accessibility.

Physical Threats

The United Nations' Global Environment Outlook (GEO) 2016 reports an overall alarming figure of 40 million people threatened by sea-level rise in India (Jolly & Ahmad, 2019, p. 3). The long 7500-kilometer coastline of India supports a large number of big cities that face ecological pressures (Pramanik, 2017, p. 1345). Sea-level rise causes concern mainly due to the potential habitat loss and the direct physical impact of inundation (Pramanik, 2017, p. 1344). The sea level has risen 8.5 centimeters in the past 50 years along the Indian coast (Panda, 2020). It is predicted that the sea level will rise by 1.4 meters in Kolkata by the end of the century. When Kolkata's average elevation of 1.5–11 meters is factored in, a compounded adverse effect occurs in which physical vulnerability is exacerbated.

Thirty-six million Indians may be at risk from flooding due to sea-level rise by 2050 (Chandrashekhar, 2019). According to a national survey conducted in India by Leiserowitz & Thaker (2012), nearly 2 out of 3 respondents indicated that it would take their household or community several months or more to recover from a severe flood or drought (p. 11). Moreover, potential monsoonal changes and sea-level rise around the low-lying coastal areas threaten India's growing coastal cities (Billett, 2010, p. 3).

Kolkata is one of the mega-cities in India that lies on the banks of the great rivers – the Ganga, Narmada, Krishna, and Kaveri – along with more than 30 other million-plus cities (Revi, 2007, p. 214). It has been ranked the fourth most vulnerable among 11 major coastal cities in Asia threatened by sea-level rise and flooding (Tikoo, 2009). Since Kolkata is situated in a river delta, it is also vulnerable to sea-level rise (Mulhern, 2020). Urbanization has been an important driver of increased flood risk in Kolkata (Dhiman et al., 2019). Typically, the urban poor reside in areas more prone to flooding and are most vulnerable to flooding and rising sea levels (Revi, 2007, p. 213). Climate change is expected to increase the severity of flooding in many Indian river basins, especially those of the Godavari and Mahanadi along the eastern coast, where Kolkata sits (Revi, 2007, p. 215). More than 50% of people living in Kolkata and its suburbs – a staggering 12 million – will be flooded out of their homes if the global community's actions under the Paris Agreement fail to limit global warming to 2°C (Mukherjee, 2015).

The Sundarbans

The Sundarbans region is one of the richest ecosystems globally and contains the world's largest continuous mangrove forest, at nearly 10,000 square kilometers. About 40% of the Sundarbans forest lies within West Bengal; the rest is in Bangladesh. The Sundarbans have been protecting the state of West Bengal from cyclones since prehistoric times (Biswas, 2021). Also, more than 3.5 million people depend on the Sundarbans for income and livelihood (Mukul et al., 2020, p. 1198).

India's Sundarbans are characterized by high poverty levels and exposure to natural hazards. Climate change is likely to exacerbate the situation (Panda, 2020). Sea-level rise, soil and water salinization, cyclones, and flooding make this area one of the most vulnerable on the Indian subcontinent. The Indian side of the Sundarbans is submerging 2–4 millimeters a year. Given that the elevation level of the Sundarbans is 0.9–2.11 meters, this alarming rate of submergence means that many parts of this region are already facing a high risk of sea-level rise and flooding (Panda, 2020).

Although managed relocation has allowed people to stay in the Sundarbans, a viable long-term solution has not yet been determined. Somini Sengupta, a reporter for the Independent, writes that the "people of Sundarbans are pouring into Kolkata" (Sengupta, 2018). Scientific evidence shows that the average yearly sea-level rise along the Sundarbans delta is much higher at 8 millimeters than the global average of 3 millimeters annually. Managing the mangroves and wetlands in the Sundarbans to preserve them for future generations would be one effective strategy for Kolkata to adapt to climate change (SNRD-Asia/Pacific, 2015–2019).

Migration

Geopolitical boundaries do not restrict climate change impacts. Every day, vulnerable people are displaced and forced to migrate due to climate change impacts. Amali Tower, Director of Climate Refugees, a nongovernmental organization, stated, "This isn't something that will happen; this is something happening now" (Tower, 2020). Tower[1] also noted that South Asia had 9.5 million disaster displacements in 2019, and India recorded the highest number of disaster displacements with five million in 2019.

At the end of 2018, there was a total of 28 million new internal displacements across 148 countries and territories, with 61% of these displacements caused by disasters (World Migration Report, p. 45). State governments and political parties in India frown upon in-migration of any kind, and politicians use it as leverage to create tension between the locals and migrants (Dasgupta, 2020). The World Bank estimates that by 2050 more than 216 million people will be forced to flee their homes globally (World Bank Report, 2021). The Climate Change, Migration and Displacement report published by the Overseas Development Institute (ODI) and the United Nations Development Programme (UNDP) states that people migrate due to social, political, economic, and other environmental reasons other than climate change (Mahapatra & Sangomla, 2020). Nevertheless, it is fair to say that climate change is at least a "threat multiplier" for the risk of migration (Huntjens & Nachbar, 2015, pp. 5–8).

Harjeet Singh, Global Lead on Climate Change ActionAid International, wrote in the India Climate Dialogue (2018) that migration is the only option left for people who permanently lose their households and livelihoods to climate change disasters. Rehabilitation will be a challenge for the Indian Federal Government in the near term because climate refugees from Bangladesh will also head to West Bengal (Mukherjee, 2015). According to Stojanov et al. (2016), migration is not a new phenomenon; therefore, it seems reasonable that migration would be part of the adaptation response to environmental threats like climate change (p. 3). Migration is an important adaptation strategy in response to climate change impacts (Jha et al., 2018, p. 122). Brown (2008) observes that forced migration in response to climate stresses can cause epidemic diseases (p. 34). Thus, proper migration policies are important for Kolkata and India as a whole.

In an interview that the authors conducted with Dr. Stellina Jolly[2] at South Asian University in New Delhi, Dr. Jolly observed that Odisha and Andhra Pradesh are the only two states taking proactive measures for early warnings and rehabilitation due to floods, thus reducing the mortality rate. Dr. Jolly also noted that gradual coastal erosion in coastal regions is becoming more prominent. She indicated that the outer islands of the Sundarbans have relocated their residents to Sagar Island, the main island of the Sundarbans. Consequently, there are now infrastructure problems and a resource drain on these islands.

Existing Legal Framework

This section addresses the absence of legal policies to protect climate refugees. The design of the National Action Plan on Climate Change (NAPCC) in India at a sub-national level focuses on eight missions: solar, enhanced energy efficiency, sustainable habitat, water, sustaining Himalayan ecosystems, "Green India," sustainable agriculture, and strategic knowledge for climate change (Pandve, 2009). However, this legal response does not include strategies to address climate change displacement. India is required to submit a National Action Plan pursuant to the Paris Agreement; however, its plan is focused exclusively on renewable energy, water resources, and agriculture (Pandve, 2009)

Research in the Global South has emphasized vulnerability and community-based adaptation issues, mostly at the local level (Gupta, 2019). Issues such as the governance of adaptation at different levels remain poorly studied in South Asia (Gupta, 2019). Most of the climate change national programs in South Asian countries identify climate change-induced displacement as a threat but do not address it significantly (Jolly & Ahmad, 2019, p. 259).

India's National Policy, Frameworks, and Action Plans

India's State Action Plans on Climate Change (SAPCC) serves as the primary policy document at the sub-national level to address vulnerabilities related to climate change and implement necessary infrastructure projects and policies; however, displacement due to climate-related threats has received very little attention. Although separate policies on coastal zone management exist at national and state levels, they barely examine the human impacts of displacement and instead largely focus on the development of climate-resilient infrastructure and other measures such as protective sea walls and dikes (Panda, 2020). There are no binding legal frameworks for climate-displaced migrants (Jolly & Ahmad, 2019). Trial courts in India have not been receptive to the plight of international climate refugees (Jolly & Ahmad, 2019, p. 206). Most refugees faced imprisonment for a specified period because free movement in India for refugees is not allowed (Jolly & Ahmad, 2019, p. 206). At a national level, however, there are few planned policies for strategic relocation for those affected by climate change (Panda, 2020).

Commenting on the status of India's response to climate refugees, Dr. Jolly stated,

> India's legal mission towards climate refugees is slow to emerge. India does not have a long-term vision for climate refugees. Most of India's initiatives are short-term, such as the temporary relocation of communities

or reconstruction of affected areas. Bangladesh has climate change included at a policy level and has some funding mechanisms in place.

Dr. Jolly further noted that there is a lack of coordination in the Indian policy system. There is no data available regarding how many climate refugees have been displaced due to climate change. India does not have a policy for climate refugees because the government looks for policies in place at the international level. The NAPCC mentions climate adaptation; however, it lacks a resettlement plan. Finally, there are no pending laws in India that address climate refugees. In 2019, the 29 states in India initiated a process to revise their five-year SAPCC, which is intended to integrate climate change concerns into mainstream government planning processes (Ginoya et al., 2019).

Dr. Deepa Badrinarayana,[3] Professor of Law at Chapman University noted, "There needs to be better leadership in the Indian Government to better communicate the risks of climate change to the poorest or vulnerable communities." She indicated that there are NGOs that communicate about climate change to vulnerable populations, but there is no direct government involvement.

State and Local Laws and Policies

Local-level policies in India have responded to climate migrants in coastal communities; however, the response has not addressed current and future impacts resulting from sea-level rise and flooding (Rajan, 2021; Rajan & Bhagat, 2019). With the help of state governments, the central government has invested resources and funding projects to control erosion by raising embankments and installing a system of river-training supports to capture sediment and slow the rivers' flow. But these measures have not proven as effective as expected. West Bengal has an existing state action plan; however, it lacks policies to address "climate refugees" or "climate migration." Addressing local concerns is also important because Indian states have primary jurisdiction over the country's water and agriculture sectors, two of the areas most impacted by climate change (Ginoya et al., 2019).

Different states have climate disaster policies in place that help vulnerable populations that other states can adopt. Odisha's government used its legal authorities to relocate residents from hazard-prone areas. Although the relocation has been praised as an effective strategy, there has been criticism about the resettlement policy, lack of fair compensation and post-settlement livelihoods, the number of houses being resettled from the original location, and migrants being sent back to their villages. The current development strategy, which encourages migration into resource-rich areas and hampers long-term vulnerability reduction and economic development, will lead to increased human vulnerability to sea-level rise (Panda, 2020). On the other hand, the State of West Bengal has prepared the State SAPCC to introduce climate adaptation programs (Government of West Bengal, 2012). Efforts are under-way to mainstream climate change concerns into development decisions and other government initiatives. However, coordination among various activities of line departments – government departments at the state level – with a common objective for combating climate change demands specific attention and dedicated efforts (Government of West Bengal, 2012). West Bengal's State Action Plan suggests adaptation strategies; however, none of them speaks to climate change-induced migration. The term "climate refugees" does not exist in the State Action Plan for West Bengal (2012).

Some states, like Kerala, are playing a vital role in protecting climate refu-gees. Kerala has a participatory democracy system and the highest literacy rate of 96.2% in the country (The Economic Times, 2020). Kerala is a decen-tralized state, the local government is strong, and the participation rate of the people is high. Similarly, West Bengal's literacy rate is 76.26% (Census India, n.d.). West Bengal has the capacity to adapt Kerala's actions to provide pro-tective measures for climate refugees. Dr. Badrinarayana observed that West Bengal has a strong sense of communism and that Kolkata has a sense of community and intellectual life. Dr. Jolly noted that there is a disaster man-agement system in place for when a storm is approaching in India. Kolkata can build on this mechanism and strive to communicate the seriousness of climate change-induced migration through a similar mechanism.

Gaps in Climate Change Communication in India

India is relatively new to climate change communication research despite its high rank in greenhouse gas emissions. Yet as one of the world's most vulner-able countries to climate change (Leiserowitz & Thaker, 2012, pg. 5), it is imperative for climate change communication to improve. This section explores the missing links in climate change communication in India. There is a pressing need to bridge the gaps between climate change communication and public awareness in India. This section discusses the need for media to use jargon-free language to communicate climate change, the need to increase vernacular journalists, and the challenge to change the Indian mindset and raise awareness of climate change as a pressing reality. It also considers India's challenges in climate change communication to set a context for the recommendations in the next section. Delays in sharing relevant information with the public on these issues can cause harsh consequences. If the appropri-ate information and warning system does not engage the public, it will strain the Indian Government's capacity to manage disasters and climate refugees.

The media spreads awareness that climate change is human-caused; how-ever, there are not many changes to recent coverage to reflect the imminent threat of climate change-induced displacement. Two English-speaking news-papers cover issues of climate change: the *Times of India* (TOI) and the *Telegraph*. TOI has covered a series of stories on climate change, climate ref-ugees, the Sundarbans, and vulnerable areas, but coverage is still limited.

Newspapers place more emphasis on air pollution than sea-level rise. Krishnendu Mukherjee,[4] TOI reporter, also mentioned that these journalists perceived that scientists do not communicate about climate change in jargon-free language, making it difficult for them to cover the issue.

English newspapers cover most articles on climate change. TOI and *Telegraph* have policies to report environmental issues, but reporting is still inadequate. Although TOI and *Telegraph* are inadequate in their climate change coverage, their climate change coverage is better than the local papers. Each state in India speaks a different language. Climate change information should be translated into regional languages to be more relevant for local populations.

According to Krishnendu Mukherjee, TOI, and Chandrima Bhattacharya,[5] *Telegraph*, there are two Bengali newspapers with very limited coverage of climate change: Ananda Bazar and Ei Somoy. In addition, there is also insufficient coverage in English newspapers. In the eastern part of India, journalists who cover environmental topics are not well versed in sea-level rise and climate change. Further, many media interviewees mentioned that editors of newspapers are not adequately aware of the issues. Chandrima Bhattacharya (see note 4) further stated that coverage of politics gets priority over everything else, and newsrooms may not be educated about the implications of climate change. Additionally, Krishnendu Mukherjee (see note 3) noted that although there is an influx of people moving from Sundarbans to West Bengal, these developments are not communicated as related in any way to climate change impacts. Indians have observed climate change in their respective areas; however, they have not understood the threats of imminent forced migration (Leiserowitz & Thaker, 2012, p. 18). The biggest challenge India faces is changing the Indian mindset. Many believe that the West has burdened them with this issue. Another misconception editors of newspapers have is that Kolkata is inland and will not be affected by sea-level rise. The environmental journalists try to convince editors about the importance of sea-level rise; however, the editors feel readers might not be interested in reading such stories.

Dr. Jagadish Thaker emphasized how only 7% of Indians say they know facts about climate change, 51% know some or little about climate change, and 40% of the people are unfamiliar with climate change (Leiserowitz & Thaker, 2012). This data shows that a majority of Indians may not know about climate change as a conceptual framework to make sense of the changes in their local area and make decisions about their future. Dr. Thaker further noted that the public wants to know more about the local impacts of climate change and is willing to pay more attention to climate change policies. The underlying problem in India is insufficient coverage of climate change and migration reaching the local communities. Creating effective communication strategies could help inform and motivate the public to petition the government to implement policies on climate migration.

Communication Theories to Improve Climate Change Communication in Kolkata

A lack of public understanding and awareness will likely result in low demand for government action on climate change in a democracy (Thaker, 2017, p. 13). This section first proposes communication theories that will help build a platform for journalists to increase awareness of climate change as an imminent threat. It then recommends various strategies to strengthen media effects to enhance climate communication to underscore the risk of forced migration.

Agenda-Setting Theory

Agenda-setting theory was developed based on three assumptions, two of which are relevant to increasing climate change communication: (1) the media establish an agenda and in so doing are not simply reflecting reality, but are shaping and filtering reality for the public, and (2) the media's concentration on the issues that comprise their agenda influence the public's agenda. Using agenda-setting theory, the media can provide a lens through which to understand and reflect on social reality (West & Turner, 2019, p.359). When the media focuses on an agenda, it influences the public's agenda and the agenda of the decision-makers (West & Turner, 2019). Persuading people about climate change when the majority of the public is preoccupied with more immediate concerns is a challenge, however.

Agenda-setting theory describes the news media's opportunity to influence the importance of topics in the public agenda. The media places emphasis on a specific topic and influences how the public perceives the news. The repetition of messages about public issues in the news day after day and the pervasiveness of the mass media in our daily lives constitute a major source of journalism's influence on the audience (McCombs & Valensuela, 2007, p. 46).

Making issues like climate change more salient in people's minds through agenda-setting theory can also shape people's responses regarding climate change. For example, during the COVID-19 pandemic, the media focused its coverage on the importance of wearing masks, social distancing, avoiding indoor spaces, and the positive impact of the vaccine. The way the media delivered these messages can also be utilized to deliver climate change messaging. When the media draws attention to climate change as a current rather than future challenge, people would respond to the short-term threats from climate change. According to Michael Phillips-Anderson,[6] Associate Professor of Communication at Monmouth University,

Agenda-setting theory provides an important lens for considering the effectiveness of climate change communication. It is difficult to get the public to attend to most issues. Suppose the media doesn't choose to make environmental issues part of the agenda. In that case, most people will not know that it is a topic of concern or consider the issue less important than if there was more attention focused on it.

Agenda-setting theory addresses an important level of media: media framing. West and Turner (2019) define media framing as how media depictions of events influence and constrain how the audience interprets events. Additionally, framing involves selecting some aspects of perceived reality to make them more salient (Weaver, 2007, p. 143). According to McCombs and Valensuela (2007), media framing is a central organizing idea for news content that supplies a context and suggests what the issue is through the use of selection, emphasis, exclusion, and elaboration (p. 47). Framing would be extremely useful to promote climate change communication in India. This way, the media's agenda to increase climate change and climate refugee awareness would interact with what the public thinks about these topics.

An interesting example of agenda-setting theory that focuses on vulnerable and poor communities through Bangladeshi radio clubs is relevant to the Kolkata and West Bengal contexts. Community Radio Clubs in Bangladesh were created to help climate change-impacted communities by inspiring girls to be a part of the solution to extreme weather impacts rather than burdens to society. Kolkata's communities can promote awareness of climate change through young women because climate change is an existential threat with the unique power to undermine all sectors of society (Climate Home News, 2019). The natives of the Sundarban Islands, West Bengal, can use a similar technique to promote awareness of the vulnerability and need to adapt to climate change through their young women and radio. West Bengal's governance framework and socioeconomic circumstances are comparable to Bangladesh's; therefore, replicating this communication technique would be beneficial. "When we empower young women in Bangladesh to play a part in building comprehensive climate solutions, they discover the agency and power to shape their own future and to improve the future for society as a whole" (Climate Home News, 2019). This strategy embraces agenda-setting theory to focus on the overall vulnerable and poor communities in Kolkata and West Bengal.

Agenda-setting theory suggests that increased media coverage increases issue salience in public. In an interview with Jayanta Basu,[7] Reporter at *Telegraph*, he observed that the media in India prioritize phenomena that have an immediate impact. He noted that

> media coverage of climate change is increasing, but not enough in mainstream media, particularly in the Indian language media, at least in Bengal. It is still often seen as a "global" phenomenon, therefore happening elsewhere in the world, and not "local" enough.

The goal of agenda-setting theory is not to overwhelm the audience with media coverage but rather to make it simple and real.

Narrative Paradigm

The narrative paradigm has three relevant assumptions: (1) humans are natural storytellers, (2) decisions about a story's worth are based on the "good reasons," and (3) rationality is based on people's judgments of a story's consistency and truthfulness (West & Turner, 2019, p. 342). People make decisions based on what makes sense to them and accept or reject the story accordingly (West & Turner, 2019). Therefore, it is vital to increase communication on climate change; however, it is equally critical to create a reliable context for the story.

For example, in India, some newspapers have a Comics section. It would be ideal for newspapers to incorporate animated stories about climate change or the potential outcomes like migration. Professor Phillips-Anderson (see note 5) observed that

> for those interested in engaging the public about the dangers of climate change, it is important that they consider the form their communication takes. Humans are persuaded by narratives and regularly use them to organize and communicate their own experiences. In order to help people understand the dangers of climate change and the impact on their lives, it will be helpful to engage the audience with stories (that are realistic and supported by credible evidence) that can help them envision what the world will be like if nothing is done to alter the trajectory of climate change.

Wade, a new animated film about the climate crisis, is a gripping nine-minute story. *Wade* imagines what will happen if a rising sea floods Kolkata, with humans and animals alike sloshing through thigh-deep, litter-strewn water. "Animation also has the honor of being a core medium for children. If they get the message early on, the future gets even more hopeful" (Taylor, 2020). This animated film targets the population in Kolkata, making it a more relevant film to watch for this region.

Increasing the number of children watching films like *Wade* would cause climate change awareness to increase dramatically in the future. This film seeks to connect people's experiences with climate change and the need to start thinking about migration. Effective climate communication requires more focus on local and regional regions. Therefore, using animated films like *Wade* that focus on the people in Kolkata is most relevant for a specific audience.

Krishnendu Kuntak,[8] a TOI reporter, mentioned that he had heard about *Wade*, and there was no coverage of the film locally in Kolkata. If the media uses this form of narrative to tell the story of what is currently happening and a depiction of what will happen, it will increase the level of awareness among the public in India. The Indian media can create powerful narratives and find themes relevant to various climate change threats in India (Mittal, 2012, p. 220). In addition, Bangladesh circulates children's newspapers, and it would be valuable for Kolkata to adopt the concept. Future generations are important to educate and inspire on climate change awareness. A short 10-page children's newspaper including coverage of climate change and migration would be ideal.

Strengthening Media Effects

Agenda-setting theory and narrative paradigm will help increase climate change communication outreach in India. However, improving smaller and simpler aspects of the media that the two communication theories cannot cover is equally essential. The different levels of the media can strengthen these adjustments.

The Indian media outlets effectively use "Climate Change," "Global Warming," "Greenhouse Gas," and "IPCC" in their climate change media analyses (Billett, 2010, p. 5). However, to improve awareness of climate refugees, the media should increase its use of related words like "migration," "climate displacement," and "climate refugees." According to Chadwick (2017), the choice of "climate change" versus "global warming" also affects public understanding. The media has spread the overall message that climate change is broad; however, they have not captured how climate change impacts will result in potential forced migration.

For example, only 108 newspaper articles directly referenced climate change and how it is having an impact on India (Billett, 2010, p. 7). Worse still, four newspapers target and reach a readership that is restricted to the socioeconomic elite (Billett, 2010, p. 15). Only a few studies have evaluated Indian media coverage of climate change, and most of these focus on English newspapers. Recent studies determined that Indian media coverage of global warming has been low relative to other countries, with only 6% of all stories published in Indian newspapers compared to 34% in the US and 19% in the UK (Thaker, 2017, p. 6). Reporter Jayanta Basu emphasizes the need for journalists to be trained to connect the science, economic, societal, and political dimensions of climate change at a local level.

In a climate-sensitive region of Bundelkhand in central India, the pilot project, "Shubh Kal" (a better tomorrow), which established community radios to bridge the knowledge gap between communities, scientists, and policymakers, was tested in the drought-prone areas. After this radio program, the community stated that it learned new information on adaptation options. Although this project focused on farming and harvesting techniques, it is a valuable example of how private, public, and community radios can establish community-wide climate change communication shows for locals in their local language with simple messages (Bisht & Ahluwalia, 2012). In addition, All India Radio is a national resource and an effective mechanism to reach rural areas in India. This service is available throughout India in multiple languages. Although this channel has not attracted the attention of urban listeners, rural populations are left with no choice but to receive news through the radio. Cricketers and Bollywood actors could also accelerate communication about the threat of climate change and related concerns like forced migration. People in India pay more attention to notable figures rather than newspapers or news channels.

The media needs to be precise and intentional about how it conveys climate change to the public. Dr. Jolly states that the media covers environmental events like the Paris Agreement; however, there is no specific coverage of climate refugees. At the Qatar Sustainability Expo, Dr. Jagadish Thaker also stated that the Indian media's coverage of climate change is sporadic; they only cover such topics when there are climate events. Additionally, Dr. Badrinarayana (see note 2) noted that the media covers floods taking place in areas but does not indicate the connection between climate change and extreme weather events. Dr. Bardinarayana (see note 2) believes climate change messages should be communicated more effectively to future generations. Some high-end schools in India teach environmental courses, but climate change education needs to be implemented on a much broader scale.

Conclusion

> News about the environment, environmental disasters, and environmental issues or problems does not happen by itself, but is … "produced", "manufactured", or "constructed."
>
> – Dr. Anders Hansen (Hansen, 2010)

Climate change communication is a fairly new concept; however, the need for effective communication is growing. Public engagement is essential to increase public awareness of climate change. People do not realize that climate change is happening at the moment and is not simply a distant threat. The poorest populations in India and other developing countries are currently witnessing and facing climate change impacts like sea-level rise and flooding. India is responsible for an abundance of internally displaced people in the world (Rajan, 2020). Consequently, it is vital to spread information and communicate about these threats and create a potential long-term solution.

Additionally, it is important to address sea-level rise and flooding because these threats are fast-growing concerns for vulnerable populations in developing countries like India. Many families are being affected due to these threats that influence their livelihood, settlements, and daily necessities. Instead of solving the climate refugee problem through short-term goals, there should be long-term policies in place to better manage millions of people migrating at the same time into Kolkata. When people are affected by the physical threats of climate change, it is important to secure adequate protection programs to respond to immediate as well as long-term requirements (International Organization for Migration, 2009).

According to the IPCC report on climate change, climate change is a complex topic that requires representing climate change to diverse groups. Through changed communication methods, the IPCC has improved its efforts to present its findings and has increased public understanding and attitudes toward climatic changes. The IPCC states that addressing climate change concerns while still maintaining scientific facts is essential to effectively communicate climate change (Pathak et al., 2021). This chapter discusses agenda-setting theory and narrative paradigm as proposed strategies to increase outreach to the most vulnerable populations. Agenda-setting theory focuses on the quantity of media produced, and narrative theory focuses on the quality of the media articles. If there is insufficient coverage of climate change or believable and relatable stories about it, people will not draw their attention to those stories and understand them. These two communication theories would also help to apply pressure on the Indian Government to take legal measures to address climate migration in India.

To help ensure that the Indian Government improves its climate change communication strategy, it is essential to alter the Indian mindset that believes the western culture has burdened developing countries with human-induced climate change threats. Another aspect that climate change communication will accelerate is the lack of climate change laws that lower the priority of effective climate change communication to the Indian people. The final strategy would be to use the state of Kerala as a good framework for climate change communication in West Bengal due to the high literacy rate in these two states.[9]

Notes

1 The authors conducted an interview via email in December 2020 with Amali Tower, Director of Climate Refugees, a research and advocacy organization that seeks to protect climate displaced peoples.

2 The authors conducted an online interview via zoom in November 2020 with Dr. Stellina Jolly, Associate Professor in the Faculty of Legal Studies at South Asian University in New Delhi, India.

3 The authors conducted an online interview via zoom in November 2020 with Dr. Deepa Badrinarayana, Professor of Law at The Dale E. Fowler School of Law at Chapman University.

4 Krishnendu Mukherjee is a reporter at the *Times of India* and reports on environmental issues in India. The authors interviewed Krishnendu in November 2020 via WhatsApp.

5 Chandrima Bhattacharya is a journalist for the *Telegraph*. The authors interviewed Chandrima Bhattacharya in November 2020 via email.

6 Michael Phillips-Anderson is an Associate Professor of Communication at Monmouth University. The interview was conducted in November 2020 via zoom.

7 Jayanta Basu is a specialized environment and climate correspondent and columnist associated with the *Telegraph*. The authors interviewed Jayanta Basu in November 2020 via zoom.

8 Krishnendu Kuntak, a journalist at *Times of India*, was interviewed by the authors in November 2020 via WhatsApp.

9 The authors gratefully acknowledge Monmouth University's Urban Coast Institute, which funded this research project in fall 2020. An interview with Ms. Samuel, in which she discusses her motivations for pursuing this project, and identifies some of the project's findings, is available on the Urban Coast Institute webpage, https://www.monmouth.edu/uci/2021/02/08/student-researcher-examines-climate-change-communication-in-india/

References

Arnold, A. (2018). *Climate change and Storytelling: Narratives and Cultural Meaning in Environmental Communication*. Palgrave Macmillan.

Ballantyne, A. G. (2016). Climate change communication: What can we learn from communication theory? *Wiley Periodicals*, 7, 329–344. https://doi.org/10.1002/wcc.392

Billett, S. Dividing (2010). Climate change: Global warming in the Indian mass media. *Climatic Change*, 99, 1–16. https://doi.org/10.1007/s10584-009-9605-3

Bisht, H. & Ahluwalia, N. (2012). Community radios and climate change communication: Mapping grassroots experiences of the 'Shubh Kal' project in Bundelkhand, Central India. Retrieved from: https://cdkn.org/wp-content/uploads/2012/11/Manuscript-Community-radios-and-climate-change-communication-Mapping-grassroots-experiences.pdf

Biswas, M. (2021). The Sundarbans: India must protect the world's largest delta from a dystopian future, and now. *Down to Earth*. Retrieved from https://www.downtoearth.org.in/blog/climate-change/the-sundarbans-india-must-protect-the-world-s-largest-delta-from-a-dystopian-future-and-now-79092

Census India. (n.d.). West Bengal literacy rate-Census 2011. Retrieved from https://www.censusindia.co.in/states/west-bengal

Chadwick, A. (2017). Climate change communication. *Oxford research encyclopedia of climate change communication*. https://doi.org/10.1093/acrefore/9780190228613.013.22

Chandrashekhar, V. (2019, Oct 31). '36m Indians at flood risk due to sea level rise by 2050' [times nation]. *The Times of India (Online)* Retrieved from https://ezproxy.monmouth.edu/login?url=https://www-proquest-com.ezproxy.monmouth.edu/newspapers/36m-indians-at-flood-risk-due-sea-level-rise-2050/docview/2310516952/se-2?accountid=12532

Climate Home News. (2019). Radio clubs help climate-hit Bangladeshi communities avoid child marriage. Retrieved from https://www.climatechangenews.com/2019/01/09/radio-clubs-help-climate-hit-bangladeshi-communities-avoid-child-marriage/

Dasgupta, K. (2020, March 02). Why climate migration doesn't have to be a crisis. *Hindustan Times*. Retrieved from https://www.hindustantimes.com/analysis/why-climate-migration-doesn-t-have-to-be-crisis-opinion/story-DxtMHrG61A7fWokM6G8U0I.html

Deweerdt, S. (2020) As climate changes, what changes people's minds? *Anthropocene* Retrieved from https://www.anthropocenemagazine.org/2020/08/as-the-climate-changes-what-changes-peoples-minds/

Dhiman, R., VishnuRadhan, R., Eldho, T. I., & Inamdar A. (2019). Flood risk and adaptation in Indian coastal cities: Recent scenarios. *Applied Water Science*. 9(5).

Ghosh, A., & Boykoff, M. (2019). Framing sustainability and climate change: Interrogating discourses in vernacular and English-language media in Sundarbans, India. *Geoforum*, 99, 142–153. https://doi.org/10.1016/j.geoforum.2018.11.014

Ginoya, N., Narayan, U., & Worker, J. (2019, July 25). As India revises state climate plan, who should have a voice? *World Resources Institute*. Retrieved from https://www.wri.org/insights/india-revises-state-climate-plans-who-should-have-voice

Government of West Bengal. (2012). West Bengal state action plan on climate change. Retrieved on November 10, 2020 from http://www.nicra-icar.in/nicrarevised/images/State%20Action%20Plan/West-Bengal-SAPCC.pdf

Gupta, J. (2019). Climate adaptation impossible without community participation. *India Climate Dialogue*. Retrieved from https://indiaclimatedialogue.net/2019/05/22/book-review-climate-adaptation-impossible-without-community-participation/

Hansen, A. (2010). *Environment, Media, and Communication*. Routledge. https://doi.org/10.4324/9780203860014

Huntjens, P. & Nachbar, K. (2015). Climate change as a threat multiplier for Human Disaster and conflict. *The Hague Institute for Global Justice*. Retrieved from https://www.thehagueinstituteforglobaljustice.org/wp-content/uploads/2015/10/working-Paper-9-climate-change-threat-multiplier.pdf

Internal Displacement Monitoring Centre. (n.d.) India. Retrieved from https://www.internal-displacement.org/countries/india

International Organization for Migration. (2009). *Migration, Climate Change, and the Environment*. International Organization for Migration (IOM). Retrieved on January 30, 2022, from https://www.iom.int/sites/g/files/tmzbdl486/files/jahia/webdav/shared/shared/mainsite/activities/env_degradation/compendium_climate_change.pdf

Jha, C. K., Gupta, V., Chattopadhyay, U., & Sreeraman, B. A. (2017). Migration as adaptation strategy to cope with climate change: A study of farmers' migration in rural India. *International Journal of Climate Change Strategies and Management*, 10(1), 121–141. https://doi.org/10.1108/IJCCSM-03-2017-0059

Johnson, B. B. (2011). Climate change communication: A provocative inquiry into motives, meanings, and means. *Risk Analysis*, 32(6), 973–991. https://doi.org/10.1111/j.1539-6924.2011.01731.x

Jolly, S. & Ahmad, N. (2019). *Climate Refugees in South Asia*. Springer.

Leiserowitz, A., & Thaker, J. (2012). Climate change in the Indian mind. *Yale Project on Cliamte Change Communication*, 1–219. Retrieved from https://climatecommunication.yale.edu/wp-content/uploads/2016/02/2012_08_Climate-Change-in-the-Indian-Mind.pdf

Mahapatra, R. & Sangomla, A. (2020, May 22). Migration out of climate change: Disaster- induced migration outstrips that due to conflicts. *Down to Earth*. Retrieved on December 22, 2020 from https://www.downtoearth.org.in/news/climate-change/migration-out-of-climate-change-71291

Maibach, E. W., Rose-Renouf, C., & Leiserowitz, A. (2008). Communication and marketing as climate change–intervention assets a public health perspective. *American Journal of Preventive Medicine*, 35(5), 488–500. https://doi.org/10.1016/j.amepre.2008.08.016

McCarthy, J. (2020, February 19). Why climate change and poverty are inextricably linked. *Global Citizen*. Retrieved from https://www.globalcitizen.org/en/content/climate-change-is-connected-to-poverty/

McCombs, M., & Valensuela, S. (2007). The agenda-setting theory. *Pontificia Universidad Católica de Chile*, 44–50. Retrieved from https://www.redalyc.org/comocitar.oa?id=97120369004

Mittal, R. (2012). Climate change coverage in Indian print media: A discourse analysis. *TheInternational Journal of Climate Change: Impacts and Responses*, 3(2), 221–231. https://doi.org/10.18848/1835-7156/CGP/v03i02/37105

Mukherjee, K. (2015, n.d.). Kolkata in vortex of climate refugee crisis. *Times of India*. Retreived from https://timesofindia.indiatimes.com/articleshow_comments/49760935.cms?from=mdr

Mukul, S. A., Huq, S., Seddon, N., & Laurance, W. F. (2020). Saving the Sundarbans from development. *Science*. Retrieved from https://doi.org/10.1126/science.abb9448

Mulhern, O. (2020, June 24). *Sea Level Projection Map-Kolkata*. Earth.org. Retreived on February 6, 2022 from https://earth.org/data_visualization/sea-level-rise-by-the-end-of-the-century-kolkata/

Panda, A. (2020, May 22). Climate change, displacement, and managed retreat in coastal India. *Migration Policy Institute*. Retrieved from https://www.migrationpolicy.org/article/climate-change-displacement-managed-retreat-india

Pandve, H. T. (2009). India's national action plan on climate change. *Indian Journal of Occupation and Environmental Medicine*, 13(1), 17–19. https://doi.org/10.4103/0019-5278.50718

Pathak, M., Roy, J., Patel, S., Some, S., Vyas, P., Das, N., & Shukla, P. (2021). Communicating climate change findings from IPCC reports: Insights from outreach events in India. *Climate Change*, 168, 23. https://doi.org/10.1007/s10584-021-03224-8

Pramanik, M. K. (2017). Impacts of predicted sea level rise on land use/land cover categories of the adjacent coastal areas of Mumbai megacity, India. *Environment Development and Sustainability*, 19(4), 1343–1366. https://doi.org/10.1007/s10668-016-9804-9

Rajan, I. S. (2019). *Climate Change, Vulnerability and Migration*. Routledge India

Rajan, I. S. (2020). *India Migration Report 2017*. Routledge India.

Rajan, I. S. (2021). *India Migration Report*. Routledge India.

Revi, A. (2008). Climate change risk: An adaptation and mitigation agenda for Indian cities. *International Instute for Evironment and Development*, 20(1), 207–229. https://doi.org/10.1177/0956247808089157

Russill, C., & Nyssa, Z. (2009). The tipping point trend in climate change communication. *Global Environmental Change*, 19(3), 336–344. https://doi.org/10.1016/j.gloenvcha.2009.04.001

Sengupta, S. (2018, August 11). Kolkata is becoming an unnecessary climate casualty–it didn't have to be. Retrieved from https://www.independent.co.uk/climate-change/news/kolkata-india-climate-change-casuality-coastal-flooding-ganges-a8479516.html

SNRD Asia. (May 2015–April 2019). *Management of the Sundarbans Mangrove Forests for Biodiversity Convervation and Increased Adaptation to Climate Change (SMP)*. SNRD Asia/Pacific. Retrieved from https://snrd-asia.org/management-of-the-sundarbans-mangrove-forests-for-biodiversity-conservation-and-increased-adaptation-to-climate-change-smp/

Stojanov, R., Kelman, I., Ullah, A., Duzi, B., Procházka, D., & Blahutová, K. (2016). Local expert perceptions of migration as a climate change adaptation in Bangladesh. *Sustainability*, 8(12), 1223. https://doi.org/10.3390/su8121223

Taylor, N. (2020, July 13). What if Kolkata was swamped in climate horror. *India Climate Dialogue*. Retrieved on December 20, 2020, from https://indiaclimatedialogue.net/2020/07/31/what-if-kolkata-was-swamped-in-climate-horror/

Thaker, J. (2017). Climate change communication in India. *Oxford Research Encyclopedia of Climate Science*. https://doi.org/10.1093/acrefore/9780190228620.013.471

The Economic Times. (2020, September 08). At 96.2%, Kerala tops literacy rate chart; Andhra Pradesh worst performer at 66.4%. *The Economic Times*. Retrieved from https://economictimes.indiatimes.com/news/politics-and-nation/at-96-2-kerala-tops-literacy-rate-chart-andhra-pradesh-worst-performer-at-66-4/articleshow/77978682.cms

Tikoo, R. (2009, November 12). *Kolkata in list of 5 most environmentally vulnerable coastal cities in Asia: WWF*. DevNews. Retreived from https://devnews.net/environment/kolkata-in-list-of-5-most-environmentally-vulnerable-coastal-cities-in-asia-wwf/

Tower, A. (2020). *Crisis Looms*. Climate Refugees. https://www.climate-refugees.org/why

Weaver, D. H. (2007). Thoughts on agenda setting, framing, priming. *Journal of Communication*, 57(1), 142–147. https://doi.org/10.1111/j.1460-2466.2006.00333.x

West, R. & Turner, L. H. (2018). *Introducing Communication Theory: Analysis and Application*. McGraw Hill Education.

Yale Program on climate change communication. (n.d.) Yale school of the environment. Retreived from https://climatecommunication.yale.edu/about/the-program/

Part III

Power and politics of climate change

Focusing on strategies and policies

11 Climate change and the government conundrum in Bhutan

Johannes Dragsbaek Schmidt

For several decades, Bhutan has been cast in international media and academic literature as a trailblazer referring to its ambitious environment-friendly and ecologically balanced policies and its protection of pristine nature with clean air, water, and primeval forest. Various democratically elected governments' reliance on the vision of Gross National Happiness (GNH) as the country's development philosophy has been coupled with the fact that the country remains exceptional due to its carbon neutrality and even negative carbon footprint (Yangka et al., 2018). Carbon sequestration from its forests explains its carbon negative status, and the country is a net sink for greenhouse gases. The goals of GNH are encapsulated in the country's overall development trajectory as environmentally friendly and a leader in nature conservation. At the same time, Bhutan is one of the smallest but fastest-growing economies globally, with annual average growth between 2013 and 2017 reaching 5.4% (4.63% in 2017), exceeding the average global growth of 4.4%. The GDP per capita increased almost tenfold since 1980, from US$332 in 1980 to US$3,438.2 in 2017 (World Bank, 2019b, p. 10).

Approximately the size of Switzerland, mountainous, but at a much higher altitude, with a small population of about 763,092 in 2019, Bhutan is a landlocked country situated between India and China. It is located at the eastern fringe of the Himalayan fragile ecosystem. The country's climate is diverse due to considerable variations in elevation and altitudes. The Duars Plain is primarily hot and humid; the Lesser Himalayas region has lower temperatures, while the areas in the Greater Himalayas are closest to that of the alpine tundra. The southern belt at the foothills of the snowcapped mountain belts (150–2,000 meters above sea level) is subtropical, with high humidity, heavy rainfall, and average temperatures of approximately 15°C–30°C. With its river valleys (2,000–4,000 meters above sea level), the central belt has cool winters, hot summers, and moderate rainfall. With snowcapped peaks and alpine meadows (4,000 meters above sea level), the Northern belt has cold winters and cool summers (ADB & WB, 2021, p. 6).

In this context, climate change is not only an environmental problem but also poses tremendous challenges to sustainability, livelihoods, and development itself. Being primarily an agrarian country, Bhutan depends on the biophysical impacts of climate change. Due to its topography, extreme weather

DOI: 10.4324/9781003216476-14

events and natural catastrophes have an ongoing impact, increasing its vulnerability in agricultural production, management of water resources, and human health. Biodiversity is also under siege from the simultaneous threats of human land use and climate change. IPCC indicates in a preliminary report snow-covered areas and snow volumes will decrease in most regions of the Himalayas during the 21st century, and snowline elevations will rise. Glacier volumes will likely decline with more significant mass loss in higher CO_2 emissions scenarios (IPCC, 2021, p. 134). The increase in temperatures and changes in precipitation patterns is a major concern for the health of the Himalayan snow cover and glaciers. The region, including Bhutan, has experienced significant melting of snow and retreat of glaciers during the past five decades. While climate change also significantly affects the environment over the high mountain regions of Asia, its impact on the Himalayan cryosphere is also a significant threat to regional water resources (Sabin et al., 2020). According to the IPCC, the sensitivity of the Himalayas to climate change is visible at higher elevations, with expected temperature increases up to three times larger than the global average. In comparison, the direct effect on the livelihoods of poor farmers at the lower elevations on the slopes bordering India will be more severe.

Climate change also threatens its vast biodiversity and increases the likelihood of natural hazards (e.g., glacier lake outbursts, flash floods, droughts, and forest fires) (Hoy et al., 2016). Bhutan is highly vulnerable to adverse impacts of climate change due to its fragile mountainous ecosystem and economic structure. The most vulnerable sectors are water resources, agriculture, forests and biodiversity, and hydropower. It is projected that the frequency and intensity of extreme climate events would increase with changing climate (National Environment Commission, 2016). The third dissemination to the United Nations Framework Convention on Climate Change (UNFCCC) discloses the impact of climate change on key sectors such as water, agriculture, energy (hydropower), human health, and glaciers. These risks and increased exposures highlight the high degree of vulnerability of the country. Furthermore, the complex, high-altitude terrain and the severe impact of the Indian summer monsoon add to a substantial exposure of the country's key economic sectors (agriculture, forestry, hydropower generation, and tourism) to climatic changes (ADB & WB, 2021).

This chapter aims to contribute to the literature about climate change and the link to governance and impacts of policies on the farmers in Bhutan. It elaborates on the visions and promising strategies from the political sphere and from the local level that could lead to a way out of the problem of balancing many development goals simultaneously. This development "trap" is quite common in the global South. The chapter is informed by a critical political ecology perspective (Busck & Schmidt, 2020), explaining the injustices of underdevelopment and poverty linking the issues of climate change, environmental change, and exploitation in terms of class, gender, or subaltern status (Bryant, 2001, p. 152; Biersack, 2006, p. 9). The accumulation of exploitation rests on the accumulation of dispossession (Harvey, 2003), which sees the peasantry and the agrarian problem through the lenses of critical agrarian studies (Bernstein, 2010). This overall structural framework is linked with policy studies encapsulating the potential problems of balancing different visions, plans, policies, and implementation (Schmidt & Christensen, 2017).

Climate change and its impact on resources affect socio-economic wellbeing. Many of these impacts are adverse, with some being irreversible, and the effects vary greatly. They are "felt more among the poorest of the poor and vulnerable communities due to limited resources and infrastructure available to these groups" (Johnson & Hutton, 2014). The Bhutanese government also recognizes that the farming and poor communities are the most vulnerable (Brassard, 2017, p. 183). This contribution aims to analyze the impact of climate change policies on agriculture and food security, water, and hydropower. The following sections analyze the role of government visions and responses toward adaptation and mitigation. The attempt is made to identify bottlenecks and contradictions within and between policies. Finally, a conclusion wraps up the findings of the chapter.

Agricultural production and food security

In 2006, the government decided to promote 100% organic agriculture. Today, 37% of farmers use agrochemicals on approximately 19% of arable land, implying that the rest is chemical-free or organic. Recent information disclosed that Bhutan has ten products certified as organic: Potato, garlic, carrot, turmeric, sea buckthorn, chamomile, mint, Camellia Sinensis, rhododendron anthology, and lemongrass (Dukpa & Wangdi, 2021, p. 20). In addition, the original intention of the government's ambitious goals to improve rural life by reducing rural-urban migration to maintain traditional practices, a sense of community, and decentralizing control over natural resources should have generated feelings of local ownership, promoted good governance, and enhanced more effective resource management (Brooks, 2013, p. 3647). However, the lack of rural income sources and employment fueled rural-to-urban migration, which is now perceived as a national challenge adversely affecting urban development, particularly youth unemployment.

Although there are success stories, climate challenges are legion. To quote the government of Bhutan's report to the UN (Royal Government of Bhutan, 2021, p. 8): The main constraints facing the agricultural sector are climate change-induced water shortages, wildlife predation of crops, pests and diseases, poor mountainous shallow soils, further worsened by increasing soil loss through surface erosion and scarcity of farm labor. The report explains the adaptation priorities and options for agriculture without elaborating on a complete implementation plan, focusing on water shortages combined with improved planning and implementation of irrigation projects. More and better storage and pumping facilities are needed, creating a database and inventory of water resources. More focus on the Human-Wildlife Conflict. Increasing knowledge, planning, and research on animal migration. Creation of an adaptation plan against soil erosion: Most smallholder farmlands sit on sloping lands where fertile topsoils are frequently lost to soil erosion. Even more crucial is the improvement of Sustainable Land Management techniques and practices and alternate crop production. Establishing farmer schools introducing post-harvest technology and techniques. Develop and apply varieties of crops and animals resilient to pests, diseases, and climate change. Exploring and continued dependence on native livestock breeds that, historically, have been able to cope with natural changes as an alternative to dependence on improved varieties. Harsh unprecedented weather conditions and other varieties of extreme climate, such as heat and cold waves, pose risks to both crops and livestock. Pasture development and providing shelters for animals in hot and cold regions are also necessary to protect them from harsh climatic conditions.

In addition, agriculture is vulnerable due to its reliance on erratic monsoon rains, brief growing periods, and climatic swings in temperature and weather in the mountains. The structure of agricultural production and concentration of farming in vulnerable areas makes it susceptible to shifting climatic weather. For example, "with 31% of agriculture on slopes, it is particularly vulnerable to landslides and soil erosion. These may be exacerbated by the projected increases in extreme precipitation intensity" (ADB & WB, 2021, p. 20).

The problem lies with the government and the much-needed policies dealing with numerous diseases and pests, erratic rainfalls, storms, droughts, flash floods, and landslides. These climate-induced problems are related to implementing resilience policies protecting Bhutan's farming. The government's task is to enhance cross-sectoral strategies like policy support, increased investment, R&D, technology generation, and disaster preparedness (Brassard, 2017; Chhogyel & Kumar, 2018).

There are real problems and not just projected increases since

> Crop damage of 1–19% due to the combined effects of extreme weather events (untimely rains, drought, and windstorms) demonstrate the urgency of charting out the most appropriate action plans for improving the resilience of Bhutanese farms. Even though there are policy initiatives for climate change mitigation and adaptation, the mainstreaming of implementable policy instruments into the development plans needs major attention. Thus, context-based efforts, in addition to being tailored, must be backed by robust policy intervention towards a resilient agro-ecological environment and livelihood program.
>
> (Chhogyel et al., 2020, p. 13)

While replacing subsistence with diversified high-value agriculture has the short-term potential to increase farmer's income, the introduction of mono-agriculture may also create vulnerabilities. Many scholars view diversity as a central determinant of food and livelihood security. Traditional multiple cropping technologies, utilized for generations in local settings, could be refined to adapt against the potential threats of global environmental and climate change and ensure food security. Replacing a diverse set of site-specific agricultural practices with intensive, market-driven mono-cropping may reduce household food self-sufficiency, increase social and ecological risk, and reduce rural households' wellbeing in the long run (Sabin et al., 2020). Nevertheless, more than 60% of the Bhutanese population are subsistence farmers. Their livelihoods are dependent on natural resources such as forests. Many depend on small plots of land, barely enough for survival.

The agricultural sector, i.e., the primary sector, contributed approximately 16% of GDP in 2019 but employed a vast majority of the working population (National Statistics Bureau, 2021, p. 9). Agriculture is vital for food security in Bhutan and is vulnerable to higher temperatures and changes in precipitation patterns (Hoy & Katel, 2019). The problem is that the impact of climate change on agricultural production and food security is one of the biggest challenges of the 21st century. The impact on various crops and land has negatively affected different kinds of farming (Chhogyel & Kumar, 2018, pp. 7–8).

What is worrying is that the mountain slopes are sensitive to changing climates due to their dependency on water. As noted by Bhutan Academican Dr. Om Katel,

> Climate change also means increasing heat stress to farmers leading to change in agricultural practices changing temperature has a direct bearing on water availability. Changing temperature also means increasing the cost of food when farmers do not produce enough to feed the families, they face food shortage and forcing them to buy foods from markets.
>
> (*Bhutan Times*, 2021)

Approximately 20% of crop losses in Bhutan include rice, maize, potatoes, apples, and mandarin. This decline in crop productivity and crop yield leads to food shortage in households, and it may lead to food insecurity if this continues. These problems are of paramount importance and are also connected with water resources and hydropower management, which the government has given top priority.

Water and hydropower

Water has immense importance in Bhutan. It supports agriculture and hydro-electric energy production, but climate change may challenge these sectors. A joint report from the Asian Development Bank and the World Bank (2021, p. 17) shows that water resources consist of glaciers, glacial and high-altitude wetlands, rivers and river basins, and groundwater reservoirs. The fragility of the Himalayan topography makes glacier lake outbursts and flooding a significant threat. Groundwater and reservoirs are limited to flat valleys, particularly in areas close to the Indian border. The country has four major river systems, which are primary water resources: The Amo Chhu (Toorsa), the

Wang Chhu (Raidak), the Punatsang Chhu (Sunkosh), and the Dangme Chhu (Manas). The Punatsang is one of the largest rivers. All rivers depend principally on glacier melt, snow, and seasonal rainfall. Although the country "has one of the highest per capita water resource availability in the world with 94,500 m³/capita/annum. However, it uses only 1% of its total generated annual freshwater due to poor water resource management and its peculiar terrain. In addition, there are increasing drought incidents all over the country" (Tariq et al., 2021).

Water is essential for irrigation farming, agro-processing, industry, households, and hydroelectric power generation. It also holds symbolic value in cultural and religious contexts. Approximately 98% of the population has access to safely managed drinking water, although this number is disputed. However, population growth and high seasonal precipitation constrain water resources at local and community levels. Various projections for river flow due to climate change show mixed results, with some showing increases and others decreasing. This way, projections in precipitation are highly variable though a slight majority of models show an increase in precipitation at the national level. Other projections show a significant rise in the southern border with India during the monsoon season when water resources are particularly abundant. Projected increases in the number of days with very heavy precipitation could further increase the risk of flooding and impact runoff, erosion, and rates of river discharge (ADB & WB, 2021, p. 17).

While the numbers show a high per capita availability of water, the reality contrasts these projections since there is already an emerging scarcity crisis at local levels. There are several reasons (Ranjan, 2018):

1   Available water and its annual supplies are unevenly distributed. Some regions receive large quantities of water, while others have limited supplies.
2   The phenomenon of climate change affects Bhutan's water resources. A study from 2016 by the National Environment Commission found that some of the country's districts, such as Thimphu, Haa, and Zhemgang, will experience water shortages by 2030. Furthermore, the study confirmed that, due to climate change, several essential water sources in the country have started to dry up.
3   In addition to the decline in water availability, the country also suffers from a lack of water supply infrastructure. It is difficult to provide a regular water supply to people scattered in different parts of the mountainous country. In addition, water supply relates to the quality of water people consume. Estimations show that more than 20% of Bhutan's population does not have access to safe drinking water, contradicting other surveys.

Water is also a blessing, especially for development purposes and wealth creation in the future. The government has cast hydropower as the panacea for green growth. Most of the hydropower projects are run-of-the-river dams that have less impact on ecosystems. Still, new reservoir dams are more damaging, and climate change may lead to unpredictability in terms of water flows that generate power. Indeed, hydropower is among the most vulnerable areas to global warming because water resource availability is closely linked to climate change. Hydroelectric plants are dependent on predictable runoff patterns. Greater unreliability of dry season flows poses potentially severe risks to water supplies in the lean season (Gautam et al., 2014, p. 7703). However, as mentioned before, modeling and various projections have different predictions. What is steadfast is that "the retreat of glaciers is intrinsically associated with changes in river water discharge; the latter is expected to fluctuate more substantially than in the past" (Hoy & Katel, 2019). Questions have also been raised about the safety of building hydropower projects in the fragile mountains of the Himalayas (Adhikari, 2020).

The main problem involved in hydropower is that India is demanding a price and payment below the market price in India itself and below world market prices. Furthermore, the loans the Bhutanese government has taken from the government in Delhi for hydropower projects "would put Bhutan under about ₹12,300 crore rupees (S$2.4 4 billion) of debt. This accounts for 77 percent of the country's total debt and is about 87 percent of its GDP" (Ranjan, 2018, p. 7). These huge debts will have repercussions and is probably a reason why the World Banknotes: "The massive scale of hydropower generation projects relative to the small size of the economy generates large fluctuations in aggregate demand, both during the construction phase and when the projects become operationalized" (World Bank 2019a, p. 11). Thimphu's aim of not relying on foreign assistance or loans may have huge implications on the government's room for maneuver in coping with climate change and environmental disasters. None of the hydropower projects went through the routine Environmental Impact Assessment test. This led to many questions about the environmental feasibility of the projects. Finally, the hydropower projects may have disrupted or affected the natural flow of water from ponds and/or spring water (Ranjan, 2018, p. 8). As Ranjan finally points out, the "prevailing undercurrent in Bhutan is that India exploits Bhutan's water sector for its benefit" (Ranjan, 2018, p. 9). This may have severe consequences for mitigation and adaptation to climate change since the Bhutanese government does not control the hydropower sector.

Nevertheless, the Bhutanese government still sees hydropower as a climate mitigation strategy. Undoubtedly, it is a high-income foreign exchange earner with approximately 45% of the country's GDP coming from this source. On the other hand, it may also escalate into conflict since the building of hydropower dams is implicitly "increasing numbers of water sources falling dry." This is "resulting in decreasing crop yields. Local-scale water conflicts between farmers and villagers in areas with high irrigation demand are likely to get accentuated in future." A majority of the local people in the comparably populated Punakha–Wangdue valley already face water shortages for irrigation and drinking. Some water shortages are also observable in high population density areas like the capital city of Thimphu (Hoy & Katel, 2019).

Government mitigation and adaptation

Bhutan has experienced significant gains in growth and standard of living. There has been a successful transition to constitutional democracy and has maintained its cultural and ecological heritage (Brooks, 2013, p. 3657). Nevertheless, climate change challenges these hard-won victories. In 1998, in "The Middle Path" outlining Bhutan's environment strategy, the government emphasized:

> In its mission to take control of development, Bhutan is blessed with four significant assets. First, because of its delayed arrival on the world stage, Bhutan — perhaps more than any other country in the world — can plan its development. Second, its cautious and incremental pursuit of foreign assisted development has meant that Bhutan is not in danger of accumulating large debts. Third, the one prerequisite upon which any truly effective development strategy depends — political will — is there.

Interestingly, the document refers to the experience from other countries, and states,

> that unless the highest authorities in a country are committed to the cause, the process of development will inevitably be co-opted by a powerful elite whose ability to profit inordinately from development activities will undermine the mechanisms of self-help and cooperation and erode both the communal and national spirit. Moreover, a powerful ally lies in the country's strong Mahayana Buddhist culture, emphasizing spiritual, as opposed to material enrichment.
>
> (National Environment Commission, 1998, p. 18)

The Constitution of the Kingdom of Bhutan reflects word-by-word the vision and philosophy of the "Middle Path." All citizens are supposed to protect the natural environment, conserve rich biodiversity, and prevent all forms of ecological degradation, including pollution. The rights over mineral resources, rivers, lakes, and forests are regulated by law. The state is assigned to ensure a safe and healthy environment, always maintaining a minimum of 60% of total land under forest cover. It must secure ecologically balanced sustainable development and sovereign rights over its biological resources, including legislation to ensure sustainable use of natural resources and intergenerational equity. "By the constitution, the Parliament can, by law, declare any part of the country to be a National Park, Wildlife Reserve, Nature Reserve, Protected Forests, Biosphere Reserve, Critical Watershed, and such other categories meriting protection" (Constitution of Bhutan, 2008, pp. 11–12).

While these documents denote a clear path and vision about societal change with emphasis on the sacred environment, climate policy is a bit more unclear. Various documents recognize the high dependence on agriculture and the significant role of hydropower for economic development but also that it increases vulnerability while being more unclear about implementation strategies.

The government's REDD+ strategy goals are straightforward and linked to the vision described above. They encompass a transition to low-emission, climate-resilient, and sustainable pathways in forestry, agriculture, energy, tourism, and other cross-cutting areas. The strategy aims to target carbon emissions from deforestation and forest degradation and target additional benefits, including enhanced livelihoods, protected biodiversity, and other ecosystem services, improved forest governance, and support for sustainable land use planning. Interestingly, policies signify a participatory and consultative process with various "empowered" stakeholders from the governmental, non-governmental organizations, and the private sector. In addition, the aim is to mobilize funding for the implementation of the proposed policies and measures and reach out to various development partners (World Bank, 2019, p. 19).

The observation is that there is a persistent quest for external funding and expertise and overlap in policies, duplication, and lack of coordination. However, the implementation has only recently been initiated.

> In order to implement climate actions, international support in the form of finance, technology, capacity building, research and awareness is required as outlined in the first NDC. Bhutan must be ready to access and receive international support while also mobilizing some amount of domestic support under the objective of achieving self-sufficiency and economic self-reliance.
>
> (GoB 2020, p. 5)

The government pursues a policy where the public sector, the private sector, and communities are seen as partners in the governance of resources. For example, sectoral policies are subject to vigorous stakeholder consultations and a policy-screening tool to ensure that the proposed policy upholds the values and principles of GNH. The GNH Commission is supposed to establish the framework for sectoral policies, which involves all social partners. Essentially, it calls for a balanced and integrated approach to socio-economic development, preservation of the environmental and cultural heritage, and promotion of good governance (Ojha et al., 2019).

However, the state is a major player through its large state-operated enterprises (SOEs). These state-owned companies operate in various commercial sectors, including the financial and energy sectors (hydropower), manufacturing, and telecommunications. This has created a small domestic market and nascent private sector; SOEs provide essential goods and services. The Ministry of Finance is the sole shareholder of all SOEs, and managers have preferential access to policymakers. In 2017, the contribution of SOEs accounted for 43% of government revenues, and gross revenues were equivalent to 30% of GDP (World Bank 2019b, p. 14).

The private sector remains underdeveloped and dominated by small and micro firms, primarily operating in the informal sector (World Bank 2019b,

p. 11). The implication is that it is difficult to involve the nascent private sector in climate adaptation and mitigation. Civil society is also weak by most standards (Schmidt, 2017), while the community level may be mobilized for specific purposes. The question is how the government and the donors cope with these issues and at the same time relate to the ambitious policies and statements about self-reliance.

The gap between good intentions and reality

Bhutan ratified the UNFCCC on 15 August 1995, and the National Environment Commission Secretariat (NECS) is the national focal point for all the multilateral environmental agreements. The National Environment Protection Act 2007, Section 20, mandates the National Environment Commission (NEC) functions as an independent authority and the highest decision-making body. The Commission is responsible for implementing the policies, regulations, and directives issued by the National Environment Commission and for all decisions regarding the management of the environment (Royal Government of Bhutan 2021, pp. 3–4).

Bhutan lacks institutional infrastructure and technical capacity in responding to changing climate and its effects, particularly in agriculture. Urgent response from the government and concerned stakeholders is crucial to undertake critical actions and minimize the effects of climate change to improve food and nutrition security and reduce poverty. The legal framework and policies incorporate international agreements, protocols, acts, policies, strategies, long-term and short-term plans, rules and regulations, and directives for responding to climate change effects adopted in Bhutan. The country has signed all multilateral agreements, including the Kyoto Protocol in 1997, Doha Amendments, and the Paris Agreement for Climate Change in 2016. There are several policies on paper, such as a green foreign direct investment policy, organic farming, biotechnology, agro-processing, and green and climate-smart practices (Tenzin et al., 2019). Other ideas note the urgency of the

> introduction of sound sustainable crop insurance policy and contingency plans to protect farmers during disasters. To protect and sustain agriculture, Bhutan would also require an effective policy to protect agricultural land. All such interventions must be backed by increased investments in critical farming areas.
>
> (Chhogyel & Kumar 2018, p. 79)

Many of these initiatives never materialized and lacked substance and context.

The enactments of state control of forests came with the Forest and Nature Conservation Act of 1969 (later revised in 1995) and the Land Act of 1979, which reduced the status of local communities from "owners" to "proprietors" (Ojha et al., 2019). Bhutan's Forest Act of 1969 has distinguished between non-sokshing and sokshing forests. For non-sokshing forests, there has been a loss of community and individual rights, leading to non-compliance, especially in rural areas because of the high costs involved in conformance to the new property rights regime (Dorji et al., 2006). For sokshings, the Forest Act reduced the sokshing "owners" to "proprietors," which lost alienation rights they exercised prior to 1969. As Brooks (2013) notes, these changes in property rights did not lead to gross mismanagement of Bhutan's forests. However, increasing economic growth may jeopardize the government's commitment to protecting 60% forest cover, although forest cover has been more than 71% for decades (Yangka et al., 2018). The forest laws were essentially a nationalization of forests, which led to the loss of community and traditional rights. The result was reduced incentives at the local level for protecting the forests and non-compliant behavior. This is a way to seek a balance between economic development with other objectives and the government's goal of maximizing GNH and ensuring the enactment of the laws for future generations.

The environmental policies and laws initially evolved under the central governments and later expanded through donor support and NGOs and stakeholders at the community level to a much lesser degree. Communities have been involved and engaged in managing natural resources for centuries. This way, "natural resources were controlled, appropriated, and managed through locally evolved norms, indigenous practices, and institutions. With the increasing involvement of the state in modern times, natural resources have increasingly been governed through policies and laws made centrally" (Ojha et al., 2019).

The challenge facing Bhutan is to achieve the right balance between economic development and carbon neutrality while increasing the degree of economic self-sufficiency. Most importantly, the country must decouple economic growth from GHG emissions, which will be possible through investments in climate-friendly technology, public awareness, international financing, and capacity building. A people-centered approach is to leverage existing networks and create connections to ensure broad-based participation, also by the poor and underprivileged, in decision-making and policy dialogue. Networks in community-based health already exist, but a lack of horizontal links across other networks, in many cases, leads to duplication and ineffectiveness in planning. A critical challenge among the agents involved in providing health, education, water, and sanitation and those involved in environment and climate change is that there are no organizational linkages amongst those stakeholders, and most work is carried out in isolation, resulting in duplication and overlap of similar programs (Brassard, 2017, p. 195; Schmidt & Christensen, 2017).

The problem is, not unlike challenges related to the implementation of GNH itself and education and health, that climate policies are genuine, ambitious, and progressive, but the lack of implementation illuminates institutional inertia, donor dependency, and asymmetry between stated goals and vertical and horizontal links between policies and implementation strategies (Schmidt, 2017; Schmidt & Christensen, 2017; Schmidt & Yezer, 2021).

Problems related to human resources and capacity are significant constraints on effective inter-agency coordination for implementing climate policies. The evidence shows that mainstreaming at the policy level has been relatively successful, but that the gaps between policy formulation, planning, and implementation occur in some cases, and there are challenges to the vertical level of the policy-making process.

Assessments were conducted to identify critical constraints and gaps in implementing climate change activities. The problems are related to a lack of financial resources, technology, research, development, and human capacity. The legal framework is insufficient, and there is a lack of coordination and integration and the absence of detailed research. These constraints and barriers connect with legal and policy enforcement and the planning of climate change actions. Lack of quality research on climate change projections, its potential impacts, and consequences of the implementation of climate change response actions. Availability of financial resources to address climate change challenges. Adequacy of technical and technological support and knowledge. Capacity-building support or climate change action. Mainstreaming and integrating climate change issues into national, subnational, and sectoral development programs and plans; Coordination among the government ministries and agencies, and the international financing and project implementing entities (Royal Government of Bhutan, 2021, pp. 121–122).

This asymmetrical distribution of power and authority across different levels of governance is quite standard in developing countries. In many, if not most cases, the government and central bureaucracy formulate policies, conduct planning, and sign laws without consulting local levels, even though decentralization has been a hallmark of Bhutan's development strategy for many years.

> The currently dominant normative paradigm of governance emphasizes the state as the legitimate site of authority; unless this authority is delegated or enacted through polycentric institutions, there is a tendency to overlook the importance of decision making at lower scales of governance. However, both local and national jurisdictions are difficult to exercise, as local- and national-level environmental challenges become increasingly intertwined.
>
> (Ojha et al., 2019, p. 568; Rosenau et al., 2004)

These tendencies are similar to the Bhutanese government's assessments.

This way, policymakers are aware of the urgency to act on adaptation but face substantive challenges. Some of those challenges are a lack of adequate data (both in terms of quantity and quality, and especially at a localized scale) about climate change impacts and lack of human and financial resources both with the data sources and the compilation team (Royal Government of Bhutan, 2021: 6 and 121), weak institutional capacity at various governance levels, social and economic barriers to intervention uptake, and poor infrastructure for development and adaptation purposes (Mishra et al., 2019).

As Ojha et al. (2019) discussed, the implementation of existing policies and legislation remains ineffective. Environmental policy implementation will improve if the government recognizes regulation and governance's multi-sectorial and cross-scalar nature. It depends on the engagement of civil society and perhaps the private sector when relevant and includes local communities. There is a persistent lack of coordination among actors/both government and non-government) and it amounts to significant challenges. As Brassard points out, "Unless communities are cooperatively empowered to be part of the governance system that can effectively address some of the negative impacts of development, their resilience will remain undermined" (Brassard, 2017, p. 171). Also, WHO is clear about the problems related to effectiveness in Bhutan: "Insufficient national capacity in terms of human and financial resources for incorporating climate change risks into all levels of health sector activities, and national climate change plans and processes" (WHO, 2021).

Another aspect of the policy implementation gap is the donor dependence of Bhutan on external assistance for its development needs. Donors are assisting these countries in developing specific policies and legal instruments. Still, they are not adequately followed through for several reasons related to relevance and inadequate institutional, human, and financial capacity (Royal Government of Bhutan, 2021, p. 55).

Concluding remarks

The World Bank is acknowledging that Bhutan has become a lower-middle-income country and has cut poverty by two-thirds in the last decade. Gross National Income (GNI) per capita, at $3,080 in 2018, is now three times the threshold for lower-middle-income countries and only 10% below the threshold for upper-middle-income countries, indicating that many advantages of attracting donor funding are disappearing (World Bank, 2021). This may be a double-edged sword for the country since Bhutan has benefitted tremendously from its LDC status and has been one of the "darlings" of the global development community.

The problem is three-fold. Now the government can attract aid and loans from abroad, not least from India due to geostrategic reasons – read China – and the international donors. Bilaterally and multilaterally. However, with continued economic growth, the country's graduation from being an LDC may endanger its ability to attract donor grants and favorable loans. This may have huge repercussions on mitigation of climate change and protecting the environment, agriculture, forests, and promoting the livelihoods of poor people.

Another conclusion relates to Bhutan's reliance on hydropower as the primary foreign exchange earner and future replacement of foreign development aid. There are several challenges ahead and indeed unknowns as well.

One is the dependency on India's benevolence and payment below world market prices and even below India's market price for electricity from Bhutan. Another is the upcoming debt trap which appears to contradict the country's major development strategies since the goal of a certain degree of self-reliance appears to be important. A third relates to the strange fact that Bhutan has not prepared an overall cost estimation necessary to preserve carbon neutrality. As the evidence suggests, there are also challenges due to lack of knowledge, human resources, and financial constraints (Yangka et al., 2018).

A final remark relates to the problem that one of the reasons why Bhutan is being regarded as a success story by the international development aid circles is because of its developmental state capacities and the political will of the governments and bureaucracy, which is a huge advantage (Schmidt & Yezer, 2021). Nevertheless, there appear to be problems with overlaps, lack of synergies, ineffectiveness in mitigation of climate, and adaptation of climate change and environmental governance. Finally, the critical political ecology perspective has been utilized to shed light on the farmers' livelihood strategies and the challenges the agricultural sector is facing, which may result in declining agricultural production, growing scarcity of water resources, and deterioration of livelihoods and human health.

References

ADB and WB. (2021). *Climate Risk*. Country Profile Bhutan, Washington DC sand Manila. https://www.adb.org/sites/default/files/publication/722636/climate-risk-country-profile-bhutan.pdf

Adhikari, I. P. (2020). Climate change impact in Bhutan. *The Bhutan Journal*, 1, 35–50.

Bernstein, H. (2010). *Class Dynamics of Agrarian Change*. Fernwood, Halifax.

Bhutan Times. (2021, 13 September) Climate change and food security, Bhutan Times. https://bhutantimes.bt/index.php/2021/09/13/climate-change-and-food-insecurity/

Biersack, A. (2006). Reimagining political ecology: Culture/power/history/nature. In A. Biersack, & J. B. Greenberg (Eds.), *Reimagining Political Ecology* (pp. 3–40). Durham: Duke University Press.

Brassard, C. (2017). Disaster governance, inequality and poverty alleviation in Bhutan: Towards integrated and preventive policies. In J. D. Schmidt (Ed.), *Development Challenges in Bhutan* (pp. 171–201). Springer, Cham.

Brooks, J. S. (2013). Avoiding the limits to growth: Gross national happiness in Bhutan as a model for sustainable development. *Sustainability*, 5(9), 3640–3664. https://doi.org/10.3390/su5093640

Bryant, R. (2001). Political ecology: A critical agenda for change? In N. Castree & B. Braun (Eds.), Social nature: Theory, practice and politics (pp. 151–169). London: Blackwell.

Busck, O., & Schmidt, J. D. (2020). Development, ecology and climate change: Resistance by the peasantry. *Journal of Development Policy and Practice*, 5(1), 9–31. https://doi.org/10.1177/2455133320909912

Chhogyel, N., & Kumar, L. (2018). Climate change and potential impacts on agriculture in Bhutan: A discussion of pertinent issues. *Agriculture & Food Security*, 7, 79. https://doi.org/10.1186/s40066-018-0229-6

Chhogyel, N., Kumar, L., & Bajgai, Y. (2020). Consequences of climate change impacts and incidences of extreme weather events in relation to crop production in Bhutan. *Sustainability*, 12(10), 4319.

Constitution of Bhutan. (2008). *The constitution of the kingdom of Bhutan*. https://www.nab.gov.bt/assets/templates/images/constitution-of-bhutan-2008.pdf

Dorji, L., Webb, E. L., & Shivakoti, G. P. (2006). Forest property rights under nationalized forest management in Bhutan. *Environmental Conservation*, 33(2), 141–147. http://dx.doi.org/10.1017/S0376892906002979

Dukpa, W., & Wangdi, J. T. (2021). Situation of family farming in Bhutan, In R. B. Shrestha, M. E. Penunia, P. Ferrand, M. Dave, & Y. Ali (Eds.), *United Nations Decade of Family Farming 2019–2028: Regional Action Plan to Implement the UNDFF and Achieve the SDGs in South Asia*. Dhaka: SAARC Agriculture Center, Dhaka, Bangladesh; Asian Farmers' Association (AFA), the Philippines; International Cooperative Alliance Asia and Pacific (ICA-AP), India; and FAO, Rome.

Gautam, B. R., Li, F., & Ru, G. (2014). Climate change risk for hydropower schemes in Himalayan Region. *Environmental Science & Technology*, 48(14), 7702–7703. https://doi.org/10.1021/es502719t

GoB. (2020). *Climate Change Policy of the Kingdom of Bhutan 2020, National Environment Commission*. Royal Government of Bhutan. https://www.gnhc.gov.bt/en/wp-content/uploads/2020/02/Climate-Change-Policy.pdf

Harvey, D. (2003). *The New Imperialism*. Oxford: Oxford University Press.

Hoy, A., & Katel, O. (2019). Status of climate change and implications to ecology and community livelihoods in the Bhutan Himalaya. In A. Saikia, & P. Thapa (Eds.), *Environmental Change in the Himalayan Region* (pp. 23–45). Springer, Cham.

Hoy, A., Katel, O., Thapa, P., Dendup, N., & Matschullat, J. (2016). Climatic changes and their impact on socio-economic sectors in the Bhutan Himalayas: An implementation strategy. *Regional Environmental Change*, 16(5), 1401–1415. https://doi.org/10.1007/s10113-015-0868-0

IPCC. (2021). First part of the Sixth Assessment Report, Climate Change 2021: The Physical Science Basis. Working Group – Contribution to the Sixth Assessment Report, https://ipcc.ch/report/sixth-assessment-report-working-group-i

Johnson, F. A., & Hutton, C. W. (2014). Dependence on agriculture and ecosystem services for livelihood in Northeast India and Bhutan: Vulnerability to climate change in the Tropical River Basins of the Upper Brahmaputra. *Climatic Change* 127, 107–121 https://doi-org.zorac.aub.aau.dk/10.1007/s10584-012-0573-7

Mishra, A., Appadurai, A. N., Choudhury, D., Regmi, B. R., Kelkar, U., Alam, M., Chaudhary, P., Mu, S. S., Ahmed, A. U., Lotia, H., & Fu, C. (2019). Adaptation to climate change in the Hindu Kush Himalaya: Stronger action urgently needed. In P. Wester, A. Mishra, A. Mukherji, & A. B. Shrestha (Eds.), *The Hindu Kush Himalaya Assessment* (pp. 457–490). Springer, Cham. https://doi.org/10.1007/978-3-319-92288-1_13

National Environment Commission. (1998). *The Middle Path*. Royal Government of Bhutan, Thimphu. https://www.thegef.org/sites/default/files/ncsa-documents/Middle_Path.pdf

National Environment Commission. (2016). *Kingdom of Bhutan Intended Nationally Determined Contribution*. Royal Government of Bhutan. https://www4.unfccc.int/sites/ndcstaging/PublishedDocuments/Bhutan%20First/Bhutan-INDC-20150930.pdf

National Statistics Bureau. (2021). *National Accounts Statistics 2020*, Thimphu. https://nsb.gov.bt/wp-content/uploads/dlm_uploads/2020/11/National-Accounts-Report-2020.pdf

Ojha, H. R., Ghate, R., Dorji, L., Shrestha, A., Paudel, D., Nightingale, A., Shrestha, K., Watto, M. A., & Kotru, R. (2019). Governance: Key for environmental sustainability in the Hindu Kush Himalaya. In P. Wester, A. Mishra, A. Mukherji, & A. B. Shrestha (Eds.), *The Hindu Kush Himalaya Assessment* (pp. 545–578). Springer, Cham.

Ranjan, A. (2018). *India-Bhutan Hydropower Projects: Cooperation and Concerns.* ISAS Working Paper, No. 309, 17 October, ISAS, NUS, Singapore https://www.isas.nus.edu.sg/wp-content/uploads/2018/10/Working-Paper-No.-309-India-Bhutan-Hydropower-Projects.pdf

Rosenau, J. N., von Weizsäcker, E., & Petschow, U. (Eds.). (2004). *Governance and Sustainability. Exploring the Roadmap to Sustainability After Johannesburg.* London: Greenleaf Publishing.

Royal Government of Bhutan. (2021). Third national communication to the UNFCCC. https://unfccc.int/sites/default/files/resource/TNC%20of%20Bhutan%202020.pdf

Sabin, T. P., Krishnan, R., Vellore, R., Priya, P., Borgaonkar, H. P., Singh, B. B., & Sagar, A. (2020). Climate change over the Himalayas. In R. Krishnan, J. Sanjay, C. Gnanaseelan, M. Mujumdar, A. Kulkarni, & S. Chakraborty (Eds.), *Assessment of Climate Change Over the Indian Region* (pp. 207–222). Springer, Singapore.

Schmidt, J. D. (2017). Donor-assisted ethno-nationalism and education policy in Bhutan. In J. D. Schmidt (Ed.), *Development Challenges in Bhutan: Perspectives on Inequality and Gross National Happiness* (pp. 29–46). Springer. Contemporary South Asian Studies, Cham. https://doi.org/10.1007/978-3-319-47925-5_3

Schmidt, J. D., & Christensen, L. K. (2017). Policy synergies in health-promoting education in Bhutan. In J. D. Schmidt (Ed.), *Development Challenges in Bhutan: Perspectives on Inequality and Gross National Happiness* (pp. 235–256). Springer. Contemporary South Asian Studies, Cham.

Schmidt, J.D., & Yezer, Y. (2021). 'Free education': Is it sustainable. In J. D. Schmidt, & Y. Yezer (Eds.), *Education Policies, the Developmental State and GNH in Bhutan: Perspectives from Government and Schools.* Studera Press. (pp. 89–105).

Tariq, M. A. U. R., Wangchuk, K., & Muttil, N. (2021). A critical review of water resources and their management in Bhutan. *Hydrology*, 8(1), 31. https://doi.org/10.3390/hydrology8010031

Tenzin, J., Phuntsho, L., & Lakey, L. (2019). Climate smart agriculture: Adaptation and mitigation strategies to climate change in Bhutan. In R. B. Shrestha, & S. M. Bokhtiar (Eds.), *Climate Smart Agriculture: Strategies to Respond Climate Change in South Asia*, SAARC Agriculture Centre, Dhaka, 180.

Tenzin, J., Phuntsho, L., & Lakey, L. (2019). Climate smart agriculture: Adaptation & mitigation strategies to climate change in Bhutan. In B. Shrestha and S. M. Bokhtiar (Eds.), *Climate Smart Agriculture: Strategies to Respond to Climate Change* (pp. 37–61). SAARC Agriculture Centre.

WHO. (2021). *Climate Change Adaptation to Protect Human Health.* WHO, UNDP & GEF https://www.who.int/globalchange/projects/adaptation/PHE-adaptation-final-Bhutan.pdf

World Bank. (2019a). *Bhutan Forest Note Pathways for Sustainable Forest Management and Socio -equitable Economic Development.* Washington, DC. https://documents1.worldbank.org/curated/en/118821562700584327/pdf/Bhutan-Forest-Note-Pathways-for-Sustainable-Forest-Management-and-Socio-equitable-Economic-Development.pdf

World Bank. (2019b). *Bhutan. Development Report. A Path to Inclusive and Sustainable Development*. Washington DC, January. https://documents1.worldbank.org/curated/en/259671548449315325/pdf/134060-WP-PUBLIC-25-1-2019-13-34-54-BDRJanuary.pdf

World Bank. (2021). *The World Bank in Bhutan. An Overview*. World Bank. https://www.worldbank.org/en/country/bhutan/overview

Yangka, D., Rauland, V., & Newman, P. (2019). Carbon neutral policy in action: The case of Bhutan. *Climate Policy*, 19(6), 672–687. https://doi.org/10.1080/14693062.2018.1551187

12 Localized policy responses for climate change and developmental complications

Discussions from three coastal regions of India

Devendraraj Madhanagopal, Lisha Samuel and Kasturi Gandhi

Introduction: the multiple vulnerabilities of coastal India

Marine fishing communities around the world are intensely subjected to the brunt effects of development-related conflicts, environmental degradation, and climate change (Garcia et al., 2018). Research across the globe, time and again, has proved that climate change has severe implications on the lives and livelihoods of small-scale marine fishing communities who are already marginalized and are at the vulnerable and deprived strata of the fishers (Barange et al., 2018; Andrews et al., 2021). India, one of the largest maritime nations in South Asia, has a coastline of around 8,118 kmthat touches nine states and four union territories, including islands. The coastal regions of India are bounded by the Bay of Bengal, the Arabian Sea, the Indian Ocean, and the union territories of Andaman and Nicobar, Lakshadweep, Daman, and Diu Islands (Nandakumar & Nalini, 2010; Gopi et al., 2019). Though the coast length of India encompasses less than 02.5% of the entire world's coast length, it is home to around 63 million people. About 47% of the Indian population resides in these coastal states and the Union territories; 73 coastal districts are home to approximately 17% of the total Indian population. Apart from that, about 250 million people live within 50 km of the coastline (ICZMPWB, n.d.-b). Based on the 2011 census, the Central Water Commission (2016) has noted that the coastal districts are home to around 188 million people, and the island territories are home to about 0.44 million people. The coasts of India have a rich history of acting as the focal points of trade and commerce (Behera, 1999), and this continues in the contemporary period. For example, around 95% of trade (in volume) and about 72% of the trade (in value) happens through these coastal regions. It shows that India, essentially a maritime nation, is heavily dependent on the sea, and by extension its coastal states, for the entire growth of the nation (Dhowan, 2021). Due to its geographical nature, cyclones frequently occur in the Bay of Bengal from October to January. Hence, the east coast is more vulnerable to extreme wave conditions due to severe tropical cyclones than the west coast

DOI: 10.4324/9781003216476-15

(Kumar et al., 2006). National Cyclone Risk Mitigation Project (NCRMP) of the Government of India has recently identified that Andhra Pradesh, Tamil Nadu, Odisha, West Bengal, and Gujarat are the five coastal states that are highly vulnerable to cyclone risks due to the frequency of cyclone occurrences, population size, and the existing institutional mechanism for disaster management (NDMA, 2015). Research shows that given the numbers of the exposed population, two cities in India, Mumbai, and Kolkata, will be highly vulnerable to coastal flooding in 2070. Both cities are ranked as the top two cities in Asia (Fuchs, 2010). In recent decades, India's coastline has experienced severe ill effects of environmentally destructive 'development' projects, tourism industries, and other industrial activities, which have done irreversible damages to the marine ecosystems, fishery livelihoods, and living spaces of marine fishing communities, especially the small-scale fishing communities of India (Down to Earth, 2008; Chouhan et al., 2016; Parthasarathy & Chouhan, 2020; Kadri, 2020; Maher, 2021). To respond to all these unprecedented climate change challenges and coastal complications, India needs effective coastal laws driven by inclusiveness. Whereas existing loopholes in the laws and actions of the state have enabled the authorities to open coastal spaces for multiple infrastructure and development-related projects without due and proper consultation of the major coastal stakeholders, who are marine fisherfolk. Such opening up of fragile inter-tidal areas and rural areas of the coastal spaces for these projects has put the entire coastal environment and millions of coastal people at risk of climate change and coastal hazards. This is particularly bringing more risks to the traditional marine fishing communities as their lives and livelihoods are entirely intertwined with coastal spaces and resources. This chapter, focusing on Tamil Nadu, West Bengal, and Maharashtra, India's three vulnerable and contested coastal states, reviews a few of those loopholes and offers critical insights into the underpinned politics of coastal laws. Using both primary and secondary data backed by two recent case studies from Maharashtra, this chapter discusses the challenges of many coastal stakeholders while seeking intervention and redressal for climate change and development-related complications in the regions they inhabit. In conclusion, this chapter advocates policy recommendations for prioritizing the most vulnerable, addressing coastal region complications better, and bolstering climate-change preparedness among local stakeholders.

Laws and policies governing coastal India: history, influence, and politics

The protection of coasts and coastal resources of India has always been multifaceted and has several interpretations from different stakeholders. Different ministries, state governments, and multiple stakeholders have been engaged in protecting coasts, and it has always been a contentious one. The Ministry of Environment, Forests and Climate Change of the Union Government of India plays one of the key roles in governing and regulating the coasts, coastal zones, and associated regions. Along with that, several international organizations, including consultancies (hired by the state), play importance roles in coastal policy making of India. Realizing the need to regulate and govern the coasts and coastal resources from multiple vulnerabilities, the Ministry of Environment and Forests (MoEF) of the Union government of India issued the first notification in 1991 (hereafter CRZ[1] 1991) under the Environmental Protection Act, 1986. This notification has classified the coastal zones into four types: CRZ – I, CRZ – II, CRZ – III, and CRZ – IV. This notification has prevented and regulated development activities, including establishing industries in the coastal zones. Given the various demands of various other stakeholders, including the state governments, union ministries, and NGOs, this 1991 notification was amended about 25 times (ICZMPWB, n.d.-b; Sundararaju, 2019). This was the first coastal law in India, and it has had a problematic existence since its implementation. As reported by Dias & Sridhar (2021), throughout the 1991–2011 period, many cases were filed against the violators of this CRZ Notification; however, there were no comprehensive actions or steps to curb the same. CRZ 1991 was superseded twice, in this 2011 notification was superseded by the and recently in 2019. The CRZ 2011 has had many objectives, geared toward, and plans for the protection of the livelihoods of coastal fisherfolk, coastal ecology, and the promotion of economic activity. In Clause 6, CRZ 2011 has instructed the state government to form the district level committees and to include the local representatives, particularly coastal fisherfolk, tribals, and other local communities, in enforcing the provisions of the CRZ 2011 Notification. Like this, it has envisaged the ways to regularize the unauthorized dwellings of the coastal fisherfolk under the permissions of CRZ 2011 so that they get formal approval from the state, given the norms and procedures of the state (CRZ, 2011, pp. 8–9). To promote sustainable use of marine resources and coastal conservation in ecologically sensitive regions such as Sunderbans, Chilika, Gulf of Khambhat, Gulf of Kachchh, Karwar, Coondapur, Vembanad, Gulf of Mannar, Bhitarkarnika, East Godavari, and Krishna, this CRZ 2011 has envisaged the participation of the local communities (particularly, fisherfolk) through a consultation process (Government of India, 2011. p. 17). CRZ 2011, however, is also not free of controversies. It has received criticism from experts, environmental groups, and civil society (Sharma, 2011; Kumar et al., 2014; Correa, 2016). Puthucherril (2011) has pointed out the flaws of this law, and highlighted that this law is too weak in confronting the challenges of climate change and sea level rise. As a result, a committee was appointed by the Ministry of Environment, Forests, and Climate Change under Dr. Shailesh Nayak to review the recommendations, suggestions, and concerns of various stakeholders of the coastal regions related to the CRZ 2011. The result of this committee was the Coastal Regulation Zone Notification of 2018, notified by the Union Government of India and hosted on the Ministry of Environment, Forest, and Climate Change website on 18 April 2018. The Union government of India considered the suggestions and objections on the CRZ 2018, and the Union Cabinet finally approved it in December 2018, and it was published in The Gazette of India on 18 January 2019 (hereafter CRZ 2019) (Press Information Bureau, 2018; The Gazette of India, 2019b). There have been widespread criticisms and protests against CRZ 2019 across coastal India, reported in national news outlets (Kukreti, 2019; Menon & Kohli, 2019). It has been reported that the state ignored 90% of the objections to the notification, which came from stakeholders including fishing collectives, environmental groups, activists, and citizens across India. Some of the prime concerns raised by them are as follows: i) The only stakeholders consulted by the Union Government of India were the state governments and the Union Territories, but not the others, especially coastal communities, including fisherfolk. It will threaten the lives of millions of poor fishers' communities who have historically resided along the coastal zones and expose them more to climate risks and coastal hazards. i) Easing strict regulations and lifting development restrictions along the coastal zones will intensify the construction and associated activities in urban coastal areas, which would be disastrous for the coastal ecology and traditional fishing communities that reside and work in those regions. iii) The new CRZ 2019 notification reduced the boundary limit in previously protected no-development zones (NDZ) in rural areas and NDZ along tidal-influenced water bodies, which will pave the way for more real estate, hotels, and beach resorts in ecologically sensitive coastal areas. (Kapoor, 2020; Dias & Sridhar, 2021).

Besides the CRZ Notifications, the Government of India launched another initiative with support from the World Bank, called the Integrated Coastal Zone Management Project (ICZMP). To effectively implement the various provisions of CRZ 2011 and IPZ 2011[2] and to protect the coasts from various environmental impacts, including climate change, and to promote the livelihoods of the coastal fishing communities, the Government of India has initiated the Integrated Coastal Zone Management Project (ICZMP) with the support from the World Bank. The total cost of this ICZMP is USD 178,929,383.50, and its listed objectives are "to assist Government of India (GoI) in building national capacity for implementation of a comprehensive coastal management approach in the country and piloting the integrated coastal zone management approach in states of Gujarat, Orissa and West Bengal". These three states were chosen for this project to represent both the west and east coasts of India. Besides that, the project sites of these states were selected based on the challenges of coastal zone management and the levels of development of the states (The World Bank, n.d.; ICZMP, 2021). Finally, another policy that has deep connections with coastal protection matters in India is the country's policy on climate change. On 13 July 2007, the Prime Minister's Council on Climate Change propounded a National Document discussing the challenges of Climate Change (NICRA, p. 12), which eventually resulted in the promulgation of the National Action Plan on Climate Change (NAPCC) on 30 June 2008. This national action plan addressed eight national missions: National Solar Mission, National Mission on Sustainable Habitat, National Mission for Enhanced Energy Efficiency, National Water Mission, National Mission for Sustaining the Himalayan Ecosystem, National Mission for a Green India, National Mission for Sustainable Agriculture, and National Mission on Strategic Knowledge for Climate Change (Government of India, 2009, p. 5–7; Press Information Bureau, 2021). The Union Government of India directed all the states to develop state action plan–on climate change, aligned with the NAPCC; state climate change action plans were to outline how they would implement the objectives of the NAPCC. As of 2019, 32 states of the Indian Union have formulated and begun implementing their own state-level climate change action plans (Saran, 2019).

Climate change and coastal complications: focus on coastal Tamil Nadu

Tamil Nadu, the southernmost state of India, regularly experiences coastal flooding and extreme storm surges. Coastal Tamil Nadu has been affected by cyclonic storms about 30 times with disastrous effects from 1900 to 2004 (Sundar & Sundaravadivelu, 2005). Tamil Nadu has a coastline of about 1,076 km, ranking the second-longest coastline among all the Indian states next to Gujarat. Tamil Nadu has 13 coastal districts and 591 marine fishing villages with a total marine fishing population of about 0.861 million. Around 0.26 million fishermen are actively engaged in fishing (Government of Tamil Nadu, 2012). Within Tamil Nadu, the southeastern coastal Tamil Nadu faces severe threats due to rapid changes in geology and geomorphology, sea-level change, tropical cyclones, and storm surges (Sheik Mujabar & Chandrasekar, 2011), and hence, the marine fishers along this coast are the direct and indirect victims of these adverse effects. A large proportion of Southeastern coastal Tamil Nadu was devastatingly affected by the 2004 Indian Ocean Tsunami disaster. Within Tamil Nadu, Nagapattinam district was the worst affected district in the Indian mainland by the 2004 Indian Ocean Tsunami. This disaster brought several losses and long-term adverse effects on the infrastructure, water resources, agricultural land, fishing assets, and coastal resources of this district (Sheth et al., 2006; Chandrasekharan et al., 2008; Kume et al., 2009). Babu (2011) has argued that in the aftermath of the tsunami disaster, private parties started playing critical roles in post-tsunami reconstruction and rehabilitation activities with the mere regulatory (weak) functions of the state. Since then, coastal Tamil Nadu has become an attractive zone for large-scale investments to construct ports, harbors, thermal power plants, and petrochemical industries (Babu, 2011; Swamy, 2011). In recent decades, the marine fisheries sector of Tamil Nadu has continuously been affected due to the discharge of industrial sewage and effluents, and it severely affects the marine resources along the coastal stretch (FIMSUL, 2011). The government of Tamil Nadu has already recognized the growing climate change threats across their coasts and needs to integrate climate change concerns with the coastal laws. Tamil Nadu is one of the frontrunners of climate change planning in the Indian subcontinent. The first draft climate change action plan was formulated in 2014 in line with the NAPCC of the Union government of India and identified seven sectors to be focused, which are as follows: 1) Sustainable Agriculture, 2) Water Resources, 3) Coastal Area Management, 4) Forest and Biodiversity, 5) Energy Efficiency, Renewable Energy and Solar Mission, 6) Sustainable Habitat, and 7) Knowledge Management. There have been numerous policies, projects, programs, and authorities at the state level to protect the coasts from climate change, coastal hazards, and other developmental and environmental complications. They are as follows: 1) Integrated Coastal Zone Management Plan for Tamil Nadu, 2) National Assessment of Shoreline Change for India, 3) Disaster Risk Management Program, 4) NCRMP, 5) National Coastal Protection Project, 6) Coastal Area Management Project of Tamil Nadu Forest Department, 7) Gulf of Mannar Biosphere Reserve (GoMBR), and 8) Biodiversity Conservation. (MoEF&CC Government of India, 2019). However, there has been substantial opposition among the traditional fisherfolk communities of Tamil Nadu toward the recent CRZ 2019 notifications and the implementation of many projects of Integrated Coastal Zone Management Plan for Tamil Nadu (The Times of India, 2019; The New Indian Express, 2022).

Dias & Sridhar (2021) have provided a detailed report on how the Coastal Zone Management Plan, which is a mandatory one for every coastal state, is loosely being prepared by the Tamil Nadu coastal authorities and how the incomplete and inaccurate maps that are prepared by those authorities are helpful to the project proponents to move ahead with their plans intact. In many cases, as the authors reported, there were no proper consultations made with the local communities, including fishers. The poorly conducted state coastal zone management authority meetings and the lack of a separate independent SCZMA website for Tamil Nadu made the required information on the violations of CRZ norms and guidelines, minutes of the conducted meetings, and the decisions taken in the meetings hard to locate (Menon et al., 2015; Dias & Sridhar, 2021). There have been various reports of media outlets on how the new CRZ 2019[3] diluted the public consultation processes, jeopardizing the coastal environment and livelihoods of the millions of marginal fishing communities and making the coastal areas as commercial and investment hubs (Ashok, 2018; India Today, 2018; Lakshmanan, 2021). The responses of the major coastal stakeholders, the fishing communities to the CRZ draft plans and CRZ 2019 notifications have been strong across coastal India, particularly, in Tamil Nadu (DNA India, 2018; The Hindu, 2018; Neelambaran, 2019). Similar to this, experts and environmental activists across various spectrums vigorously objected to the CRZ regulation on public platforms, including courts (India Today, 2018; Krishna Chaitanya, 2018; Dodhiya, 2021).

Climate change and coastal complications: focus on coastal Maharashtra with case studies

Located on India's western coast, Maharashtra is India's third largest state area wise, geographically characterized by plateaus, islands, marine and inland forests, harbors, creeks, and a 720-km-long coastline (Maps of India, n.d.). Among its total population of over 124 million are indigenous and migrant people. Due to its identity as one of India's foremost ports, the state has always attracted a high volume of cross-border labor, trade, and industries (Government of Maharashtra, 2019). Similarly, owing to its location—flanked by the Arabian Sea to the west—generations of fisherfolk communities are settled along the region's coast and are considered among the state's earliest inhabitants (Hegde, 2015). There is great disparity in the region's socio-economic conditions (Hatekar & Raju, 2013), and so too in its climate experiences; while parts close to the coast experience annual torrential flooding, those further inland battle with recurring and devastating drought (Ranjan, 2020). When it comes to climate change issues in the state, there is ample research evidence supporting the fact that marine fisherfolk populations are among those most disproportionately affected in Maharashtra, and have been for a long time (Salagrama, 2012a; Salagrama, 2012b; Kupekar & Kulkarni, 2013; Yadav et al., 2020) Marine fisherfolk of Maharashtra are chiefly constituted of the indigenous Koli communities, i.e., the Agris, Son, Malhar, Mahadev, and Dhor Koli subsects (Hegde, 2015). Owing to their long-standing connection with the landscape, they have a socially and politically distinct identity as a notified scheduled Tribe. Many of them still follow their traditional manner of fishing, which has earned them the title 'artisanal fishers' (Chouhan et al., 2016). There are over 2 million marine fishers in Maharashtra (Government of India, 2020) and out of these, about half are Kolis—accounting for under 1% of the state's total population (Census India, 2011b). Their physical location along the coast, however, makes them the most vulnerable to both natural and manmade disasters, which we will explore in this section through case studies from Maharashtra's Raigad district. The rationale for choosing this coastal district for this chapter is that it has been a venue for climate change issues exacerbated by extreme changes in land use, seasonality in employment, reduction in fish catch and daily income, food insecurity, dispossession, migration, and policy shortcomings. On the frontier of these issues are Raigad district's Koliwadas—villages and traditional fish processing grounds of the Koli communities (CMFRI, 2020). The Kolis' plight offers a valuable and human-centric lens to understand the state's climate change issues and policy shortcomings. Selected case studies from these regions seek to shed light on what's missing from the state's climate change policy portfolio, comprising the Coastal Regulation Zone Notifications (1991, 2011, 2019) and the Maharashtra State Adaptation Action Plan for Climate Change (MSAAPC).

Case 1:   Raigad District in Maharashtra

Spanning 7,152 square km and with a population of over 26 million, Raigad district is located along the state's western coast (Census India, 2011b), bordering the Arabian sea and several creeks. It is home to over

60,000 fisherfolk across 16,619 households, with most belonging to the Agri-Koli subsect (Singh et al., 2010). Between 2020 and 2021, in addition to the havoc wreaked by the Covid-19 pandemic[4] two cyclones—Nisarga (June 2020) and Tauktae (May 2021)—made landfall in Raigad, ravaging its coastal landscape (Phadke, 2020; The Times of India, 2021). Soon after in July 2021, the district received torrential rain for weeks; flood levels reached up to 25 feet, leading to widespread death and destruction (India.Com, 2021). These recent disasters added greatly to the coastal district's festering ecological and economic wounds caused over the past many decades, starting in the 1970s.

Displacement and dispossession

In 1971, the northern region of Raigad was chosen for the construction of a new city, Navi Mumbai, built to redistribute some of Mumbai city's population and industrial infrastructure further inland. The new city, of about 100 square km, was replete with a new terminal port, townships, railway lines, highways, and coastal roads. 15,000 hectares consisting of fields, villages, creeks, wetlands, and mangroves—landscapes which were erstwhile fishing grounds and residences for the fisherfolk—were reclaimed to build it (Shaw, 2004).[5] Later in 1988, 2,584 hectares of Raigad's Sheva island on the southern tip of Navi Mumbai were acquired to build a terminal port owned by the Jawaharlal Nehru Port Trust (JNPT). JNPT is owned by the Government of India, this port is India's largest container handling port and manages around 55% of the country's containerized cargo (Mazumdar, 2015). The Environmental Impact Analysis (EIA) carried out for the project stated the following. I) The ecological balance of the region had already been disturbed by the agricultural and salt harvesting techniques of the local population, and the pollution caused by the development activities of Navi Mumbai. Ii) Commercial fishing in the harbor was 'non-existent, impractical' and there was 'not much potential' for preserving the area for fishing. If one goes by oral testimonies[6] of the fisherfolk populations of Uran, however, there was in place a thriving fishing economy—a good fishery for up to 30 species of fish, eels, oysters, mussels, prawns, crabs, and lobsters—used and inhabited over generations by Hanuman Koliwada, Panje Koliwada, and residents of Nhava Island.

Owing to the construction of the port, 256 fisherfolk families from Hanuman Koliwada at Sheva were relocated to a transit camp further south in the village of Uran. As Navi Mumbai and the JNPT properties expanded over the years (called Special Economic Zones [SEZs]),[7] so did property values for those who could afford them; but for various indigenous populations, their villages and transit camps became lone clusters running on bare minimum welfare[8]—detached from coastal ecosystems of work due to new roads, highways, and townships, and confined to their villages by concrete walls built to mark property boundaries. Additionally, JNPT guards began to regulate the waters around the port, often hindering access for local fishers.

These problems bear heavily on those project-affected persons (PAPs) relocated from Sheva Island even today. At present, there are around 600 Koli families of Hanuman Koliwada living in 227 homes in the transit camp. Although they had been promised 17 hectares by the JNPT during the relocation, they ended up getting only 2 hectares to build their homes, fish-drying grounds, and boat mending quarters. Due to tidal surges, the remaining 15 hectares are now covered with mangroves (Janwalkar, 2021). These mangroves are, in turn, protected from being cleared by the Coastal Regulation Zone Notification. The catch here is that in the neighboring Koliwadas of Uran, CRZ rules have been circumvented, and hectares of coastal ecosystems (including ancient mangrove forests and wetlands) have been reclaimed for the expansion of Navi Mumbai and its SEZs. Fisherfolk at the Hanuman Koliwada transit camp have been fighting a 30-year-old resettlement and rehabilitation battle with the JNPT. These fisherfolk once lived and worked at the tip of Nhava, where the JNPT is now located. Not only do the fisherfolk struggle to live in the new village assigned to them due to its small size, but they also find it hard to use the waters now patrolled by JNPT guards. Dwindling fish catch, water pollution, and displacement woes have made life difficult for them, and even recently issued policies for climate change mitigation (covered in the following sections) do not show adequate solutions.[9]

Case 2: Beach Erosion in Nagaon of Maharashtra

Down south from the Hanuman Koliwada transit camp is a coastal fishing village called Nagaon, situated along a sandy beach; it is spread over 280 hectares and has a population of 3,837 people—mostly marine fisherfolk. In 1998, Nagaon started experiencing intense beach erosion. What began with a bit of sand 30 years ago, continues to the present day with larger repercussions for the locals. Rows of coconut trees have collapsed into the tide, and wells and fields are turning saline. Homes that were once built at a safe distance from the high tide line are now mere meters from the Arabian Sea. With cyclones, storm surges, and rising sea levels, the locals face an extremely uncertain future. Not every stakeholder's experience of the erosion, however, has been the same. The government's Oil and Natural Gas Corporation (ONGC) and the Naval Base, situated on this beach, have built heavy breakwater boundaries around their properties, causing intense and continuous erosion for the remaining properties. Generations of affected residents have written to municipal authorities. Letters written by them to the Public Works Department, the Mumbai Maritime Board, and the Office of the Collector of Raigad have been circulated within state government offices but have not helped bring recourse.[10] Policy-wise, one tool that could've helped mitigate the crisis has been neglected since its very inception—the Coastal Zone Management Plan (CZMP). According to the CRZ Notification of 2011, the CZMPs were to classify beaches as 'high eroding', 'medium eroding', or 'low/stable'; the CZMPs prepared in the past two decades, however, bear no such classification. A review of the three sets of CZMPs prepared for the region in the past three decades will, therefore, indicate nothing about the erosion in Nagaon. From the perspective of the people living and working here, the situation is grave.

Projects of national interest

Existing laws pertaining to coastal environments prohibit the scale of reclamation being undertaken in Maharashtra, and there are central and local authorities who share the responsibility of enforcing these mandates (CRZ, 2011). Yet, whenever infrastructure projects in the 'national public interest' are proposed—including the Coastal Road, Shivaji Statue, Navi Mumbai International Airport, Mumbai Trans Harbor Link, Dronagiri, and JNPT in Raigad—the rules aren't merely circumvented but rewritten upon recommendation by these very authorities. More often than not, they are not in favor of the 'public' closest to the project site but are hailed for their benefit to the 'larger public'. For example, in 1997, when the Indian Ministry of Civil Aviation sought a site for a second airport for Mumbai, a piece of land in the Kopar-Panvel area near Navi Mumbai was proposed (Ministry of Civil Aviation, 2007). This site was listed as a wetland in the Maharashtra Wetlands Atlas, and was home to the Gadhi, Kalundri, Taloja, and Kasadi rivers, and the Panvel and Ulwe creeks. As per MRSAC (1994) and MRSAC (1997), it had dense mangroves, creeks, and vegetated mudflats. Kopar-Panvel was home to a robust fishing economy too, spanning over one thousand hectares, five villages, and home to over 5,000 people—many of them Koli fisherfolk (Bhatia, 2014).

In 2007, the Ministry of Civil Aviation granted Navi Mumbai's town planning authority—the City and Industrial Development Corporation (CIDCO)—an in-principle approval to set up a greenfield airport through a Public Private Partnership (PPP) (Narayan, 2007). The only hurdle to this plan was CRZ 1991 (which prohibited the reclamation of ecologically sensitive coastal areas), so CIDCO approached the National Coastal Zone Management Authority (NCZMA) to amend the notification in favor of the airport[11]. In October 2007, the NCZMA obliged, albeit with a list of conditions: CIDCO would have to carry out detailed hydrodynamic, geological, and Environment Impact Assessment (EIA) studies covering issues of sea level rise, loss of mangroves, and deforestation, and conduct public consultations with the Kopar-Panvel locals[12]. CIDCO commissioned the studies to Centre for Environmental Science and Engineering at the Indian Institute of Technology (IIT) Mumbai (Jamwal, 2010). In July 2008, the MoEF approved this course, and on 15 May 2009, amended CRZ 1991 to permit reclamation of CRZ I areas of Navi Mumbai for the airport. It should be noted that public consultations were never held prior to this, and it was merely stated by the ministry that in view of the special circumstances, it is in the public interest to dispense with the notice under Rule 5(3)(a). The terms 'special circumstances' or 'public interest' in the statement issued were neither explained nor justified[13]. It was as late as May 2010 that the Maharashtra Pollution Control Board (MPCB) finally announced a public hearing for the proposed airport. Most locals boycotted it because they hadn't been provided any draft EIA studies for consideration. The MPCB conducted the 'public hearing' with CIDCO officers, consultants, and press representatives, without any 'public' present. After adding observations from the purported 'public hearing', CIDCO handed over the EIA studies to the MoEF in June 2010 (Jamwal, 2010). The studies failed on several counts. I) They did not include the cumulative impacts of the project and its associated activities on the Navi Mumbai coastline. ii) Social Impact Studies were not carried out. iii) They stated that there were only 'degraded mangroves' on site, which was proven false via site visits by the Expert Appraisal Committee (EAC)[14]. The EAC also observed that quarrying the 97-meter-high hill on site would change the hydrology and drainage pattern of the entire sub-region, including the Jawaharlal Nehru Port in the south, and areas in Panvel in the east, and may cause flooding. Against all such observations, the EAC itself recommended the project to the MoEF for an Environmental Clearance (EC) and CRZ Clearance in November 2010[15]. These economic, political, and social sources of India's coastal woes may be effectively addressed by policy, which is not merely geared toward conservation from above, but is able to function effectively at the human level too (Gadgil, 2019). Unfortunately, however, that level of rootedness is rarely seen in policy concerning environmental protection. Before reviewing why these policies fail at the human level, it is important to briefly visit some specific outcomes of Maharashtra's climate change related policies, which span the past three decades.

CRZ, maps, and Maharashtra

As discussed earlier, the foremost policy for coastal regulation dates to 1991; this is the Coastal Regulation Zone Notification, first issued in 1991 and then superseded twice, in 2011 and 2019. A critical outcome of this Notification was to be the Coastal Zone Management Plan, or CZMP—a map meant to demarcate land use along the coast. This included urban or rural areas, ecologically sensitive areas, tide lines, habitats of marine species, and so on. The CZMPs were (in theory) to also classify critically vulnerable coastal areas and eroding stretches of beaches (The Gazette of India, 1991; The Gazette of India, 2011). These maps, it was stated, would be used by project proponents seeking permission/approval to build in coastal areas. According to the third author's field research, und 1991 to 2021, even though the enforcement of the CRZ Notifications has become decentralized, delays (five years for CRZ 1991 and seven for CRZ 2011), flawed data, and inadequate public participation continue to be a norm. Most importantly, Cadastral Maps (in addition to the CZMPs) on a 1:4,000 scale mandated by CRZ 2011, showing detailed land use (which would benefit the Koliwadas and other stakeholders alike), have never been shared with the public. The latest CZMPs (under CRZ 2019), although prepared relatively fast (in February 2020 by the National Centre for Sustainable Coastal Management), were uploaded on the Mumbai Metropolitan Region (MMR) CZMA website for public review but were poorly disseminated among rural coastal stakeholders. In the drafts for the MMR, the following general issues emerged: i) The Hazard Line for several areas in Mumbai was missing. ii) Existing sea links and bridges over water bodies, wetlands, koliwadas, nesting grounds of birds were not demarcated. iii) Bunds and embankments were demarcated as the High Tide Line. iv) Instead of demarcating a 50-meter buffer zone around Mangroves, the buffer zone was demarcated within the mangrove area itself.

At the public hearings for these draft CZMPs under CRZ 2019, held from February to March 2020, a common thread emerged—several stakeholders attending the consultations had not had timely access to the CZMPs, and therefore could not deliver proper feedback. Moreover, the hearings themselves were conducted poorly; delays of up to two hours were common, and representatives of the coastal authorities were not always present to answer questions from stakeholders; this meant that whoever was conducting the meeting instead would merely 'note down' feedback to convey to those in charge. The burden of keeping feedback concise in interest of correct transmission, therefore, fell on the public. For instance, at the public hearing for the 2019 Draft CZMPs for Raigad, a small room was decided upon as the venue, an odd choice against the fact that coastal communities of the entire district would be present, a district undergoing severe coastal issues both ecological and social. Since the size of this room was inadequate, the participants were shifted to a larger room midway through the hearing. Unfortunately, the introduction to the hearing and the presentation explaining the Notification and the CZMPs, took place in the smaller room with half of the people crowding the corridors outside and unable to participate. At the hearings, several indigenous coastal stakeholders demanded to see cadastral maps, which had been promised by CRZ 2011, but were yet absent from visual discourse even in 2020. They also invited the authorities present to visit their villages and collaboratively rectify the inaccuracies of the CZMPs. Whereas comments were noted down at the hearings, whether changes would be made in the final CZMPs was not guaranteed. Later that month on March 24, the state of Maharashtra went into lockdown owing to the Covid-19 pandemic, and till date, the CZMPs under CRZ 2019 remain to be finalized.

It is not only the inaccuracies and silences on the maps themselves, but also the delays, piecemeal dissemination, approvals, amendments, and modifications relating to them that tell us who has territorial power over coastal zones, both in practical and symbolic senses. For the present context of CZMPs and the CRZ Notifications, Harley's idea of 'cartographic silence' may be merged with the similar notion of 'the deafening silence of state' forwarded by Raymond L. Bryant and Sinéad Bailey in their book *Third World Political Ecology* (1997) concerning the economic and political sources of environmental issues in developing countries (Bryant & Bailey, 1997). The "silence", for example, lies in CRZ 1991 spotlighting issues of coastal pollution from industries, yet mentioning nothing of the displacement of indigenous coastal populations in urbanizing Indian geographies like Raigad through the 1970s and 1980s. For example, the first set of CZMPs circulated between 1996 and 1997 showed various landward stretches of the MMR coastline as "blank" and only seaward or waterbody side of the High Tide Line as coastal zones, bore inconsistencies in matters of cartography, scale, and legends across various sheets, and failed to show vital features like roads, highways, housing colonies, commercial sectors, industries, etc. Logically, if such mistakes had been corrected, the CRZ Notification might have been more successful and the CZMPs more reliable. This trajectory, however, is simplistic and disregards underpinnings of the larger political system in which the CRZ, CZMPs, and their stakeholders operate. Given the power disparities among the coastal stakeholders of this region, what sticks in the geography's visual documentation will likely depend upon who 'constructs' it and, furthermore, to what degree rules are followed beyond the documentation will depend on who can garner authority. The flawed and inconsistent visual discourse of the CZMPs then, is not merely a technical glitch, but an intentional practice of cartographic silence present through data gathering, compilation, editing, drafting, printing, and publication. Whereas numerous technical constraints to cartography and representation exist in any general circumstance, in the case of the CZMPs for MMR, their silences have been repeatedly antithetical to the interests of conservation, indigenous communities, and genuine climate change mitigation. To understand this better, we need to look toward what the CZMPs do not demarcate, as 'that which is absent from maps is as much a proper field of enquiry as that which is present' (Harley, 1988[16]). If we take a few steps back and regard the generationally flawed coastal land use maps, it becomes evident that the foundation for coastal planning and protection policy is fundamentally flawed. Until these flaws are corrected, future policies may never achieve desired outcomes.

Maharashtra's climate action plan

The second influential policy for the regulation and planning of the state's coastal spaces is its climate action plan, or the MSAAPC. Taking cue from the NAPCC 2008 (Press Information Bureau, 2021), in 2010, the Government of Maharashtra (GoM) appointed The Energy and Resources Institute in Delhi (TERI) to prepare a plan that would 'address the urgent need to integrate climate change concerns into the State's overall development strategy'. TERI's report included climate projections for three future time slices: the 2030s, 2050s, and 2070s, with the model baseline being the region's average climate between the years 1970 and 2000. To help contextualize these projections against issues specific to Maharashtra, the study prioritized four areas, i.e., hydrology and fresh water sources, agriculture and food systems, coastal areas (marine ecosystem and biodiversity), and livelihood (including migration and conflict).

As opposed to a micro-level analysis of the state's pain points, TERI's reports used a Macro Level Vulnerability Index (MLVI) based on 19 indicators relating to exposure, sensitivity, and adaptive capacity, to identify districts most vulnerable to climate change in Maharashtra. This index yielded the following results: Nandurbar was the most climate change vulnerable district, followed by Dhule and Buldhana districts (TERI, 2014). Using the same scale, Raigad was not identified as vulnerable, and should invite critical analysis on whether our parameters are able to recognize human vulnerabilities adequately.

Environmentalists and meteorologists who read the report and consequent policy further criticized both on several fronts. For starters, the MSAAPC was drafted over eight years, a delay against back-to-back ecological upheavals on the ground, including landslides, thunderstorms, and cyclones (see Government of Maharashtra, 2016). Secondly, it failed to go in depth into the climate hazards acutely felt by coastal districts, together known as the Konkan Region. One of those issues was thunderstorms. From 1979 to 2011, casualties due to lightning strikes have been steadily increasing in the state (especially among male agricultural laborers in the Konkan region), with about 55%, 67%, 60%, and 65% in lightning events, fatalities, injuries, and casualties, respectively, have been recorded in the past ten years (Singh et al., 2017). Yet, the MSAAPC does not even mention lightning. According to Ashok Jaswal, former scientist with the India Meteorological Department (IMD) Pune, "Rising temperature means more evaporation and high moisture content in the atmosphere, which leads to more thunderstorm activity and an increased incidence of lightning"(Jamwal, 2019). The MSAAPC also fails to describe the preparedness and mitigation measures for extreme weather events. According to Parineeta Dandekar, associate coordinator of the South Asian Network on Dams, Rivers and People (SANDRP), "Increased rainfall will lead to heavy flooding, which will have a direct bearing on the state's water infrastructure. But the action plan fails to elaborate upon ways to manage the water infrastructure in times of climate change" (The Wire, 2019). Most importantly, the MSAAPC does not delve into the inequalities which end up exacerbating the climate change effects on already vulnerable populations. While climate change issues seem to be a future concern, several areas in the state battle them at present. As we have seen in this chapter's case studies themselves, for several residents of the Raigad district, the future time slices of the MSAAPC are irrelevant, and the onset of climate change chaos is a present-day reality. Unfortunately, people's experiences from this time—day-to-day dysfunctions relating to ecology and politics—have not found much representation in the region's climate change policy.

West Bengal

West Bengal is a maritime state located in the eastern part of India, stretching from the Himalayas in the north to the Bay of Bengal in the south. In the North, West Bengal has three main rivers—Teesta, Torsa, and Jaldhak, which are tributaries to the Brahmaputra. Two additional main rivers are passing through the state: Ganga and Hooghly. Due to these, major rivers flowing through the state are projected to have a high sea-level rise and flood risk. (WBSAPCC, 2017-2020; Basu, 2020). As per the 2011 census, the population of West Bengal is around 91.3 million, which accounts for about 7.54% of the total population of India (Census India, 2011a). The coastal length of West Bengal is about 220 km, with a coastal zone of about 9,630 square km. The coastal zone of West Bengal supports an approximate population of 7 million (ICZMPWB, n.d.-a).

Climate risks and developmental complications in West Bengal: double burden

Tropical storms are major hazards for coastal regions. Many severe local storms occur over the Gangetic plain in relation to West Bengal. Warming of the sea in the Bay of Bengal causes intense cyclones that regularly hit the coasts—when storms with such intensity and speed hit, the coastline destroys coastal resources (Vellore et al., 2020; Pardikar, 2021). Recently, in April and May 2019, 'Fani', a cyclone that developed in the Bay of Bengal, resulted in the high-speed wind causing territorial rain and excessive flooding. It devastated human lives and made huge property losses in the eastern India (Kumar et al., 2020). Like this, in May 2020, the super cyclone 'Amphan' hit the Bay of Bengal, costing human life (72 people in India and 12 in Bangladesh) and high destruction of property. Amphan also affected the largest mangrove forest in the world (Kumar et al., 2021). As a result of the recent cyclones coupled with sea-level rise and human interventions, coastal erosion is rising (Padma, 2022). Diamond Harbor, a coastal port in West Bengal, recorded the maximum sea-level rise in 2019 (Singh, 2019). The Sea Level Research Group at the University of Colorado found that the sea level off the coast of Bengal is rising at 4.04 ± 0.44 mm/year (Environment Department/ Government of West Bengal, 2021). Further, the impact of sea-level rise in Sundarbans is accelerated due to the slow subsidence of land at 2.9 mm/year (Environment Department/Government of West Bengal, 2021, p. 12). In addition to sea-level rise, coastal erosion, a natural phenomenon that is carried out by waves, tidal and littoral currents, and deflation are getting exacerbated by activities like land reclamation, dredging of harbors, navigational channels and tidal inlets, construction of groins, jetties, and other structures on the coast (Kukreti, 2019). The Sundarbans are a very important region in West Bengal that is highly impacted by storms. The Sundarbans is around 170 square km of coastal land and is in a net erosional state, the delta front undergoing net erosion. Sundarbans, the world's largest tract of contiguous mangrove forest, is inhabited by large flora and fauna, including the Bengal Tiger, various species of fish, Indian python, and many more (Everard et al., 2019). It is also the World's largest coastal wetland, covering about 1 million hectares in the delta of the rivers Ganga, Brahmaputra, and Meghna. The Sundarbans region is shared between Bangladesh (~60%) and India (~40%) (Gopal & Chauhan, 2006). Indian Sundarbans is located within the north and south 24 Parganas districts of the state of West Bengal. The total area of the Sundarbans region in India is 9630 square km. As of 2011, the population of Indian Sundarbans is around 4.37 million, with a high density of 957people/square km. These regions are one of the most backward regions in India due to their remote and hostile terrain conditions and extreme poverty of the majority (around 43.5%) of people of these regions (Dasgupta & Shaw, 2014). Also, these regions have been the hotspots of climate change risks and developmental complications for the past few decades. Considering all the risks to the Sundarbans, it has been declared a natural 'World Heritage Site' and the declared sanctuary of the dwindling Royal Bengal Tiger. Large parts of it also fall under the regulatory regime of the Coastal Regulation Zone. The high dependency on natural capital, less involvement in social networks, the salinity of water, inaccessibility to healthcare facilities, and lack of perception of climatic events are responsible for the vulnerabilities of coastal fishing communities and other coastal communities (Basu & Basu, 2020, p. 770). In recent decades and years, the Sundarbans have been increasingly affected by cyclones, sea-level rise, and coastal erosion. New research has found that the Sundarbans in India and Bangladesh have lost 24.55% of their mangroves due to soil erosion over the past three decades (Ghosh, 2020; Bhargava et al., 2021). As a result of these physical climate threats, people have started flooding from Sundarbans and Bangladesh into West Bengal in search of new livelihoods and homes, causing a double burden on the states' resources.

Fishers of the Sundarbans in West Bengal are dependent on fishing, crab catching, and coastal ecosystems for their livelihoods. These livelihoods in the Sundarbans are threatened by the high levels of risk from cyclones, sea-level rise, embankment failures, salinity intrusion, and loss of mangrove forests (Basu & Basu, 2020, p. 766). The degradation of mangrove forests has put local communities and stakeholders at risk. To respond to such threats, the state government of West Bengal has arranged an Early Warning System (EWS) in the coastal Sundarbans to combat the stress of cyclones and storm surges under Disaster Risk Reduction. It has taken various initiatives to expand crop insurance packages for small and marginal farmers' security against crop loss during floods or cyclones in the state (Basu & Basu, 2020, 770). In 2021, the State Government had not yet effectively made studies available to the public on the growing risks of cyclones and sea-level rise. However, the Government has secured warning systems when a cyclone is to hit a certain area that led to moderate death during 'Amphan' amidst the pandemic (Kumar et al., 2021).

Small-scale fishing communities of West Bengal have already started migrating in search of new livelihoods—fishing being a major means of livelihood for coastal communities. Promoting fisheries as an industry is one of the adaptation strategies under the West Bengal SAPCC; however, the marine fisheries sector is also at high risk of climate change and developmental complications. In contrast to these adaptation strategies, as reported in the local news outlets, going against these regulations, 'development' projects such as

Merchant Shipping Corridor adversely affected fishers of West Bengal. The National Fishworkers' Forum has stated that these multiple industrial projects destroy marine life and sabotage the livelihood of approximately 80% of fishermen in the state (The Statesman, 2018). Even though a letter has been sent to the respective authority, the State Government denied getting any formal intimation on this issue. There is a big list of violations of CRZs across the coasts of West Bengal. For instance, the coastal area of Mandarmani sea beach, West Bengal, has 130 resort hotels operating without considering CRZs as against the regulations set by CRZs (Basu, 2019). Another example of CRZ violation would be in Purba Medinipur, a 41-km-long coastal stretch, an important fishing harbor for local fisherfolk (Environment Department/Government of West Bengal, 2021). The dunes and the beach at Mandarmani have been flattened due to Hoteliers violating CRZ regulation. Many hotels are either within the inter-tidal space or in 'no construction zone' as per the CRZ of 2019 (Environment Department/ Government of West Bengal, 2021, p. 37). These hoteliers obstruct the lives of about 0.2 million fishermen dependent on the coastline to make a living (Environment Department/Government of West Bengal, 2021). The West Bengal State Coastal Zone Management Authority (WBSCZMA) imposed a processing fee to the hoteliers at the end of the appeal and were suggested to re-issue a CRZ clearance. However, the outcomes of this appeal were not beneficial for local stakeholders as the damage had already been done. Aggarwal (2018) states that environmentalists feel that the 2018 CRZ draft was not prepared by consulting the public; it was only to benefit the industries and economic output. This draft opens the coastline for industry, real estate, and tourism rather than protecting the wellbeing of the coastline. In 2019, it was stated that 90% of objections received from fishing communities, environmental groups, and stakeholders were being disregarded while a decision was being made on the CRZ Notification (Anand, 2021). As per the CRZ of 2019, most changes were made for industries or companies to actively seek permits to regulate projects within the CRZs (Environment Department/Government of West Bengal, 2021). However, none of these changes directly protect the wellbeing of the coastal communities. In May 2021, the government attempted to facilitate the legitimization of illegal industries operating in CRZs. The recent amendments to the CRZ 2019 further dilute sensitive coastal groups and put communities and the ecology at risk (Anand, 2021). Tajpur and Haldia's coastal development project in 2019 is another example of how the government oversees the fishing communities' concerns. These communities have not been informed about this deep-sea port project or have been officially visited by the government to guide them on their fishermen's livelihoods will be impacted. The community of Haldia— an industrialized fishing town—states that their income has been reduced by one-third since these projects have taken over (Chakraborty, 2019). Even though CRZs are meant to eventually protect local stakeholders, communities, and fisherfolk, these regulations are not entirely considering the adverse effects these industries cause. In 2018, an organization of fishermen in West

Bengal wrote to the central government to scrap the draft CRZ Notification 2018, alleging the adverse effects it would cause to the local communities' livelihoods. This notification would cause a pollution load in the sea and encroachment in the CRZ (Sundararaju, 2019). The communities that live in coastal resources consistently feel neglected considering that much of the livelihoods of these communities are dependent on the coasts. The National Fishworkers Forum has constantly been in turmoil with the government on implementing and diluting the CRZ rules. To protect the livelihoods and households of these local coastal communities, it is necessary that these coasts need a science-based policy for research, conservation, and management (Warrier, 2018).

West Bengal State Action Plan on Climate Change

In 2010, the State Action Plan on Climate Change (SAPCC) preparation began its workshop and committee (Kumar, 2018). The goal of the SAPCC is to provide adaptation strategies, a region-based approach, for resource management in West Bengal. According to the West Bengal's State Action Plan on Climate Change, the major concerns in the Sundarbans are the intensity of cyclones and constant breaching of embankments, changing water mass properties, and impacts on mangrove species. In the last decade, the run-off in the eastern rivers of Sundarbans has decreased, leading to higher salinity and greater seawater sulfate amount (Mahadevia &Vikas, 2012). However, the world does not realize the everyday threats of the inhabitants of Sundarbans. The sixth IPCC report on Climate Change projected that relative sea-level rise is likely to inundate most of the Sundarbans by the mid-21st century. Those that remain could be threatened by saltwater incursion. The world only pays attention to the Sundarbans Islands when a disaster hits (Biswas, 2021). The climate refugees of West Bengal have already started migrating in search of new livelihoods. The WBSAPCC includes adaptation strategies for the Sundarbans, where two out of 11 strategies are planned toward the community. One strategy is to raise existing houses in these vulnerable areas on stilts, and the second is to put forth rehabilitation strategies and alternative livelihoods (WBSAPCC, 2012). However, these adaptation strategies are only meant for a short-term solution and may not be effective, just like the EWS for cyclones. The state needs long-term solutions or policies to handle the repercussions of climate disasters. The West Bengal Pollution Control Board (WBPCB) states that the mangrove plantations in the Sundarban Islands would not be able to reverse the effects of erosion; it will only work in protected areas where the offshore gradient is gentle (Environment Department/Government of West Bengal, 2021, p. 29). The WBPCB plans to reinforce the existing 30-km-long seafront with concrete and their protection with planned vegetative buffers (Environment Department/Government of West Bengal, 2021). The Government also plans to build two parallel dykes along the coast, declaring them as a 'no construction zone' in compliance with the CRZ (The Gazette of India, 2019a). Even though the Government of West Bengal plans to protect the mangroves and its coastal communities through these planned processes, how long will these strategies take to be put into effect? Although this plan is not set in place and is currently being considered by the Government of West Bengal, it may require the relocation and rehabilitation of 630 households in Sagar Island and 605 households in Moushuni Island (Environment Department/ Government of West Bengal, 2021). In reality, this project will take immense time, capital, and labor to achieve its desired output. Rehabilitation of these households could cause utter chaos as these coastal communities have lived in these areas for a long time. These people would have to give up their livelihood and homes for this project to take into effect. At the same time, the West Bengal SAPCC has featured adaptation strategies based on agriculture, water resources, forests and biodiversity, human health, and renewable energy; these strategies fail to mention any people-centric short-term or long-term solutions like migration or rehabilitation facilities due to natural disasters or climate change. 'At the end of the day, only states with climate-literate policy planners will be able to prepare their constituencies with a better set of options for the future' (Dasgupta, 2018).

Conclusion

This chapter provides an overview of the ongoing challenges in India's coastal regions as a result of developmental complications, CRZs, and climate change. It provides a summary of those conflicts that arise from coastal laws through secondary data (in the case of coastal Tamil Nadu) and primary data that are packed with field research (in the case of coastal Maharashtra), and secondary data that are packed with online interviews with the key informants (in the case of coastal West Bengal). Overall, taking a step back from these detailed observations about cartography, gaps in visual discourse, and developmental complications across coastal India, one needs to reckon with the fact that contemporary climate change policy cannot repeat the mistakes of previous environmental policies. Any contemporary climate change policy which does not effectively address local level long-standing issues is not human-centric enough. Climate change is not a process that will happen soon, but something has been affecting local fisherfolk for several years already. For example, the MSAAPC itself, drafted over eight years, has a future focus: the 2030s, 2050s, and 2070s, while the ongoing issues faced in places like Raigad are not listed. To fully critique climate change policy responses, it is important to take a longitudinal look at previous environmental policies of the state. More importantly, if we are to ensure the protection of ecological landscapes and affected populations, climate change policy needs to shift gears and address the needs of local communities urgently.

In the case of West Bengal, we have also provided more emphasis on the Sundarbans, how Indian Sundarbans have become a prone region to multiple climate associated risks, and how the existing coastal laws and climate action plans are inadequate to address the multiple vulnerabilities of these coastal stakeholders (Mehta et al., 2019). It is obvious that the future of fisheries will be shaped by cross-sectoral solutions to the current problems (Rao et al., 2020). NGOs in Kolkata can adapt short-term as well as long-term climate change goals like monitoring coastal regulation zones, creating opportunities for climate change communication, and setting up policies for proper management of climate change disasters. In the years to come, developing countries can take the opportunity to increase research techniques for deep-sea storms and invest in technology that can reduce seawater temperatures to avoid intensive cyclones from occurring. Coastal Regulation Zones (CRZ) are governed in place to protect the fragile ecosystems near the sea from human and industrial activity close to the sea. West Bengal lacks strong climate change communication and research programs to raise public understanding.

In India, even though coastal spaces have been governed by a variety of laws, institutions, and authorities, in this chapter, in the context of coastal law, we have focused on Coastal Regulation Zones Notifications and Climate Change Policy and Actions of the states (to some extent) as both plays influential roles in governing the coastal spaces and provides an authoritative action map to the authorities to govern them. Nevertheless, as this chapter shows, both have developed with little emphasis on people's participation since the beginning. In particular, the recent Coastal Regulation Zone Notification 2019 has further diluted the concept of local people's participation and is primarily driven by a top-down approach. As Basu and Mandal (2022) write, current coastal laws and policies lack clear insights and visionary pathways, resulting in more complications in governing the oceans and coastal populations. Such complications increased the vulnerability of the coastal stakeholders, mainly traditional marine fishing communities, and exacerbated people's resistance to the flawed coastal laws. This is evident in the case of coastal Tamil Nadu and Maharashtra. In coastal Tamil Nadu, in particular, such increasing resistances and conflicts between the people and the state in terms of CRZs have further worsened the livelihoods of the traditional fishing communities (Rodriguez, 2010; Kumar et al., 2014) and put them more at risk of climate change and risks. This may further bring more complications not just to the coastal spaces but also to the development agenda for the entire nation, and it has to be addressed through local-centric, human, and inclusive policy-oriented approaches and actions.

Notes

1 Though we agree that the term 'coastal laws' is an umbrella term that denotes many laws, regulations, and even institutions, however, in this chapter, we mainly place emphasis on the Coastal Regulation Zones Notifications, and a limited emphasis on climate change policy. Hence, throughout this chapter, we mean that

the term 'coastal law' is mainly used to denote 'CRZ' (Coastal Regulation Zones) notifications and climate change policy.

2 The Island Protection Zone notification 2011 (IPZ 2011): This notification applies only to Andaman and Nicobar and Lakshadweep islands. IPZ 2011 aims to protect coastal resources and restrict activities and industries, including mining activities in the island regions. For more details: Dakshin (2018). Recently, the Government has enforced the Island Coastal Regulation Zone (ICRZ) plan in 2019 (hereafter ICRZ 2019). The preamble of this notification has noted the following: To conserve and protect coastal biodiversity from unrestricted industrial activities and to promote the livelihoods security of the fisherfolk and other communities of the island, and to promote sustainable development based on scientific evidence, and to protect from the dangers of global warming, ICRZ 2019 declares the eight bigger oceanic islands in Andaman and Nicobar, and the water area up to territorial water limits of the country, as the Island Coastal Regulation Zone (The Gazette of India, 2019a).

3 Coastal Regulation Zone Notification 2019 (The Gazette of India, 2019b).

4 The pandemic hit the Koli community particularly hard due to the country-wide lockdown and multiple fishing bans; fisherfolk were left without work for months, and deep-sea fishers were left stranded at sea (Nidhi, 2020).

5 In the book *Making of Navi Mumbai*, author Annapurna Shaw documents how there were several local protests against land acquisition and inadequate compensation. These protests were met with brute force by the State, and several protesting locals were shot dead (Shaw, 2004).

6 The third author of this chapter undertook fieldwork in Maharashtra in 2019 as part of her master's thesis. She documented these testimonies.

7 A special economic zone (SEZ) is an area in which the business and trade laws are different from the rest of the country. SEZs are located within a country's national borders. Their aims include increased trade balance, employment, increased investment, job creation, and effective administration.

8 The Kolis, in exchange for their traditional land, were promised ration (monthly grains and gas to cook), and displaced Kolis were promised jobs at the Port.

9 As noted in the footnote 7, the third author conducted fieldwork in Maharashtra in 2019. Her primary interviewees were residents of coastal villages in Raigad district.

10 As per interviews conducted by the third author of this chapter with Nagaon locals in March 2019.

11 Notice of Motion on PIL 87/2006 of Bombay High Court issued by City Industrial and Development Corporation (CIDCO) to Bombay High Court, April 29, 2009.

12 Coastal Regulation Zone Notification 1991 inserted vide Coastal Regulation Zone Notification Amendment (May 15, 2009).

13 Conservation Action Trust & Anr. Versus Union of India & Ors. Public Interest Litigation no. 57 of 2019, High Court of Bombay (2019).

14 Environment Impact Assessment Studies for NMIA, Centre for Environmental Science and Engineering at the Indian Institute of Technology (IIT) Mumbai, June, 2010.

15 89th Expert Appraisal Committee Meeting, Minutes of Meeting, July 21-23, 2010
90th Expert Appraisal Committee Meeting, Minutes of Meeting, August 18-20, 2010.

16 J.B. Harley's collected essays *"The new nature of maps: Essays in the history of cartography"* explains the ideological dimensions of cartography (Harley, 2001).

References

Aggarwal, M. (2018, March 12) Government amends coastal zone rules to give relief to projects. *LiveMint*. Retrieved August 12, 2019 from https://www.livemint.com/Politics/aTFWIrny7VVoNqKg8eyshN/Government-amends-coastal-zone-rules-to-give-relief-to-proje.html

Anand, G. (2021, December 24). The proposed changes to the Coastal Regulation Zone Notification of 2019 further dilute protective provisions, putting the ecology and vulnerable communities in ecologically sensitive zones at risk. *The Leaflet*. https://theleaflet.in/proposed-amendments-to-the-crz-notification-2019-and-its-implications/

Andrews, N., Bennett, N. J., Le Billon, P., Green, S. J., Cisneros-Montemayor, A. M., Amongin, S., ... & Sumaila, U. R. (2021). Oil, fisheries and coastal communities: A review of impacts on the environment, livelihoods, space and governance. *Energy Research & Social Science*, 75, 102009. https://doi.org/10.1016/j.erss.2021.102009

Ashok, S. (2018, April 23). What the new Coastal Regulation Zone draft says, how it differs from the earlier version. *The New Indian Express*. https://indianexpress.com/article/explained/what-the-new-coastal-regulation-zone-draft-says-how-it-differs-from-the-earlier-version-5147723/

Babu, S. (2011). Coastal accumulation in Tamil Nadu. *Economic and Political Weekly*, 46 (48), 12–13.

Barange, M., Bahri, T., Beveridge, M. C., Cochrane, K. L., Funge-Smith, S., & Poulain, F. (2018). *Impacts of climate change on fisheries and aquaculture: Synthesis of undarb knowledge, adaptation and mitigation options.* FAO Fisheries and Aquaculture Technical Paper No. 627. Rome: FAO.

Basu, J. (2020). Bengal most vulnerable to climate risk, flags India's first assessment report. *Down to Earth*. https://www.downtoearth.org.in/news/climate-change/bengal-most-vulnerable-to-climate-risk-flags-india-s-first-assessment-report-72117

Basu, J. P. & Basu, A. (2020) Climate change vulnerability and resource dependent communities: An empirical study in coastal Sunderban, West Bengal, India. *Ecology, Environment, & Conservation Journal*, 26 (2), 766–772.

Basu, A., & Mandal, S. (2022). Protecting coastal environment in India: Reading laws in the context of climate change. *Asian Journal of Legal Education*. https://doi.org/10.1177/23220058221111299

Behera, K. S. (Ed.). (1999). *Maritime Heritage of India*. New Delhi: Aryan Books International.

Bhargava, R., Sarkar, D., & Friess, D. A. (2021). A cloud computing-based approach to mapping mangrove erosion and progradation: Case studies from the Sundarbans and French Guiana. *Estuarine, Coastal and Shelf Science*, 248, 106798. https://doi.org/10.1016/j.ecss.2020.106798

Bhatia, R. (2014). *The Noble Mansion – Navi Mumbai Airport Influence Notified Area (NAINA) Series*. Peepli. https://www.peepli.org/stories/about-naina/

Biswas, M. (2021). The Sundarbans: India must protect the world's largest delta from dystopian future, and now. *Down to Earth*. https://www.downtoearth.org.in/blog/climate-change/the-sundarbans-india-must-protect-the-world-s-largest-delta-from-a-dystopian-future-and-now-79092

Bryant, R. L., & Bailey, S. (1997). *Third world political ecology*. London: Routledge.

Census India (2011a). *West Bengal population census 2011 West Bengal religion, caste data – census 2011*. Census India. Retrieved December 12, 2021, from https://www.censusindia.co.in/states/west-bengal

Census India (2011b). *Primary Census Abstracts*. Registrar General of India, Ministry of Home Affairs, Government of. India. http://www.censusindia.gov

CMFRI (2020). *Annual Report 2019*. Central Marine Fisheries Research Institute, Kochi. 284 p.

Chakraborty, S. (2019, February 12). Tajpur and Haldia: How coastal development is impacting fishing in West Bengal. *Down to Earth*. Retrieved January 22, 2022, from, https://www.downtoearth.org.in/news/governance/tajpur-and-haldia-how-coastal-development-is-impacting-fishing-in-west-bengal-63212

Chandrasekharan, H., Sarangi, A., Nagarajan, M., Singh, V. P., Rao, D. U. M., Stalin, P., ... & Anbazhagan, S. (2008). Variability of soil–water quality due to Tsunami-2004 in the coastal belt of Nagapattinam district, Tamilnadu. *Journal of Environmental Management*, 89(1), 63–72. https://doi.org/10.1016/j.jenvman.2007.01.051

Chouhan, H. A., Parthasarathy, D., & Pattanaik, S. (2016). Coastal ecology and fishing community in Mumbai: CRZ Policy, sustainability and livelihoods. *Economic & Political Weekly*, 51(39), 48–57.

Correa, M. (2016). India: fishing communities, hemmed in by development. *SAMUDRA Report*, 74, 12 – 17. http://hdl.handle.net/1834/36023

CRZ (2011). *Coastal regulation zone notification – the 6th of January 2011*. Ministry of Environment and Forests (Department of Environment, Forests and Wildlife) New Delhi, Retrieved January 21, 2022, from, http://www.environmentwb.gov.in/pdf/CRZ-Notification-2011.pdf

Dakshin (2018). *The Island protection zone notification 2011*. Dakshin Foundation. Retrieved May 13, 2021, from https://www.dakshin.org/wp-content/uploads/2018/06/IPZ-handbook.compressed.pdf

Dasgupta, D. (2018). *West Bengal updates state climate action plan*. India Climate Dialogue. Retrieved May 13, 2021, from https://indiaclimatedialogue.net/2018/07/06/west-bengal-updates-state-climate-action-plan/

Dasgupta, R., & Shaw, R. (2014). *Participatory planning for enhancing community resilience in mangrove rich Indian Sundarbans*. 42 pages, Kyoto University, Kyoto, Japan.

Dhowan, R. K. (2021). India: A Resurgent Maritime Nation, Harnessing the Blue Economy. In R. Jha (Ed.), *Twenty K.R. Narayanan Orations: Essays by Eminent Persons on the Rapidly Transforming Indian Economy* 1st ed. (pp. 387–396). Acton ACT: ANU Press. https://doi.org/10.2307/j.ctv1prsr3r.40

Dias, L., & Sridhar, A. (2021). How CRZ violations are being regularized instead of regulated across India's shoreline. *The Bastion*. https://thebastion.co.in/politics-and/environment/conservation-and-development/how-crz-violations-are-being-regularized-instead-of-regulated-across-indias-shoreline/

DNA India (2018, December 04). Fishermen call Tamil Nadu Coastal zone management plan map 'incomplete'. *DNA*. https://www.dnaindia.com/india/report-fishermen-call-tamil-nadu-coastal-zone-management-plan-map-incomplete-2691988

Dodhiya, K. (2021, April 02). PIL challenging CRZ 2019 notification filed in Bombay HC. *The Hindustan Times*. https://www.hindustantimes.com/cities/mumbai-news/pil-challenging-crz-2019-notification-filed-in-bombay-hc-101617303179190.html

Down to Earth (2008, December 31). Fishers at bay. *Down to Earth*. https://www.downtoearth.org.in/coverage/fishers-at-bay-2811

Environment Department/ Government of West Bengal (2021). *Protection of coastal areas and earthen embankment through vegetative solutions*. West Bengal Pollution Control Board. Retrieved January 21, 2022, from, https://www.wbpcb.gov.in/files/Tu-08-2021-08-23-18Report_Final.pdf

Everard, M., Kangabam, R., Tiwari, M.K., McInnes, R., Kumar, R., Talukdar, G.H., Dixon, H., Joshi, P., Allan, R., Joshi, D., & Das, L. (2019). Ecosystem service assessment of selected wetlands of Kolkata and the Indian Gangetic Delta: Multi-beneficial systems under differentiated management stress. *Wetlands Ecology and management*, 27(2), 405–426. https://doi.org/10.1007/s11273-019-09668-1

FIMSUL (2011). *Fisheries Management Options for Tamil Nadu & Puducherry*. (Authors: V. Vivekanandan and H.M. Kasim). A Report prepared for the Fisheries Management for Sustainable Livelihoods (FIMSUL) Project, undertaken by the UN FAO in association with the World Bank, the Government of Tamil Nadu and the Government of Puducherry. Report No. FIMSUL/R20.FAO/UTF/IND/180/IND. New Delhi, Chennai and Puducherry, India

Fuchs, R. J. (2010). Cities at risk: Asia's coastal cities in an age of climate change. *Asia Pacific Issues*, 96, 1–12. Retrieved September 1, 2021, from https://scholarspace.manoa.hawaii.edu/bitstream/10125/17646/1/api096.pdf

Gadgil, M. (2019, December 23). Ecology is for the people. *The India Forum. A Journal-Magazine on Contemporary Issues*. https://www.theindiaforum.in/article/ecology-people

Garcia, S.M., Ye, Y., Rice, J. & Charles, A., eds. (2018). Rebuilding of marine fisheries. Part 1: Global review. FAO Fisheries and Aquaculture Technical Paper No. 630/1. Rome, FAO. 294 pp. https://www.fao.org/3/ca0161en/ca0161en.pdf

Ghosh, S. (2020). Erosion, an important cause of mangrove loss in the Sundarbans. *Mongabay*. https://india.mongabay.com/2020/05/erosion-an-important-cause-of-mangrove-loss-in-the-sundarbans/

Gopal, B., & Chauhan, M. (2006). Biodiversity and its conservation in the Sundarban mangrove ecosystem. *Aquatic Sciences*, 68(3), 338–354. https://doi.org/10.1007/s00027-006-0868-8

Gopi, M., Kumar, M.P., Jeevamani, J.J.J., Raja, S., Muruganandam, R., Samuel, V.D., Simon, N.T., Viswanathan, C., Abhilash, K.R., Krishnan, P., & Purvaja, R. (2019). Distribution and biodiversity of tropical saltmarshes: Tamil Nadu and Puducherry, Southeast coast of India. *Estuarine, Coastal and Shelf Science*, 229, 106393. https://doi.org/10.1016/j.ecss.2019.106393

Government of India (2009). National Action Plan on Climate Change. Prime Minister's Council on Climate Change. Retrieved May 20, 2021, from http://www.nicra-icar.in/nicrarevised/images/Mission%20Documents/National-Action-Plan-on-Climate-Change.pdf

Government of India (2011). *Coastal regulation zone notification - The 6th of January 2011*. Ministry of Environment and Forests, Government of India. Retrieved May 23, 2021, from http://www.environmentwb.gov.in/pdf/CRZ-Notification-2011.pdf

Government of India (2020). *Handbook on Fisheries Statistics 2020*. Department of Fisheries, Ministry of Fisheries, Animal Husbandry & Dairying, Government of India, New Delhi. https://dof.gov.in/sites/default/files/2021-02/Final_Book.pdf

Government of Maharashtra (2016). *Maharashtra State Disaster Management Plan*. State disaster management authority, Mumbai. Retrieved May 23, 2021, from https://rfd.maharashtra.gov.in/sites/default/files/DM%20Plan%20final_State.pdf

Government of Maharashtra (2019). *Economic survey of Maharashtra 2019–20.* Directorate of Economic and Statistics, Planning Department, Government of Maharashtra. Retrieved December 18, 2021, from https://maharashtra.gov.in/Site/upload/WhatsNew/ESM_2019_20_Eng_Book.pdf

Government of Tamil Nadu (2012). *State planning commission 2011 – 2012.* Tamil Nadu State Planning Commission. Retrieved April 18, 2021, from http://www.spc.tn.gov.in/annualplan/ap2011-12/chapter.2.5.pdf

Government of West Bengal (2012). *West Bengal state action plan on climate change.* Government of West Bengal. Retrieved December 2020, http://www.nicra-icar.in/nicrarevised/images/State%20Action%20Plan/West-Bengal-SAPCC.pdf

Harley, J. B. (1988). Silences and secrecy: The hidden agenda of cartography in early modern Europe. *Imago Mundi*, 40, 57–76.

Harley, J. B. (2001). *The new nature of maps: Essays in the history of cartography.* Baltimore: Johns Hopkins University Press.

Hatekar, N., & Raju, S. (2013). Inequality, income distribution and growth in Maharashtra. *Economic and Political Weekly*, 48 (39), 75–81.

Hegde, S. (2015). Son Kolis – the aboriginal inhabitants of Bombay (now Mumbai) in transition. *International Letters of Social and Humanistic Sciences*, 62, 140–146. https://doi.org/10.18052/www.scipress.com/ilshs.62.140

ICZMP (2021). *Implementation Completion Report (ICR) Review IN: Integrated Coastal Zone Mgmt Project (P097985).* The World Bank. Retrieved May 23, 2021, from https://documents1.worldbank.org/curated/en/404141622855878378/pdf/India-IN-Integrated-Coastal-Zone-Mgmt-Project.pdf

ICZMPWB (n.d.-a). *Coastal Zones.* Integrated Coastal Zone Management Project, Government of West Bengal. Retrieved December 12, 2021, from https://www.iczmpwb.org/main/coastal_zones.php

ICZMPWB (n.d.-b). *ICZM Background.* Integrated Coastal Zone Management Project, Government of West Bengal. Retrieved May 23, 2021, from http://www.iczmpwb.org/main/iczm_background.php

India Today (2018, April 22). Draft CRZ rules: Opening fragile coastal areas for real estate will affect ecology, fishing, says experts. *India Today.* https://www.indiatoday.in/pti-feed/story/draft-crz-rules-opening-fragile-coastal-areas-for-real-estate-will-affect-ecology-fishing-says-experts-1217450-2018-04-22

India.com (2021, July 23). Maharashtra Rains: Red alert for six rain-hit districts, extremely heavy showers expected; NDRF deployed. India News, Breaking News | India.com. Retrieved November 13, 2021, from https://www.india.com/maharashtra/mumbai-rains-live-updates-23-july-2021-local-train-status-in-mumbai-central-railway-western-railway-main-line-harbour-line-uddhav-thackeray-raigad-pune-thane-rainfall-4835628/

Janwalkar, M. (2021, January 21). Mumbai: Villagers seeking rehabilitation for 34 years call off protest, give govt time to respond. *The Indian Express.* Retrieved November 13, 2021, from https://indianexpress.com/article/cities/mumbai/villagers-seeking-rehabilitation-for-34-years-call-off-protest-give-govt-time-to-respond-7155320/

Jamwal, N. (2010) Backdoor Democracy – Central Directives on Coastal Zones could be used to Circumvent Rules, *Down to Earth.* https://www.downtoearth.org.in/news/backdoor-democracy-285

Jamwal (2019, November 11). Maharashtra's Climate Action Plan Comes up Short. *The Wire.* https://thewire.in/environment/maharashtras-climate-action-plan-comes-up-short

Kadri (2020, December 11). Coastal road projects don't just damage the environment – they are also outdated. *The Wire.* https://thewire.in/environment/coastal-road-project-damage-environment-outdated

Kapoor, M. (2020, February 26). Govt disregarded 90% objections to 2019 Coastal Zone Law: Investigation. *IndiaSpend.* https://www.indiaspend.com/govt-disregarded-90-objections-to-2019-coastal-zone-law-investigation/

Krishna Chaitanya, S. V. (2018, April 02). Tamil Nadu pushes 'incomplete' Coastal Zone Management Plan, calls for hearing in 12 districts. *The New Indian Express.* https://www.newindianexpress.com/states/tamil-nadu/2018/apr/02/tamil-nadu-pushes-incomplete-coastal-zone-management-plan-calls-for-hearing-in-12-districts-1795775.html

Kukreti, I. (2019). Coastal regulation zone notification: What development are we clearing our coasts for. *Down to Earth.* https://www.downtoearth.org.in/coverage/governance/coastal-regulation-zone-notification-what-development-are-we-clearing-our-coasts-for-63061

Kumar, S., Lal, P., & Kumar, A. (2020). Turbulence of tropical cyclone 'Fani'in the Bay of Bengal and Indian subcontinent. *Natural Hazards, 103*(1), 1613–1622. https://doi.org/10.1007/s11069-020-04033-5

Kumar, S., Lal, P., & Kumar, A. (2021). Influence of super cyclone "Amphan" in the Indian subcontinent amid COVID-19 pandemic. *Remote Sensing in Earth Systems Sciences,* 1–8. https://doi.org/10.1007/s41976-021-00048-z

Kumar, V. (2018). *Coping with climate change: An analysis of India's state action plans on climate change,* New Delhi: Centre for Science and Environment. http://cdn.cseindia.org/attachments/0.40897700_1519110602_coping-climate-change-volII.pdf

Kumar, V. S., Pathak, K. C., Pednekar, P. N. S. N., Raju, N. S. N., & Gowthaman, R. (2006). Coastal processes along the Indian coastline. *Current Science,* 91 (4), 530–536. http://drs.nio.org/drs/handle/2264/350

Kumar, M., Saravanan, K., & Jayaraman, N. (2014). Mapping the coastal commons: Fisherfolk and the politics of coastal urbanisation in Chennai. *Economic and Political Weekly,* 49 (48), 46–53.

Kume, T., Umetsu, C., & Palanisami, K. (2009). Impact of the December 2004 tsunami on soil, groundwater and vegetation in the Nagapattinam district, India. *Journal of Environmental Management,* 90(10), 3147–3154. https://doi.org/10.1016/j.jenvman.2009.05.027

Kupekar, S., & Kulkarni, B. (2013). Climate change and fishermen in and around Uran. Dist Raigad. (Maharashtra). *IOSR Journal of Environmental Science, Toxicology and Food Technology,* 4(1), 52–57.

Lakshmanan, S (2021, April 23). Coastal violators push projects in Tamil Nadu after centre weakens CRZ norms. Science*e The Wire.* https://science.thewire.in/environment/coastal-violators-push-projects-in-tamil-nadu-after-centre-weakens-crz-norms/

Mahadevia, K. & Vikas, M. (2012). Climate change – impact on the Sundarbans: A case study. *International Scientific Journal Environmental Science, 2*(1), 7–15.

Maher, A. (2021). Troubles of the coast: Industrialization, climate change, marginality, and collective action among fishing communities in Kerala, India. *The Yale Undergraduate Research Journal,* 2(1), 19.

Maps of India (n.d.). *Maharashtra map: Map of Maharashtra – state, Districts Information and facts.* Maps of India. Retrieved September 13, 2021, from https://www.mapsofindia.com/maps/maharashtra/

Mehta, L., Srivastava, S., Adam, H. N., Bose, S., Ghosh, U., & Kumar, V. V. (2019). Climate change and uncertainty from 'above'and 'below': perspectives from India. *Regional Environmental Change, 19*(6), 1533–1547. https://doi.org/10.1007/s10113-019-01479-7

MoEF&CC Government of India (2019). Tamil Nadu State Action Plan for Climate Change. Ministry of Environment, Forest and Climate Change, Government of India. May 13, 2022, from https://moef.gov.in/wp-content/uploads/2017/09/Tamilnadu-Final-report.pdf

Mazumdar, A. (2015). Anchoring the past. *Economic and Political Weekly,* 50 (6), 78–80.

MRSAC, (1997). *Mangrove Status Maps.* Maharashtra Remote Sensing Applications Centre, Pune.

MRSAC (1994). *Wetlands maps.* Maharashtra Remote Sensing Applications Centre, Pune.

Menon, M., Kapoor, M., Venkatram, P., Kohli, K., & Kaur, S. (2015). *CZMAs and coastal environments: Two decades of regulating land use change on India's coastline.* New Delhi: Centre for Policy Research-Namati Environmental Justice Program. https://namati.org/wp-content/uploads/2015/06/CZMA-Study-Executive-Summary-final-15.6.2015.pdf

Menon, M., & Kohli, K. (2019, January 17). The coast is unclear: On the 2018 CRZ notification. *The Hindu.* https://www.thehindu.com/opinion/op-ed/the-coast-is-unclear/article26006723.ece

Nandakumar, D., & Nalini, N. (2010). Coastal fisheries in India: Current scenario, contradictions, and community responses. In R. Q. Grafton., R. Hilborn., D. Squires., M. Tait., & M. J. Williams (Eds.), *Handbook of Marine Fisheries Conservation and Management* (pp. 274–286). New York. Oxford University Press.

Narayan (2007, April 26). Govt puts Navi Mumbai airport plan on fast track. *The Economic Times.* https://economictimes.indiatimes.com/industry/transportation/airlines-/-aviation/govt-puts-navi-mumbai-airport-plan-on-fast-track/articleshow/1955250.cms?from=mdr

NDMA (2015). *Environment and social management framework. National cyclone risk mitigation project II (A world bank funded project).* National Disaster Management Authority Ministry of Home Affairs, Government of India. Retrieved May 23, 2021, from http://ncrmp.gov.in/wp-content/uploads/2019/GD/NCRMP-II.pdf

Neelambaran, A. (2019, November 02). Fisherfolk, residents oppose Adani port expansion. *NEWS Click.* https://www.newsclick.in/fisherfolk-residents-oppose-adani-port-expansion

Nidhi, J. (2020, April 2). Lockdown enforced when they were at sea - so more than a lakh of fishers now wait in Deep Waters - Gaon Connection: Your connection with rural india. *Gaon Connection.* Retrieved November 13, 2021, from https://en.gaonconnection.com/lockdown-enforced-when-they-were-at-sea-so-lakhs-of-fishers-now-wait-in-deep-waters/

Padma, T. (2022, February 22). Scientists raise caution as mangroves erode faster along Bengal coastline. *Mongabay.* Retrieved from https://india.mongabay.com/2022/02/scientists-raise-caution-as-mangroves-erode-faster-along-the-bengal-coastline/

Pardikar, R. (2021). Climate change is making India's west coast more vulnerable to cyclones. *Eos,* 102. https://doi.org/10.1029/2021EO163064. Published on 13 September 2021.

Parthasarathy, D., & Chouhan, H. A. (2020). New coastal claims and socio-legal contestations in Mumbai: Artisanal fishers and the problematic of the urban environment. In Rao, M. (Ed.), *Reframing the environment* (pp. 100–110). London: Routledge India.

Phadke, M. (2020, June 3). Cyclone Nisarga makes landfall close to Alibaug, to weaken as it reaches Mumbai. *The Print.* https://theprint.in/india/cyclone-nisarga-makes-landfall-close-to-alibaug-to-weaken-as-it-reaches-mumbai/434837/

Press Information Bureau (2018). *Cabinet approves Coastal Regulation Zone (CRZ) notification 2018.* Press Information Bureau. Ministry of Environment, Forest and Climate Change, Government of India. Retrieved May 23, 2021, from https://pib.gov.in/Pressreleaseshare.aspx?PRID=1557595

Press Information Bureau (2021). National Action Plan on Climate Change (NAPCC) (Ministry of Environment, Forest and Climate Change) December 01, 2021. Press Information Bureau (Research Unit), Ministry of Information and Broadcasting, Government of India. https://static.pib.gov.in/WriteReadData/specificdocs/documents/2021/dec/doc202112101.pdf

Puthucherril, T. G. (2011). Operationalising integrated coastal zone management and adapting to sea level rise through coastal law: where does India stand?. *TheInternational Journal of Marine and Coastal Law*, 26(4), 569–612. doi: https://doi.org/10.1163/157180811X593407.

Ranjan, A. (2020, January 22). *The Climate Emergency Situation in Maharashtra: A Big Challenge for Uddhav Thackeray.* ISAS.NUS.EDU.SG. Retrieved August 13, 2021, from https://www.isas.nus.edu.sg/papers/the-climate-emergency-situation-in-maharashtra-a-big-challenge-for-uddhav-thackeray/

Rao, A. D., Upadhaya, P., Pandey, S., & Poulose, J, (2020). Simulation of extreme water levels in response to tropical cyclones along the Indian coast: A climate change perspective. *Natural Hazards*, 100, 151–172. https://doi.org/10.1007/s11069-019-03804-z

Rodriguez, S. (2010). *Claims for survival: Coastal land rights of fishing communities.* Dakshin Foundation, Bangalore, 42. Retrieved on January 22, 2022, from https://www.dakshin.org/wp-content/uploads/2013/04/Claims-for-Survival.pdf

Salagrama, V. (2012a). Climate change and livelihoods: Perspectives from small-scale fishing communities in India. *Food Chain*, 3(1–2), 32–47. https://doi.org/10.3362/2046-1887.2013.004

Salagrama, V. (2012b). *Climate change and fisheries: Perspectives from small-scale fishing communities in India on measures to protect life and livelihood.* SAMUDRA Monograph. Chennai: International Collective in Support of Fishworkers. http://hdl.handle.net/1834/27415

Saran, S. (2019). *India's climate change policy: Towards a better future.* Media Center, Ministry of External Affairs, Government of India. Retrieved August 13, 2021, from https://mea.gov.in/articles-in-indian-media.htm?dtl/32018/Indias_Climate_Change_Policy_Towards_a_Better_Future

Sharma, C. (2011). CRZ Notification 2011: Not the end of the road. *Economic and Political Weekly*, 46 (7), 31–35.

Shaw, A. (2004). *The making of Navi Mumbai.* New Delhi: Orient Longman.

Sheik Mujabar, P., & Chandrasekar, N. (2011). Coastal erosion hazard and vulnerability assessment for southern coastal Tamil Nadu of India by using remote sensing and GIS. *Natural Hazards*, 69(3), 1295–1314. https://doi.org/10.1007/s11069-011-9962-x

Sheth, A., Sanyal, S., Jaiswal, A., & Gandhi, P. (2006). Effects of the December 2004 Indian ocean Tsunami on the Indian mainland. *Earthquake Spectra*, 22(S3), 435–473. https://doi.org/10.1193/1.2208562

Singh, O., Bhardwaj, P., & Singh, J. (2017). Distribution of lightning casualties over Maharashtra, India. *Journal of Indian Geophysical Union*, 21(5), 415–424.

Singh, R., Pandey, P. K., Sharma, R., & Ojha, S. N. (2010). Fisheries profile mapping of coastal districts in Maharashtra state through GIS. *Journal of the Indian Fisheries Association*, 37, 45–50.

Singh, S. S. (2019, July 10). Bengal port records country's highest sea level rise in 50 years Subcontinent amid COVID-19 pandemic. *The Hindu.* https://www.thehindu.com/news/national/bengal-port-records-countrys-highest-sea-level-rise-in-50-years/article28364149.ece#:~:text=in%20coastal%20areas.-,Four%20ports%20in%20India%20recorded%20a%20higher%20sea%2Dlevel%20rise,the%20Ministry%20of%20Earth%20Sciences

Sundar, V., & Sundaravadivelu, R. (2005). *Protection measures for Tamil Nadu coast, final report submitted to public works department, GoTN.* Department of Ocean Engineering, Indian Institute of Technology Madras. Retrieved April 12, 2021, from https://tnsdma.tn.gov.in/app/webroot/img/document/library/15-Protection-Measures-For-Tamil-nadu-cost.pdf

Sundararaju, V. (2019, January 18). Why we need a coastal zone protection act. *Down To Earth.* https://www.downtoearth.org.in/blog/environment/why-we-need-a-coastal-zone-protection-act-62876

Swamy, R. H. (2011). *Disaster capitalism: Tsunami reconstruction and neoliberalism in Nagapattinam, South India.* (Doctoral dissertation). University of Texas at Austin. Retrieved May 20, 2021, from http://hdl.handle.net/2152/ETD-UT-2011-05-3461

TERI (2014). *Assessing climate change vulnerability and adaptation strategies for Maharashtra: Maharashtra State Adaptation Action Plan on Climate Change (MSAAPC).* New Delhi: The Energy and Resources Institute. 302 pp. [Project Report No. 2010GW01]. Retrieved November 13, 2021, from http://moef.gov.in/wp-content/uploads/2017/09/Maharashtra-Climate-Change-Final-Report.pdf

The Gazette of India (1991). *CRZ Notification 1991.* The Gazette of India Extraordinary. PART II—Section 3—Sub-Section (ii). Published on February 20, 1991. Retrieved December 14, 2021, from http://environmentclearance.nic.in/writereaddata/SCZMADocument/CRZ%20Notification,%201991.pdf

The Gazette of India (2011). CRZ notification 2011. The gazette of India extraordinary. PART II—Section 3—Sub-section (i). Published on JANUARY 18, 2019 Retrieved December 14, 2021, from http://environmentclearance.nic.in/writereaddata/SCZMADocument/CRZ_Notification2019.pdf

The Gazette of India (2019a). *Ministry of Environment, Forests and Climate Change Notification. New Delhi, the 8th March, 2019.* Retrieved May 23, 2021, from http://environmentclearance.nic.in/writereaddata/SCZMADocument/ICRZ_Notification2019.pdf

The Gazette of India (2019b, January 18). *Coastal Regulation Zone (CRZ) notification 2019. National Centre for Sustainable Coastal Management.* Retrieved January 2022 from http://www.sczmaodisha.org/pdf/CRZ%20Notification%202019%20English.pdf

The Hindu (2018, April 28). Fishermen want draft CRZ maps to be cancelled. *The Hindu.* https://www.thehindu.com/news/national/tamil-nadu/fishermen-want-draft-crz-maps-to-be-cancelled/article23714170.ece

The New Indian Express (2022, January 6). Fishermen oppose amendments to Coastal Regulation Zone notification, cite damage to ecosystem, community displacement. *The New Indian Express.* https://www.newindianexpress.com/states/tamil-nadu/2022/jan/06/fishermen-oppose-amendments-to-coastal-regulation-zone-notificationcite-damage-to-ecosystem-commu-2403722.html

The Times of India (2019, January 28). Fishermen's body threaten stir against new rules. *The Times of India.* https://timesofindia.indiatimes.com/city/goa/fishermens-body-threaten-stir-against-new-crz-rules/articleshow/67716187.cms

The Statesman (2018, October 31). West Bengal: Fishermen protest merchant shipping corridor. *The Statesman.* https://www.thestatesman.com/cities/fishermen-protest-merchant-shipping-corridor-1502703043.html

The Times of India. (2021, May 18). After 'Nisarga' in 2020, cyclone 'Tauktae' batters Raigad again: Mumbai News. *The Times of India.* Retrieved November 13, 2021, from https://timesofindia.indiatimes.com/city/mumbai/after-nisarga-in-2020-cyclone-tauktae-batters-raigad-again/articleshow/82736783.cms

The Wire (2019). Maharashtra's climate action plan comes up short. The Wire. https://thewire.in/environment/maharashtras-climate-action-plan-comes-up-short

The World Bank (n.d.). *Integrated coastal zone management.* The World Bank. Retrieved May 23, 2021, from https://projects.worldbank.org/en/projects-operations/project-detail/P097985

Vellore, R.K., Deshpande, N., Priya, P., Singh, B.B., Bisht, J., & Ghosh, S. (2020). Extreme Storms. In: R. Krishnan, J. Sanjay, C. Gnanaseelan, M. Mujumdar, A. Kulkarni, & S. Chakraborty. (Eds), *Assessment of Climate Change over the Indian Region* (pp. 155–173). Singapore: Springer. https://doi.org/10.1007/978-981-15-4327-2_8

Warrier (2018). The coasts need science-based policy action. *Mongabay.* Retrieved November 23, 2021, from, https://india.mongabay.com/2018/05/commentary-the-coasts-need-science-based-policy-action/

WBSAPCC (2017-2020). *West Bengal state action plan climate change.* Government of West Bengal. Retrieved December 21, 2021, from http://www.environmentwb.gov.in/pdf/WBSAPCC_2017_20.pdf

WBSAPCC (2012). *West Bengal State Action Plan on Climate Change.* Government of West Bengal. Retrieved December 21, 2021 from https://www.theclimategroup.org/sites/default/files/2020-10/west_bengal_future_fund_report.pdf

Yadav, B. M., Wasave, S. M., Mandavkar, S. S., Patil, S. V., Shirdhankar, M. M., & Chaudhari, K. J. (2020). Factors contributing for vulnerability on livelihood of fishers of Ratnagiri, Maharashtra State, India. *Asian Journal of Agricultural Extension, Economics & Sociology*, 91–96. https://doi.org/10.9734/ajaees/2020/v38i1130456

13 A "Coral State." Socio-political implications of the reefs' crises in the Maldives

*Marcella Schmidt di Friedberg and
Stefano Malatesta*

Corals, islands, and society: background and chapter's structure

Nasser declared in 1997, "The Maldives is entirely made up of atolls and associated coral structures." This chapter focuses on corals as the living and cultural heritage of the Maldives. The essay is essentially theoretical in nature and, through a systemic and critical reading of international literature and national reports, acts, and frameworks, it provides an overview of the links between coral reefs' health, climate change, and other elements of contemporary Maldivian societies. In the first paragraph, we introduce the geographical importance of coral reefs, to be considered as a synecdoche of the changes in progress, in fact, corals, as a system, both gather and represent the huge spectrum of socio-environmental interaction shaping Maldivian human ecology; in the second, the main anthropogenic threats to the bioecology of these marine habitats. In the following part, we discuss the exposure of the coral islands to the effects of climate change and the predominant position that adaptation and mitigation policies currently have in the national environmental agenda. The impact of climate change on reefs generates relevant feedback to island geographies. Due to this background, in the final part of the chapter, we present a few considerations on the cultural value of reefs and the implicit link between the aesthetic value of reefs and tourist attractiveness. We develop these arguments by addressing three socio-cultural dimensions: the link between corals and human activities, corals as a national heritage, and reefs as the major source of the international tourist market.

A Coral State?

The Republic of Maldives is an island state (298 sq. km), formed by a double chain of atolls stretching over 750 km in the Indian Ocean, between the islands of Lakshadweep in the north and Chagos in the south, southwest of the Deccan Peninsula. Of the 1,192 coral islands, divided into 20 natural atolls, 188 are inhabited (2017) (National Bureau of Statistics, 2018), about 160–170 (2020) host tourist resorts, while others are devoted to rural and productive activities. One-third of the resident population (total of 378,114) lives in the capital Malé, covering an area of about 5.8 sq. km. With an

DOI: 10.4324/9781003216476-16

extension of approximately 21,500 km² (Gischler et al., 2014), the coral reefs of the archipelago are among the largest in the world:

> The islands themselves and their beaches are made of corals, and other calcifying organisms and their subsistence is strictly linked to the health of corals. Maldives host some 300 species of corals even though the presence of other species and cryptic species (species morphologically indistinguishable that can be told apart using DNA) cannot be excluded. Many of these species grow large, contributing substantially to the formation of the reef.
>
> (Galli et al., 2021, p. 196)

From a geological point of view, the islands

> are accumulation of sand produced by the coral reef community, healthy coral reefs keeping up with a good carbonate production are of paramount importance to the Maldives. Large sediment supply after coral mortality events is followed by a dramatic reduction in sediment production, consequent starving of shallow carbonate platform, and potentially may lead to reef island erosion and final drowning.
>
> (Basso, Savini, 2021, p. 189)

The importance of corals in the country is that the Maldivian word "atoll" has acquired an international dimension. The etymological roots of the term can be traced back to the Dhivehi word *atholhu*. Charles Darwin, in his book *The Structure and Distribution of Coral Reefs* (1842), recognizes the indigenous origin of the word and its established use in Europe: well did François Pyrard de Laval, in the year 1605, exclaim, "C'est une merveille de voir chacun de ces atollons, environné d'un grand banc de pierre tout autour, n'y ayant point d'artifice humain." I have invariably used in this volume the term "atoll," which is the name given to these circular groups of coral islets by their inhabitants in the Indian Ocean and is synonymous with lagoon-island (Malatesta et al., 2021, pp. 3–4). Ibn Battuta, in turn, in the 16th chapter of his work *The Journeys* (الرحلة, *Rihla*, 1327–1330), described the atolls of the Maldives as rings with entrances like gateways.

From *The Death of a Nation* (the former President Mohamed Gayoom's pioneering statement at the *Small State Conference on the Sea Level Rise* in 1989) until the *Second National Communication of Maldives to the UNFCCC* (2016), the Maldivian Governments and Ministries set mitigation of climate change causes, and adaptation to climate change impacts as national priorities (Malatesta & Schmidt di Friedberg, 2017). Management of coral reefs has indubitably been a key component of this political program. The Nation, corals, and climate change compose an indivisible trio of contemporary political discourses. Scientific literature (Ban et al., 2011; Cauchi et al., 2021), tourist operators, and international agencies (https://www.iucn.org/regions/asia/countries/maldives) frequently adopt terms such as "coral nation" or "atolls nations," generally referring to small island states and jurisdictions, distributed along with the equatorial and tropical belts of the Indian Ocean and the Asia Pacific (e.g., the Republic of the Marshal Islands, Kiribati, Seychelles, French Polynesia). Bremner, recalling the famous Hayward theorization, defines the Maldives as a "coral aquapelago" to be understood as "a space performed into being by the interactions between many different species of terra-aqueous actants" (Bremner, 2017, p. 19). Our essay does not reduce the complex relationships of the human geography of the Maldivian islands (recently invested in a rapid process of cultural, social, political, and economic transformation) to the interaction between reefs and local communities alone. However, we have chosen to focus on coral reefs as a key element of the climate change emergency. In this sense, in this chapter, we define the Maldives as a "Coral State," stressing the multi-layered importance of coral reef ecosystems for the human geography of the archipelago and focusing on the importance of coral reef protection as a key goal of national environmental policies. Our perspective does not exclude the social and cultural values of other insular ecosystems (such as seagrass meadows or palm forests). However,

> For the Maldives, where the whole population practically lives on reefs, it would not be erred in stating that their livelihoods are totally dependent on the reefs and services obtained from them. In addition to their directly obvious economic significance, the coral reefs are also responsible for the protection of the coast from the open sea and storms and even from the obliteration of these low-lying islands.
>
> (Jaleel, 2013, p. 104)

According to the interpretation here proposed, corals act as a prism connecting reefs and islands' biogeography, the cultural meanings of the environment (Knoll, 2018; Schmidt di Friedberg & Abdulla, 2021), indigenous knowledge, heritage and adaptation, and mitigation policies.

Literature in the fields of bioecology and earth sciences has shown direct links between epiphenomena of climate change in tropical regions and the health of coral reefs. In the Indian Ocean Region, corals are "very vulnerable to climate change, and evidence suggests they face rapid degradation because of this, compounded with other disturbances" (Jaleel, 2013, p. 106). Moreover, the spatial patterns of Maldivian reef islands are influenced by seasonal climatic variations (Kench & Brander, 2006).

Coral reefs are the most important biological system of the Maldivian archipelago, one of the key elements of its attractiveness in the tourist market, and a habitat dramatically threatened by the consequences of climate change. They are *de facto* the dominant ecosystem of these islands; moreover, the physical geography of the archipelago is a direct outcome of coral reefs' bioconstruction activity. Abdulla Nasser explained,

> The physical setting of the Maldivian atolls varies from open structures with numerous *islands, faros* (ring-shaped reefs) *patches* and *knolls* in the atoll lagoon [...]. *Faros* is a ring-shaped reef emerging during low tidal

water, each with its own sandy lagoon, and is separated by deep channels. They generally have a rim of living coral consisting of branched and massive types. [...] In geological time, the filling up of the lagoons of faros by reef sediments has resulted in the formation of *coral islands*. [...] The islands are made up of coralline sand and have a very low elevation (on average, they are no more than 2 meters above sea level). The soil is highly alkaline, the water table is high, and the vegetation is sparse.

(Nasser, 1997)

Galli et al. (2021) underlined the irrevocability of reefs' biodiversity conservation to balance the multiple sources of biotic, abiotic, and anthropic stressors that currently affect corals: namely, abnormally elevated ocean temperatures, severe change in salinity, pollution, sea-level rise, increased concentration of dissolved CO_2, unregulated fishing practices, chemical contamination from agriculture outputs, coastal development, sand dredging, and land reclamation. Basso and Savini (2021) discussed the responses of Maldivian coral reefs to climate change effects focusing on three ongoing processes: sea-level rise, water warming, and ocean acidification. They claimed that "it is important to take into consideration the particular response of Maldivian area [where] sea-level rise is accompanied by an increase in temperature and a decrease in ocean pH" (p. 187). They concluded that "healthy corals still have the potential to survive the ongoing climate change and grow at a rate that can keep up the rate and magnitude of the expected sea-level rise" (p. 188). However, the authors added, "the pattern of their survival will probably depend on a combination of other pressures" (p. 189) mainly related to anthropogenic factors and impacts.

Shakeela, Becken, and Johnston added:

Coral reefs are also important beyond their biodiversity and recreational values because they provide a natural defense against storms and flooding (Becken & Hay, 2012). However, a wide range of stress factors, in addition to higher sea surface temperature and sea level, are affecting coral reefs and marine biodiversity. Damage from tropical cyclones and possible decreases in growth rates due to the increasing acidification (i.e., effects of higher CO_2 concentrations on ocean chemistry) will likely reduce the resilience of reefs. In addition, human activities, such as sewage discharge, waste pollution, physical damage, and disturbance of marine organisms, are adding to local pressures.

(Shakeela et al., 2015, p. 33)

Corals: between death and protection

Due to the high sensitivity of corals to sea surface temperature and acidification patterns, these organisms act as "the first signs of ecological stress from global warming" (Mohamed et al., 2016, p. 75). Coral bleaching is probably the most visible and disruptive phenomenon produced by combining the natural and anthropic threads mentioned above. There is scientific evidence

(Ghina, 2003, McNamara et al., 2019) of the association between *El-Niño* and large-scale and supraregional bleaching events in the Indian Ocean Region; however, the consequences of climate change, above all on sea temperature and ocean acidification (Jaleel, 2013), directly influence events' frequency: a condition directly determining corals' recovering opportunities. Furthermore, it has been shown that the severity and magnitude of bleaching events may increase significantly with projected climate change (Mohamed et al., 2016). After the massive crises in 1997–1998, with a mortality of over 90% of shallow-water coral colonies (Id.), in a very short space of time, the archipelago faced a chain of events: in 2010, 2012, 2015, and 2016. In 2016, from late April to May, "the overall percentage of bleached corals recorded was around 75 percent" (Galli et al., 2021, p. 202). Furthermore, "additional stress is exerted on the coral reefs due to channel building, land reclamation, sand mining, sea wall building, harbor construction, and dredging" (MEE, 2015b, p. 17). Land reclamation and harbor construction, largely promoted and encouraged as infrastructural adaptation measures in the last decade (Schmidt di Friedberg et al., 2020), may be considered strategies that seriously impact the health of the Maldivian coral reefs. These infrastructures are built by removing massive amounts of sand and corals from shallow waters generating consequences on the bioecology and hydrodynamics of lagoons. Therefore, it is a measure to contain the effects of climate change on the geography of the atolls (especially coastal erosion) and, on the other, an accelerator of coral community disintegration and an anthropic disturbance for the biology of the reefs.

The recent *Rapid Assessment of Natural Environment in the Maldives* (Dryden et al., 2020), starting from damage caused by the 2016 massive bleaching event, confirmed the centrality of coral reefs as the most important marine habitat and the key socio-economical relevance of the Maldivian marine environment. Therefore, the document prioritized the protection of corals and endangered species. Mohamed (2021) identified three founding texts setting coral reefs and biodiversity protection in the Maldives[1]: The 1987 *Fishery Law (5/87)*, the 1993 *Environmental Protection and Preservation Act (EPPA 4/93)*, and the ongoing *National Biodiversity Strategy and Action Plan 2016–2025 (NBSAP)*. The 1987 *Fishery Law* banned the export of pipe and black corals. Art no. 1 of the *EPPA* declared that "the natural environment and its resources are a national heritage that needs to be protected," explicitly mentioning coral reefs within the list of vital environmental resources. The *NBSAP* expanded the value of corals by stating that they

> act as a natural defense against the surging seas. They also provide natural replenishment of the sand by which the islands are formed. [furthermore, they are] a source of food, income, as well as a place for many tourist and recreational activities.
>
> (MEE, 2015b, p. 6)

Therefore, the protection of marine biodiversity is presented by combining an ecosystem service approach and the declaration of the cultural and social extrinsic values of corals. *NBSAP* identified seven main goals to target by 2025 (MEE, 2015b):

- Minimizing the pressure on coral reefs due to anthropogenic activities and climate change.
- Establishing some forms of protection on at least 10% of reef areas.
- Restoring impacted ecosystems.
- Getting close to the rate of zero natural habitat loss.
- Preventing the extinction of locally known threatened species (by 2020).
- Eliminating the illegal trade of protected species (by 2018).
- Minimizing the impact of waste and sewage management on coral reefs (by 2020).

Climate change policies and environmental protection (mainly through the institution of Marine Protection Areas) are both recommended as fundamental assets of *NBSAP*. Ban et al. (2011) discussed the multi-layered and multi-scale positive socio-economical feedback emerging from the institution and implementation of Marine Protected Areas in coral reefs nations. According to Mohamed (2021), in the Maldives, the establishment of MPAs is still one of the most efficient strategies to protect coral reefs. She reported, "since 1995, 61 Protected Areas have been established, including MPAs" (p. 168), showing a valuable (and constantly growing) level of public and institutional commitment. In the meanwhile, the Environmental Protection Agency (an office directly linked to the Ministry) prepared a list of more than 250 sensitive sites, including mangroves and reef areas. However, many sites still "lack clear management objective, plans, and resources" (Id.), running the risk of acting as mere paper parks.

The impact of climate change

"Climate change is a cross-cutting development issue as it affects every aspect of Maldivian ways and livelihoods" (MEE, 2015b, p. 10): due to the high level of exposure of coral reefs to abiotic and anthropic stressors directly or indirectly associated with climate change, the implementation of climate change mitigation and adaptation measures have been the priority of environmental policies at the national level (Malatesta & Schmidt di Friedberg, 2017). The 2016 *Second National Communication of Maldives to the UNFCCC* listed coral reefs (risk assessment and risk management) among the eight main assets that define the country's response to climate change effects. In addition, the impact of climate change on reefs is mentioned in the presentation of other key assets directly related to coral bioecology, such as fisheries or tourism.[2] This text highlighted the close links between adaptation policies, mitigation strategies, and reef protection. In the following paragraph, as stated above, we point to three pivotal socio-cultural dimensions of corals: the implication on human activities, the use of reefs as a national heritage, and the aesthetic value of these ecosystems.

Broadly speaking, the impact of climate change on reefs is, more generally, part of the discourse on tropical and equatorial island regions' exposure levels. This discourse should not be oversimplified in the discussion of coral states' policies. The vulnerability of atoll islands has been widely investigated and deconstructed (Barnett, 2001, Moore, 2010; Arnall, Kothari & Kelman, 2014; Kelman, 2014; Petzold, & Ratter, 2015; Cauchi et al., 2021; Moncada et al., 2021). Experts in different fields stressed the importance of considering the geography, social structures, and politics of the archipelago in discussing the Maldivian response to climate change effects (Sovacool, 2012; Malatesta & Schmidt di Friedberg, 2017; McNamara et al., 2019). In particular, McNamara et al. (2019) pointed out the complex set of spatial patterns and variables, such as marginality and connectivity, influencing vulnerability to climate change in Laamu Atoll (in the Southern region), while Malatesta and Schmidt di Friedberg (2017) stressed the constant recourse to climate change as the background to the entire corpus constituting contemporary environmental policies. Kelman recently used the expression "manufactured islands of vulnerability" (2020, p. 9), recalling the popular and interpretive stereotypes associating small islands with environmental vulnerability, while McNamara et al. (2019) claimed further understanding of climate change vulnerability in the Maldives. The authors proposed a study of the complex set of social and spatial variables defining local vulnerability to climate change. As discussed in this chapter, climate change is both the priority in national discourses and narratives and the key issue running through the entire policy framework for environmental laws. Fighting against climate change is both a political and a social issue, and the study of the complex links between island geography and reef ecology is instructive in overcoming oversimplifications such as those often applied in defining tropical island regions as "prone to climate change." According to Moore (2010), the predominance and institutionalization of climate change act as two main characters in the political arena of island states. In the case of the Maldives, this dominance is clearly visible when reading the body of legislation produced from the presidency of Mohamed Nasheed (2008–2012) until the *Second National Communication of Maldives to the UNFCCC* (2016).

Socio-environmental issues

Nasser stated,

> Being a country with the more territorial sea than dry land, Maldivians depend on resources almost entirely from the sea. The coral reefs, which built the country, play a vital role in the economic and social well-being of the country.

(Nasser, 1997, p. 1)

The social implications of marine habitat protection are crucial for environmental science and policies. Very often, the establishment of MPAs, the introduction of limits to human activities, and the application of standards and controls to fishing or tourism (e.g., the introduction of fees or special permits) generate conflicts between users and stakeholders. On the other hand, traditional Maldivian knowledge, particularly the body of narratives and myths related to nature and the ocean, can help address the current environmental crisis. Drawing on traditional knowledge can also be useful for conservation management and the creation of marine and terrestrial protected areas and encouraging a new focus on traditional resource use. This section does not discuss environmental conflict or negotiation; rather, we try to show the benefits of interpreting coral reefs as service providers.

As we underlined, the national regulatory body presented coral reefs as valuable ecosystem services providers: they assure food security, commercial opportunities, coastal protection, and strong cultural values, as stressed by Hein et al. (2019). In their study on the socio-ecological benefits of coral restoration in tropical regions,[3] including a case in the Maldives, the authors extended this interpretation to the protection and restoration of reefs by claiming, "socio-cultural benefits of coral restoration are likely to include the spectrum of ecosystem services provided by coral reefs such as the provision of alternative livelihood opportunities, increased educational opportunities and building stewardship" (2019, p. 15). Ban et al. (2011) illustrated the social and economic benefits (for coral states) linked to marine environment protection by including fisheries, tourism, and aesthetic richness. As for MPAs'

Figure 13.1 Coral nurseries in Faaf-Magoodhoo (Marine Research and High Education Centre, University of Milano-Bicocca).

Photograph by Luca Saponari – MaRHE Center.

establishment and management, coral restoration, "used as a management tool to minimize accelerating coral reef degradation resulting from climate change" (Hein et al., 2019, p. 14), has been widely discussed (Gibbs et al., 2021), animating an ongoing debate on its ecological effectiveness.

McNamara et al. (2019) defined the Maldivian socio-economic structure, like other "Small Island States" economies, as specialized in climate-sensitive sectors like tourism and fisheries. This description is reflected in national plans, laws, and reports (MEE, 2015a, 2015b; Mohamed et al., 2016, ADB, 2020; Dryden et al., 2020). The economic value of corals also involves other different production sectors:

> Nowadays, most corals have commercial value. They are being used in the aquarium business and in the pharmaceutical industry, manufacturing ophthalmic lenses, bone marrow transplantation, and many others (Ellis and Sharron, 1999). In medicine, chemical components of coral are used in treating diseases such as cancer, AIDS, pain and other disease.
>
> (Moradi, 2016, p. 125)

The protection of corals may positively contribute to the development of these sectors by generating direct benefits for a wide spectrum of stakeholders: fishers, tourist operators, visitors, and local administrators. Ban et al. (2011) supported this vision by underlying economic and ecosystem values of corals and reef biodiversity, suggesting that fishers directly or indirectly benefit from health and reach coral reef systems. They proposed a socio-political view on coral protection and support this idea by claiming that "the incorporation of broader social value and potential benefits of MPAs [...] in many countries with coral reefs, community priorities concern consistent flows of ecosystem goods and services" (2019, p. 26). Barnes et al. (2019) discussed the role, and direct involvement, of fishers' networks as key actors in coral reef management, emphasizing the reciprocal benefits (for human activities and reef biodiversity) emerging from this approach. Moreover,

> the human [socio-political] dimension is particularly relevant to coral restoration [and more generally to coral protection] since people are involved in all stages of the process [...] involvement of volunteers, and citizen scientists in restoration efforts have the potential to improve local and global stewardship of reef resources.
>
> (Hein et al., 2019, p. 15)

The involvement of locals, the relevance of socio-political issues in environmental governance, and the role of MPAs in disseminating environmental information and awareness are among the key components Mohamed targeted in her review of Maldivian environmental policies (2021).

Corals as heritage

From the earliest times, corals were used by inhabitants of coastal areas as a source of support for various activities:

> Coral reef biota is important sources of food and of reef limestone, sands, rubble and blocks for use as building materials. The physical barriers provided by coral reefs protect coasts from erosion by storm waves. Tourism associated with coral reefs provides many countries with significant foreign exchange earnings. Beyond these perhaps obvious benefits, reefs have been effective tools in art and architecture of many nations.
>
> (Moradi, 2016, p. 125)

In ancient cultures, corals were considered as plants, minerals, or mythological beings. In the 1700s, according to naturalists, they were plants that turned into rocks as they emerged from the water. (Id.). The Persian scholar and polymath Al Biruni, who visited the Maldives on his travels in 1017, described the red coral as a "stone tree." Corals were used as building materials, to make jewelry and decorative objects in medicine, and as amulets to ward off the evil eye (Id.). Coral stone structures are present in different parts of the world. According to UNESCO, "Coral has been used as a building material throughout the coastal settlements of the tropical belt in the Indian Ocean region, in Arabian/Persian Gulf, in the Red Sea and the Central American region" (https://whc.unesco.org/en/tentativelists/5812/).

In the Maldives, buildings depended primarily on the local availability of materials during the middle and modern ages. Coral stone and wood were the only durable materials, and coral stone became the primary building material for monumental architecture and houses. Archaeological studies reported the use of coral stone in buildings as far back as 165 CE during the Hindu and Buddhist periods. Over forty mosques built with coral stones have been documented, dating back to between the 14th and early 20th centuries. Figure 13.2 shows a coral stone mosque in Gaaf Dhaal.

Three styles can be identified in Maldivian construction techniques: *Veligaa* or coral sandstone, *Hirigaa* or coral stone, and Thelhigaa or coral rubble with lime masonry. Among these, the "Hirigaa" or "coral stone carpentry" technique is unique in the Maldives:

> Porite corals, called Hirigaa, from the stony coral family Scleractinia, were the main type of coral stone mined for construction in the Maldives. These boulder-like are slow-growing, and coral heads bigger than one meter in diameter are commonly found among the reefs. [...] Live coral boulders were extracted manually from the shallow reef, where it was available in abundance, using rods and ropes. It was then put into a raft or boat and taken to land, chipped or shaped, and dried before use.

Figure 13.2 Coral stone Maalande Miskiy (Flower Mosque) in GDh.Vaadhoo.
Photograph by Federica Adamoli – MaRHE Center.

> Techniques used in coral stone construction varied during different historical periods of Maldives.
>
> (Ahmad, Jameel, 2016, p. 52)

Live reef corals were very suitable for architectural and sculptural works, so the surfaces of the facades were finely decorated. Presently, six of these mosques – defined by Jameel and Ahmad as the "Vanishing Legacy" of the Indian Ocean – have been nominated as the first Maldivian site on the UNESCO World Heritage Tentative List (2013). We read in "Justification of Outstanding Universal Value":

> The Coral Stone Mosques of the Maldives represent a unique example in the Indian Ocean of an outstanding form of fusion coral stone architecture. They have Outstanding Universal Value as an example of a type of coral stone architecture with coral carvings and detailed lacquer work quality not seen in any part of the world. The architecture, construction and accompanying artistry are in themselves a work of human creative achievement.
>
> (UNESCO, https://whc.unesco.org/en/tentativelists/5812/)

And a little further on, in Criterion (ii):

> The building method comes directly from the Indian subcontinent but is applied to coral stone – the only example in the world of this particular material being used in this manner. [...] The marriage of process, materials, and techniques from other Indian Ocean cultures into a homogenous, consistent form in itself represents a masterpiece of creative genius in the coral stone mosques.
>
> (Id.)

Corals have been traditionally used on the islands also in home architecture, banned by law and regulated by the *Coral and Sand Mining Regulation*, public spaces (e.g., the curved coral stone walls), and fishing. The mining practice made it necessary to harvest living coral stone even from the "faros" of the outer atolls, which actually protect the islands' integrity from erosion and monsoon storms. Until it was banned, coal mining resulted in a dramatic reduction in coral diversity and abundance and, in some cases, almost total depletion of live coral after extraction (Brown and Dunne, 1988). Despite the normative body, NBSAP reported that corals still continue to be used as core components in construction, contributing to the exploitation of this natural resource. Shakeela et al. add:

> Coral stones from the lagoons, there is inconsistency within the regulations. The RPCETI [Regulation on the Protection and Conservation of Environment in the Tourism Industry] prohibits the extraction of coral stones for any purpose relating to the leasing of an island for tourism development or the renovation or repair of jetties. The same legislation, as well as the Regulation on the Mining of Stone, Sand, and Coral, and Law on Stone, Sand and Coral Mining in Inhabited Islands (Law no. 77/78), however, provides that stones can be mined with permission obtained from the local government authority. Thus, there are clear inconsistencies between regulations.
>
> (Shakeela et al., 2015, p. 81)

Abdulla Naseer (2020) investigated the visible clues of abandoned stone weirs in Faafu and Dhaalu Atolls (in the South-Central region), "the weirs are dug in the coral, to convey and catch *Mushimas* (mackerel) schools" (Malatesta et al., 2021, p. 5). These techniques indubitably incorporate and express a rich set of traditional practices, now at risk of disappearance, and

> in some cases, owe tourism to their survival [...] the recently established National Centre for Cultural Heritage (2019), under the mandate of the Ministry of Arts, Culture and Heritage, started an impressive survey of the country's material and intangible heritage.
>
> (Malatesta et al., 2021, p. 7)

Tourism, corals, and climate change: beyond the enchantment of Maldivian reefs?

Tourism and fisheries (the two economic pillars of Maldives) are intimately correlated to reef bioecology. They both directly depend on the health of reef ecosystems:

> With the introduction of tourism in the Maldives, the coral reefs gained a major economic significance. Tourism in the Maldives is centered around coral reefs and relies on these rich and healthy reefs for the

industry's well-being. Diving, snorkeling, water-sports, sun, and white sandy beaches are the major products sold for tourists in the Maldives.

(Nasser, 1997, p. 4)

In this essay, we focus on luxury tourism, particularly the accommodation facilities defined as the "Maldivian resorts" and their famous "one-island-one-resort" spatial pattern. According to official data and reports (MoT, 2019a, 2019b),[4] resorts still keep a dominant role as the favorite accommodation structures with a medium average of around 77% of tourists (with remarkable differences between nationalities). From a spatial point of view, state's control over luxury tourism and the reproduction of touristic enclaves are two basic elements shaping the geography of Maldivian resorts (Domroes, 2001, dell'Agnese, 2018). The narratives supporting the success of these facilities in the international market have been based on statements sounding like the "tourism sector in the Maldives has developed around the natural beauty of the islands like white sandy beaches and the rich coral reef ecosystem" (Shareef, et al., 2015, p. 4), or "other than the sandy beaches of the Maldives, the most valuable assets are the reefs" (Shareef et al., 2015, p. 9). These images have been proposed and reproduced by annual surveys indicating pristine beaches and enchanting submarine landscapes as the main attractions for international visitors (MoT, 2019b). Beyond the aesthetic value of coral reefs, however, the protection of marine habitats is linked to a set of key factors for tourism development.

The role of environmental management in the consolidation of resorts' attractiveness has been discussed by international scholars (Zubair, Bowen & Elwin, 2011; de-Miguel-Molina, de-Miguel-Molina, & Rumiche-Sosa, 2014; dell'Agnese, 2021, Basaglia et al., 2022). Zubair, Bowen, and Elwin (2011) have shown that the construction of structures along the coastline (jetties, harbors, walkways) and of the iconic water villas are highly disturbing actions toward the reefs. Moreover, mechanical damage, the removal of entire patches, and reef cover by sand are locally significant stressors. The health of coral communities is directly linked to the survival of many charismatic species, often used as a brand of the Maldivian tourist experience, such as sharks, turtles, and manta rays. At the same time, human-made structures, such as water villas, piers, or banks, are vital to maintaining the balance of the highly humanized geography of Maldivian resorts, so much so that Shareef et al. defined beach erosion as "the biggest environmental issue" (2015, p. 8) for tourism.

Coral health has recently become a strategic goal of some resorts' environmental plans. In addition to the awareness of the aesthetic and biological value of these structures and of the feedback they generate on the attractiveness of their business, some operators have included awareness-raising activities, coral restoration workshops, environmental information sessions, and lectures within what they offer to their guests (dell'Agnese, 2021). This type of strategy may provide educational and informational value, may act as a marketing plan (de-Miguel-Molina, de-Miguel-Molina, & Rumiche-Sosa, 2014), and, in some cases, is offered to visitors as an opportunity to be involved in citizen science projects (Basaglia et al., 2022). These proposals may include the establishment of a Marine Lab, a focal point informing guests about the reef ecosystems and environmental best practices, weekly meetings, and events designed to generate dialogue between experts from a range of fields (biologists, geographers, marine scientists) and guests, beach cleaning events, or coral frame workshops. In this perspective, the initiative of private operators may facilitate the integration between the iconic role of corals as a symbol of marine life vulnerability and the involvement of tourists in environmental protection policies and strategies.

The protection of natural and cultural heritage has recently become a national priority, including in relation to tourism, as part of the country's rapid transformation process. While the continued expansion of luxury tourism leaves intact the western myth of the Maldives as a paradise destination for its white beaches and coral reefs, alternative representations are being sought to revitalize the heritage: culture, monuments, and crafts have the potential to strengthen the country's national identity. The expansion of new non-European markets (China, India, Russia, and the Arab world), moreover, proposes a new model of tourism that does not necessarily draw on the Western imaginary:

> The Maldives is now a well-known brand to tourists, but the country needs to revive its arts and crafts, as there are new types of tourists arriving from Russia, China and the Middle East who are less attracted by a pristine beach holiday and more interested in local culture and the purchase of products.
>
> (Kothari, Arnall, 2017, p. 996)

Environmental emergencies – climate change, coral bleaching, sea-level rise, water acidification – contribute to the new focus toward heritage: on the one hand, the fact that the Maldives may be at risk of "disappearing" and is, therefore "to be seen now, before it is too late" increases the country's tourism appeal. On the other hand, plans for new infrastructures, land reclamation, ports, artificial islands, new settlements, and barriers against rising seas, as well as bleaching and the possible death of corals due to global warming, may lead to totally new scenarios for the future of Maldivian tourism. Processes of change offer an opportunity to rethink heritage as a cross-cutting priority in the country's governance and sustainable development. The new and active Ministry of Arts, Culture, and Heritage is committed to safeguarding objects and sites of historical importance for future generations and to ensuring the documentation and protection of cultural heritage. Among ongoing initiatives, the Maldives Heritage Survey (2018) has been launched to systematically catalog and document endangered assets and create a database of them, thereby integrating the needs of development, sustainability, conservation, and new technologies.

Local tourism, as a means of promoting culture, protecting heritage, and bringing economic opportunities to the inhabited islands, however, has been developing very slowly, and the government is well aware of this. As pointed out in the fourth TMP: "Maldives tourists come and will continue to come primarily for beach and marine activities. Taking visitors to see historical monuments will never be a prime motivator of a visit to the Maldives" (MOTAC, 2012, p. 124).

Specialized studies are still lacking, but

> It would be interesting to investigate to what extent the tourist system is dependent on the presence of a healthy coral reef or how much the Maldivian luxury tourist model can do without it. The resilience of the marine ecosystem and coral restoration projects could also give rise to the hope of restoring coral systems in various atolls, also suggesting a possible tourist redevelopment of areas further away from the capital in relation to the attraction exerted by the quality of corals.
>
> (Schmidt di Friedberg, 2019, p. 12)

Conclusion

In calling the Maldives a "Coral State," we have chosen to consider coral – both as a living organism and as a monumental heritage – as a key element in interpreting climate change and addressing its consequences. We discussed the value and function of corals by proposing an overview of their socio-political meaning. As has been shown, corals play a key function in the complex relationships between human systems and the bioecology of the Maldivian islands. At the same time, these living structures are a threatened habitat and a crucial resource for the country's major industries. We define the Maldives as a "Coral State," stressing the multi-layered importance of coral reef ecosystems for the human geography of the archipelago and focusing on the importance of coral reef protection as a key goal of national environmental policies.[5] On top of this, the link between corals and heritage has recently become spotted in the political arena. This chapter concludes with a statement by the current president reminding us of the social, political, and cultural meanings of the term Coral State. Recently, in his presentation of the upcoming fifth TMP (2020–2025), President Solih emphasized the importance of linking the country's nature and culture, and heritage more closely with the tourism industry with a view to strengthening an already established tourism identity. The new plan is inspired by the principles and practices of sustainable tourism and reflects a commitment to seek out new ways to conserve Maldivian cultural and natural heritage:

> The Maldives Fifth Sustainable Tourism Master Plan will define strategies and activities and provide directions for further development of Maldives' tourism sector to promote development based on sustainable tourism principles and practices. The objectives being the development of environmentally sensitive business operations, support for the protection of cultural and natural heritage, and tangible economic and social benefits to locals.
>
> (MoT, 2019, p. 31)

Notes

1 Jaleel (2013) and Shadiya (2021) analytically presented the environmental governance of the Maldives. Readers may refer to their works.
2 The impact of climate change on islands and reefs systems may shape a multi-hazard scenario (ADB, 2020) affecting all socio-environmental relations (water management, coastal erosion, fishing, waste management, lagoon pollution, coastal
infrastructures, and food production). This complex frame has been effectively
presented by Mohamed et al. (2016) and MEE (2015a).
3 Figure 13.1 shows coral nurseries providing the restoration programs developed
in Faaf Atoll by MaRHE Center.
4 We refer to 2019 MoT's reports considering the potential variation and anomalies
related to Covid-19 influence on tourism.
5 On 11 February 2022, a contract was signed to start work on the first Maldives
Floating City. The eco-friendly structure of the city has been designed to resemble
the hexagonal structure of brain coral.

References

ADB. (2020). *Multihazard Risk Atlas of Maldives: Summary - Volume V: Vol. IV*
(Issue March). https://www.adb.org/publications/multihazard-risk-atlas-maldives

Ahmad J., & Jameel M. (2016). *Coral Stone Mosques of Maldives: The Vanishing
Legacy of the Indian Ocean*, Cairo, Gulf Pacific Press.

Arnall, A., Kothari, U., & Kelman, I. (2014). Introduction to politics of climate
change: Discourses of policy and practice in developing countries. *Geographical
Journal, 180*(2), 98–101. https://doi.org/10.1111/geoj.12054

Ban, N. C., Adams, V. M., Almany, G. R., Ban, S., Cinner, J. E., McCook, L. J., Mills,
M., Pressey, R. L., & White, A. (2011). Designing, implementing and managing
marine protected areas: Emerging trends and opportunities for coral reef nations.
In *Journal of Experimental Marine Biology and Ecology, 408*(1–2). https://doi.
org/10.1016/j.jembe.2011.07.023

Barnes, M. L., Bodin, Ö., McClanahan, T. R., Kittinger, J. N., Hoey, A. S., Gaoue,
O. G., & Graham, N. A. J. (2019). Social-ecological alignment and ecological conditions in coral reefs. *Nature Communications, 10*(1). https://doi.org/10.1038/
s41467-019-09994-1

Barnett, J. (2001). Adapting to climate change in Pacific Island countries: The problem of uncertainty. *World Development, 29*(6), 977–993. https://doi.org/10.1016/
S0305-750X(01)00022-5

Basaglia, L., Pecorelli, V., Pepe, A., Saponari, L. & Malatesta, S. (2022, in press),
Luxury tourism and environmental awareness: A case study in Alifu Dhaalu Atoll,
Maldives, *Geotema.*

Basso, D. & Savini, A. (2021). Sea-level changes and the reefs of the Maldivian archi-
pelago. in Malatesta, S., Schmidt di Friedberg, M., Zubair, S., Mohamed, M., &
Bowen, D. (Eds.), *Atolls of the Maldives. Nissology and Geography.* Rowman and
Littlefield, Lanham, pp. 176–195.

Becken & Hay (2012). *Climate Change and Tourism.* From Policy to Practice, London:
Routledge.

Bremner, L. (2017). Observations on the concept of the aquapelago occasioned by
researching the Maldives. *Shima, 11*(1), 18–29. https://doi.org/10.21463/shima.
11.1.05

Brown, B., Dunne, R. P. (1988). The environmental impact of coral mining on coral
reefs in the Maldives. *Environmental Conservation, 15*(02), 159–165.

Cauchi, J. P., Moncada, S., Bambrick, H., & Correa-Velez, I. (2021). Coping with environmental hazards and shocks in Kiribati: Experiences of climate change by atoll communities in the Equatorial Pacific. *Environmental Development, 37.* https://doi.org/10.1016/j.envdev.2020.100549

Darwin, C. R. (1842). The Structure and Distribution of Coral Reefs: Being the First Part of the Geology of the Voyage of the Beagle, under the Command of Capt. Fitzroy, R.N. during the Years 1832 to 1836. London: Smith Elder and Co.

dell'Agnese, E. (2018). Islands within Islands? The Maldivian resort, between segregation and integration. *Tourism Geographies, 21*(5), 1–17.

dell'Agnese, E. (2021). Greening the resort, de-bordering the enclave. In Malatesta, S., Schmidt di Friedberg, M., Zubair, S., Mohamed, M. & Bowen, D. (Eds.), *Atolls of the Maldives. Nissology and Geography.* Rowman and Littlefield, Lanham, pp. 106–124.

de-Miguel-Molina, B., de-Miguel-Molina, M., & Rumiche-Sosa, M. E. (2014). Luxury sustainable tourism in small island developing states surrounded by coral reefs. *Ocean and Coastal Management, 98,* 86–94. https://doi.org/10.1016/j.ocecoaman.2014.06.017

Domroes, M. (2001). Conceptualising state-controlled resort islands for an environment-friendly development of tourism: The Maldivian experience. *Singapore Journal of Tropical Geography, 22*(2), 122–137. https://doi.org/10.1111/1467-9493.00098

Dryden et al. (2020). *Rapid Assessment of Natural Environment in the Maldives.* IUCN Maldives, Malé

Galli, P., Montano, S., Seveso, D., Maggioni, D. (2021). Coral reef biodiversity of the Maldives. in Malatesta, S., Schmidt di Friedberg, M., Zubair, S., Mohamed, M., & Bowen, D. (Eds.), *Atolls of the Maldives. Nissology and Geography.* Rowman and Littlefield, Lanham, pp. 196–212.

Ghina, F. (2003). Sustainable development in small island developing states. *Environment, Development and Sustainability, 5*(1), 139–165.

Gibbs, M. T., Gibbs, B. L., Newlands, M., & Ivey, J. (2021). Scaling up the global reef restoration activity: Avoiding ecological imperialism and ongoing colonialism. *PLoS ONE, 16*(5 May). https://doi.org/10.1371/journal.pone.0250870

Hein, M. Y., Birtles, A., Willis, B. L., Gardiner, N., Beeden, R., & Marshall, N. A. (2019). Coral restoration: Socio-ecological perspectives of benefits and limitations. *Biological Conservation, 229.* https://doi.org/10.1016/j.biocon.2018.11.014

Jaleel, A. (2013). The status of the coral reefs and the management approaches: The case of the Maldives. *Ocean and Coastal Management, 82,* 104–118. https://doi.org/10.1016/j.ocecoaman.2013.05.009

Jameel, M. M., & Ahmad, Y. (2015). "Architectural heritage of Maldives and its revival through tourism", Conference Paper, Conference *Islam and Multiculturalism: Islam in Global Perspective,* January, NYU Abu Dhabi.

Kelman, I. (2014). Climate change and other catastrophes: Lessons from Island vulnerability and resilience. *Moving Worlds: A Journal for Transcultural Writings, 14*(2), 127–140.

Kelman, I. (2020). Islands of vulnerability and resilience: Manufactured stereotypes? *Area, 52*(1), 6–13. https://doi.org/10.1111/area.12457

Kench, P. S., & Brander, R. W. (2006). Response of reef island shorelines to seasonal climate oscillations: South Maalhosmadulu atoll, Maldives. *Journal of Geophysical Research: Earth Surface, 111*(1), 1–12. https://doi.org/10.1029/2005JF000323

Knoll, E.-M. (2018). The Maldives as an Indian Ocean crossroads. In *Oxford Research Encyclopedia of Asian History* (Issue December). https://doi.org/10.1093/acrefore/9780190277727.013.327

Kothari U. & Arnall, A. (2017). Contestation over an island imaginary landscape: The management and maintenance of touristic nature, *Environment and Planning A, 49*(5), pp. 980–998.

Malatesta, S., & Schmidt Di Friedberg, M. (2017). Environmental policy and climate change vulnerability in the Maldives: From the 'lexicon of risk' to social response to change. *Island Studies Journal, 12*(1). https://doi.org/10.24043/isj.5

Malatesta, S., Schmidt di Friedberg, M., Zubair, S., Mohamed, M. & Bowen, D. (Eds) (2021). *Atolls of the Maldives. Nissology and Geography*. Rowman and Littlefield, Lanham.

McNamara, K. E., Clissold, R., Piggott-Mckellar, A., Buggy, L., & Azfa, A. (2019). What is shaping vulnerability to climate change? The case of Laamu atoll, Maldives. *Island Studies Journal, 14*(1), 81–100. https://doi.org/10.24043/isj.67

MEE. (2015a). Guidance manual for climate risk resilient coastal protection in the Maldives, https://www.environment.gov.mv/v2/en/download/13722

MEE. (2015b). *National biodiversity strategic and action plan 2016–2025*, https://www.environment.gov.mv/v2/en/download/4318

Ministry of Tourism Arts and Culture MoTAC (2012). *Fourth Tourism Master Plan 2013–2017*. Malé: Ministry of Tourism Arts and Culture Others erased or reframed in the text.

Mohamed, M. (2021). Environmental and ecosystem protection. in Malatesta, S., Schmidt di Friedberg, M., Zubair, S., Mohamed, M., & Bowen, D. (Eds.), *Atolls of the Maldives. Nissology and Geography*. Rowman and Littlefield, Lanham, pp. 159–175.

Mohamed, M. et al. (2016). *Second National Communication of Maldives to the UNFCCC*: Ministry of Environment and Energy, Malé.

Moncada, S. et al. (2021). *Small Island Developing States. Vulnerability and Resilience Under Climate Change*. Springer International Publishing, Cham.

Moore, A. (2010). Climate changing small islands. considering social science and the production of Island vulnerability and opportunity. *Environment and Society: Advances in Research, 1*, 116–131.

Moradi, Z. (2016). The role of coral in art and architecture: An overview, *International Journal of Aquatic Biology, 4*(2), 125–142.

MoT. (2019a). Tourism Yearbook 2019, https://www.tourism.gov.mv/en/downloads/publications

MoT. (2019b). Visitors Survey 2019, https://www.tourism.gov.mv/en/downloads/visitor_survey

Naseer, A. (2020). *Fishing for Bigeye Sad (Selar Crumenophthalamus). Using Fish Weirs on Coral Reefs*, Oral Lecture Workshop "Places and Local Community" MaRHE Center, Faafu-Magoodhoo.

Nasser, A. (1997). "Profile and status of coral reefs in maldives and approaches to its Managemen" in Vineeta Hoon *Proceedings of the Regional Workshop on the Conservation and Sustainable Management of Coral Reefs*. Proceedings No. 22, CRSARD, Madras http://www.fao.org/3/x5627e/x5627e0a.htm#6%20profile%20and%20status%20of%20coral%20reefs%20in%20maldives%20and%20approaches%20to%20its%20management

National Bureau of Statistics. (2018). *Statistical pocketbook of Maldives 2018. 2.* http://statisticsmaldives.gov.mv/nbs/wp-content/uploads/2019/01/Statistical-Pocketbook-of-Maldives-2018-Printing.pdf

Petzold, J., & Ratter, B. M. W. (2015). Climate change adaptation under a social capital approach - An analytical framework for small islands. *Ocean and Coastal Management, 112*, 36–43. https://doi.org/10.1016/j.ocecoaman.2015.05.003

Schmidt di Friedberg, M. (2019). History of the landscape gaze. The image of tourism in the Maldives. in Eduardo Martínez de Pisón, E., & Ortega Cantero, N. (Eds.), *Paisaje y turismo.* Madrid, Instituto del Paisaje de la Fundaciòn Duques de Soria, Ediciones de la Universidad Autónoma de Madrid, Madrid pp. 97–120.

Schmidt di Friedberg, M, & Abdulla, A. (2021). The gender dimension of environment in the Maldives. in Malatesta, S., Schmidt di Friedberg, M., Zubair, S., Mohamed, M., & Bowen, D. (Eds.), *Atolls of the Maldives. Nissology and Geography.* Rowman and Littlefield, Lanham, pp. 45–63.

Schmidt di Friedberg, M., Malatesta, S., & dell'Agnese, E. (2020). Hazard, resilience and development: The case of two Maldivian islands. *Bollettino della Società Geografica Italiana* 14, 3(2): 11–24. https://doi.org/10.36253/bsgi-1087

Shadiya, F. (2021). Governance, activism and environment in the Maldives. in Malatesta, S., Schmidt di Friedberg, M., Zubair, S., Mohamed, M., & Bowen, D. (Eds.), *Atolls of the Maldives. Nissology and Geography.* Rowman and Littlefield, Lanham, pp. 145–158.

Shakeela, A., Becken, S., & Johnston, N. (2015). *Gaps and Disincentives that Exist in the Policies, Laws and Regulations which Act as Barriers to Investing in Climate Change Adaptation in the Tourism Sector of the Maldives,* Griffith University, Ministry of Tourism, Malé.

Shareef, A. et al. (2015). *Baseline Analysis of Adaptation Capacity and Climate Change Vulnerability Impacts in the Tourism Sector,* Ministry of Tourism, Malé.

Sovacool, B. K. (2012). Perceptions of climate change risks and resilient island planning in the Maldives. *Mitigation and Adaptation Strategies for Global Change, 17*(7), 731–752. https://doi.org/10.1007/s11027-011-9341-7

UNDP Maldives. (2012). Increasing climate change resilience of Maldives through adaptation in the tourism sector. *Tourism. Gov. Mv, June,* 1–88. http://www.tourism.gov.mv/downloads/tap/2012/reports/Report_on_Vulnerability.pdf

Zubair, S., Bowen, D., & Elwin, J. (2011). Not quite paradise: Inadequacies of environmental impact assessment in the Maldives. *Tourism Management, 32*(2), 225–234. https://doi.org/10.1016/j.tourman.2009.12.007

14 Polycentricity as a Framework for Understanding India's Climate Policy

Derek Kauneckis and Juhi Huda

Introduction

While much of the media's attention has been on international negotiations around greenhouse gas (GHG) emissions and the increasingly dramatic manifestations of climate change, less attention has been paid to how subnational governments are already responding to the dual needs of mitigation and adaptation. While nations like India have historically contributed very little to the global atmospheric accumulation of GHGs, there have been local and regional initiatives to monitor, sequester, and attempt to reduce GHGs. There is even more activity in terms of adaptation as communities, and political jurisdictions attempt to cope with climate change impacts (Atteridge et al., 2012; Jogesh & Dubash, 2015; Wei et al., 2012). India's climate policy operates in a complex environment. As a nation, it is highly vulnerable, is a major emitter already, and has ample coal supplies. Nevertheless, it has historically contributed less GHG emissions than many countries of its size (Shukla et al., 2003). Multiple drivers within India's policy environment led to a complex array of actors, interactions, and outcomes around both mitigation and adaptation policies.

Understanding the emerging climate policy and actions necessitates using a framework that can look past a reductionist top-down model of the central government as the key actor formulating climate policy. India's policymaking environment is dominated by a strong central government and a large bureaucratic administrative structure (Mitra, 2012). The limits of central governments in responding to new challenges, developing innovative policies, and avoiding capture by industry groups are well documented. Focusing climate policy analysis on national governments only both fails to explain much of the policy change occurring as well as limits proscriptive actions for more effective mitigation and adaptation actions. There is a profound organizational complexity of Indian civic life that is glossed over when attention is only paid to the actions and decisions of the central government and international actors.

This chapter discusses how the diversity of India's governance structure relates to emerging climate policy. We first discuss the remarkable organizational and social diversity in India and forms of political decentralization.

DOI: 10.4324/9781003216476-17

Next, we examine the types of climate risks, hazards, and exposure across different regions. The theoretical approach of polycentricity is introduced as a framework to better understand interactions across and between different civic organizations, communities, cities, and regions and how they respond to climate challenges. The current state of climate policy is then reviewed, paying particular attention to various subnational efforts across both adaptation and mitigation. It concludes with a discussion of how the polycentricity approach can provide novel insight into emerging climate policies.

The Complexity of Indian Governance

India's governance structure is a result of its history as an administrative state, colonialism, an independence movement, as well as its social and geographic heterogeneity. In terms of societal diversity, India is not only the largest democracy in the world but also the most diverse. It has the fourth-highest number of languages in the world with 122 major languages, 1,599 other languages, and 22 official languages (Government of India, 2001). Religious heterogeneity is equally diverse, with six major religions and over another dozen subdivisions. This high degree of social diversity has resulted in a highly decentralized governance structure, although with a strong central government at the national level. Following independence, India formally adopted a socialist model for its economy and social policies (Harmel, 2016; Mitra, 2012). This is most evident in the central planning and the cycle of national Five-Year Plans, the size of the federal bureaucracy, and the continued strong hand of the central government across all aspects of policymaking (Jalan, 2000; Murali, 2017). Even with this high degree of centralization, public services delivery responsibilities are spread across over 255,000 village-level local governments (Gram Panchayat), 6,635 intermediary level governments (Panchayat Samiti) 653 districts (Zila Parishad), 4,449 local urban government bodies, and 28 state governments. Many of the explicit responsibilities of these various levels of government are directly impacted by climate change, from providing drinking water, managing community forests and common grazing lands, and disaster preparation and recovery, among others within India's federal system (Fadia, 1984; Jörgensen, 2020; Parikh and Weingast, 1997).

The important role that social movements played in Indian independence resulted in a robust civic culture and extensive proliferation of non-governmental organizations (Chandra et al., 2016; Sen, 1992). With one of the largest and most diverse third sectors in the world, at 3.3 million registered organizations, the role that NGOs play in political and social systems is complex (International Center for Not-For-Profit Law, 2021). The organization of civil society into formal entities can serve multiple functions, from the self-organization of citizens around common interests to more explicit political organizations. NGOs have complex relationships with formal government, sometimes complementing and being the implementation agency for policy, in others offering services when government agencies do not (Banks and Hulme, 2012; Batley and Rose, 2011; Beer et al., 2012). The NGO sector in India has increasingly provided oversight and can have a contested relationship with formal government actors as they advocate for inclusion in governance and function as watchdogs of the policy processes (Sen, 1999).

India's economy is similarly diverse, with traditional subsistence agricultural villages, a large manufacturing sector, and global leadership in information technology services and biotechnology. It is the sixth-largest consumer market globally and from 2013 to 2018 was the world's fastest-growing economy. Agricultural production remains a critical component of its economy, accounting for about 50% of GDP and employing 66% of India's rural population. Poverty remains high, and India has greater economic inequality than most industrial economies (IMF, 2020; OECD, 2019). This has important implications for climate policy. As the negative impact of a rapidly changing climate disproportionally falls on those relying directly on natural resources for food and household economic activities, the agricultural sector and subsistence farmers will feel the greatest impacts. Poverty is a leading indicator of climate vulnerability. Given a large number of poor relying on natural resources, those already vulnerable to environmental and economic shocks are likely to face increased risks (Singh et al., 2017). There are risks associated with climate change at the other end of the economic spectrum as well. India's embeddedness in global supply chains means that shocks from natural disasters abroad can impact the national and local economies, from spikes in food and fuel prices to financial markets responding to risks. The disparity in economic growth across the population is also reflected in underinvestment in both urban and rural infrastructure and the presence of large informal housing on untitled and public lands. The compound effect of a fragile infrastructure, informal housing, and poverty adds additional risk to those populations that do not have a resilient infrastructure that can buffer against extreme weather, flooding, extreme heat, droughts, and other climate impacts (Freedman, 2000; Pandey and Jha, 2012).

This socially and organizationally diverse environment and complex decentralized and mixed governance structure in Indian policymaking mean it is difficult to find a single source of decision-making around policies intended to reduce carbon emissions and adapt to a changing climate. The national government is a critical factor, although policies are more an outcome of interactions across political executives at multiple levels, senior administrators competing for bureaucratic agencies, and government and non-governmental actors involved in implementation than any single legislative or executive body (Jain, 1987). In order to shift perspectives from looking at climate policy as the only output of formal national political organizations, a framework is needed to unpack this complexity better both to guide research and improve the design and implementation of effective climate action. A growing body of work shows that subnational states and provinces play an essential role in environmental performance and climate action (Fernandes et al., 2019). The next stage is grappling with the inherent complexity of the policymaking system and how it interacts with climate risks.

India's climate policymaking has been characterized by "periods of activity, inactivity, and institutional restructuring," with an initial focus on the management of climate diplomacy during 1992–2007 (Pillai & Dubash, 2021, S97). India's climate policymaking has always favored a centralized, top-down approach even though the country underwent a process of liberalization and decentralization in the 1990s. Increasing international pressure in the late 2000s resulted in two important developments: the National Action Plan on Climate Change (NAPCC) and a Prime Minister's Council on Climate Change (PMCCC) with a Special Envoy on Climate Change, which was "a one-person office that relied on the entrepreneurial abilities of the incumbent to work the governmental system" (Jörgensen et al., 2015; Pillai & Dubash, 2021). Both were at the national government level and largely emphasized the primacy of development over climate action. NAPCC was criticized for being overly focused on the central government, given its focus on eight sectoral missions related to mitigation and adaptation that required climate targets to become a part of the daily functions of the bureaucracy. Through a narrative of "co-benefits" NAPCC brought together climate and development opportunities under one umbrella while staying faithful to the principle of common but differentiated responsibility (Dubash, 2021). In 2009, a new environment minister, Jairam Ramesh, attempted to reframe India's climate policy to build a more robust institutional base for climate policy but faced considerable pushback from bureaucratic, political, and civil society actors, thereby stifling any institutional reform.

In 2009, the national government directed all 29 states and 7 union territories to prepare State Action Plans on Climate Change (SAPCCs) (Gogoi, 2015). The spaces are the primary formal policy document at the subnational level for addressing climate vulnerabilities and increasing climate resilience (Chaturvedi et al., 2019). These are supposed to lay a framework for climate change action. The plans are structured uniformly with an outline of the unique vulnerabilities of each state in relation to climate change and each state government's climate adaptation approach for current and future impacts. Most SAPCCs focus primarily on adaptation, with some focus on mitigations of GHGs. The focus remains on sectors crucial to the economy, with possible actions covering a one-to-five-year period. SAPCCs have the potential to enable an institutional platform for concerns related to environmental sustainability within developmental planning. However, the plans have been criticized for shortcomings in the "approach, process, formulation of outcomes, and implementation efforts" (Dubash & Jogesh, 2014, n.p.). SAPCCs are characterized by "[t]hin conceptual frameworks, processes that provide no space for generating a vision of change, limited state capacity, and truncated time frames" (Dubash & Jogesh, 2014, n.p.). There is also a lack of consistency in the methodology used (Gogoi, 2015). For SAPCCs to be effective, there needs to be upscaling and capacity enhancement strategies and a localization process to downscale to the district level and lower (Chaturvedi et al., 2019).

While an Executive Committee on Climate Change (ECCC) was established at the National level in 2013 to coordinate mission implementation, its role was limited. A new government elected in 2014 emphasized development and further stymied national climate activity. Despite weakening national level attention, climate policy has spread across different levels of government even though the institutional bodies that were central to NAPCC are no longer functional. India lists 35 national mitigation actions in its 2018 Biennial Update Report to the UNFCCC. However, 25 of these are beyond the scope of the eight sectoral missions listed by the NAPCC (Government of India, 2018). Climate change was not explicitly on the agenda of the newly elected Bharatiya Janata Party's (BJP) election agenda. In keeping with its rebranding strategy, the Ministry of Environment and Forests (MoEF) became the Ministry of Environment, Forests, and Climate Change (MoEFCC), and there was an increased focus on renewable energy, most notably, solar energy. The Prime Minister's Office (PMO) still took a leading role in climate policymaking initiatives. The ECCC limited itself to the eight sectoral missions with a scant focus on cross-sectoral climate policy (Pillai & Dubash, 2021).

India's Climate Risk

The Global Climate Risk Index 2021 ranks India as the seventh most-affected country regarding climate change impacts (Eckstein et al., 2021; Nandi, 2021). The latest IPCC report projected an increase in extreme weather events in South Asia (IPCC, 2021). With its diverse landscape, it is no surprise that it experiences a wide range of impacts. While the large-scale regional effects are understood, it is less certain how these will manifest at the local level given the wide disparities in adaptive capacity and diversity of urban and rural areas.

Temperatures already widely vary, with northern India averaging as low as 2°C with extremes as low as −45°C in parts of the Himalayas. The Himalayan range provides a barrier for the rest of Asia, allowing warmer temperatures to prevail in the rest of India, with averages in some states as warm as 29°C (Picciariello et al., 2021). India has experienced an increasing trend in surface temperature in the last century (IPCC, 2021; Krishnan et al., 2020; Mall et al., 2006). By the end of the twenty-first century, the national average is projected to increase by 4.4°C with an increase in the frequency of warm days and nights, assuming the high concentration RCP8.5 scenario (Krishnan et al., 2020, p xiv). The frequency and severity of heatwaves have already increased across India, with southern and western India experiencing 50% more heatwave events between 1985 and 2009 compared to the previous 25 years. Heatwaves in 2013 resulted in more than 1,500 deaths, while the 2015 heatwaves caused more than 2,000 deaths (Mazdiyasni et al., 2017; Picciariello et al., 2021). The frequency of summer heatwaves is expected to increase, with the average expected to approximately double (IPCC, 2021; Krishnan et al., 2020).

The propensity for droughts has increased all over India with a decline in the seasonal summer monsoon rainfall. Between 1951 and 2016, the frequency and spatial extent of droughts have increased, and climate projections estimate a further increase in drought frequency, severity, and spatial extent by the end of the twenty-first century. While there has been a reduction in the annual frequency of tropical cyclones between 1951 and 2018, the frequency of severe cyclonic storms has increased (Krishnan et al., 2020). India experienced one of its most active cyclone seasons ever in 2019. Cyclone Fani, the worst of the eight cyclones, affected a total of 28 million people, with economic losses estimated at US$8.1 billion (Eckstein et al., 2021). Rainfall distribution varies widely across the country, with the northeastern state of Meghalaya getting over 4,000 mm while the Thar Desert in the west is getting less than 100 mm of annual rainfall. India's around 7,500 km long coastline also significantly influences local climate patterns. The southwest monsoons are responsible for most rainfall, whereas the northeast monsoons bring rain to southern India. The variation in temperature and rainfall distribution has resulted in diverse ecological zones spanning alpine, desert, humid subtropical, and both wet and dry tropical ecosystems (Picciariello et al., 2021).

Flooding events are projected to increase further, with more glaciers and snowmelt compounding flood risk in the Himalayan region, and major river basins (Indus, Ganga, Brahmaputra) are at considerable risk if no additional adaptation and risk mitigation measures are taken. Extreme precipitation events have increased, even while total rainfall has declined. Central India experienced three times more precipitation between 1950 and 2015, resulting in thousands of deaths and millions displaced (Singh et al., 2014; Roxy et al., 2017). In 2019, monsoon conditions persisted for a month longer than normal, with 110% of the normal rainfall from June to September 2019. The floods from the heavy rains in 2019 resulted in 1,800 deaths, displaced 1.8 million people, affected 11.8 million people, and caused economic damage of approximately US$10 billion (Eckstein et al., 2021).

Current changes to precipitation have not impacted freshwater supplies since the decline in average precipitation (Krishnan et al., 2020) has been offset by an increase in runoff from melting snow and glaciers in the Hindu-Kush Himalaya. Glacial melt will have wide-ranging impacts on future water supplies as the Hindu-Kush Himalayan region has an estimated 50% of all glaciers outside the polar regions. These have been retreating rapidly (Singh et al., 2016). Retreating glaciers and diminishing snow are exacerbating flooding, with 2013 floods in Uttarakhand killing over 4,000 people. The 2021 collapse of a Himalayan glacier caused considerable flooding and resulted in the death of over 100 people (Arcanjo, 2019; Doman & Shatoba, 2021). The decline in rainfall is beginning to adversely impact groundwater since aquifers are not recharging adequately as Indian agriculture is becoming increasingly dependent on groundwater supplies due to precipitation reductions. Water security is already at risk, with a billion people in India facing severe water scarcity at least one month a year, while 180 million face severe water scarcity at some point during the year (Mekonnen & Hoekstra, 2016; Picciariello et al., 2021).

Health-related issues such as heat stroke, stress-related disorders, neurological diseases as well as vector-borne diseases like malaria and dengue fever due to higher and more moist temperatures may also be on the rise. Due to sea-level rise, the densely populated coastal regions have increased exposure to a range of climate risks, from extreme weather and flooding to the prevalence of tropical disease vectors (IPCC, 2021; Krishnan et al., 2020). Central and northwestern India is more susceptible to extreme heat, though southern and coastal parts will also be affected to some extent (Rohini et al., 2019). With declining rainfall, diminishing glaciers, and reduction in precipitation, water scarcity will negatively impact food production in the river basins and threaten the livelihoods of "209 million people in the Indus basin and 62 million in the Brahmaputra basin" (Picciariello et al., 2021, p. 5). Increasing sea levels will adversely affect infrastructure and property in low-lying, densely populated coastal cities, including Mumbai, Chennai, and Kolkata. Low-income rural communities dependent on the coastal ecosystem for their livelihood will also be negatively impacted. Furthermore, storms are expected to increase in frequency and severity, threatening low-income urban households located in densely populated settlements that lack essential services and infrastructure (Swapna et al., 2020; Sarthi et al., 2014; Satterthwaite et al., 2020). Without considerable adoption of mitigation and adaptation measures, climate change impacts will likely pose a serious threat to India's economic and sustainable development.

Polycentricity as a Framework to Understanding Climate Policy Complexity

Polycentricity provides an analytical framework for linking climate risks to the response within a complex decentralized policy system. Polycentricity emerged in response to public administration research that was focused on central governments and the associated administrative agencies as the source of policy outputs (Thiel, et al., 2019). Polycentricity was developed as a counterargument to the emphasis at the time that saw governments as the only provider of public goods and added analytic space for civic, private, and other forms of co-production. The premise is that in many federal systems, policy outputs are likely to be from systems of multilayered and overlapping centers of authority interacting across different policy domains. Polycentricity is based on the idea that groups of individuals and networks of organizations can often produce shared public goods.[1] Independently without relying on public authorities (McGinnis, 1999). It proposes highlighting how self-governing groups can organize to produce public goods at smaller scales, leading to greater efficiency and higher levels of provision. It builds on the local production of public goods as communities attempt to respond to local problems. Polycentricity allows examining how co-production in conjunction with other government efforts can occur and when external efforts might be counterproductive, and even crowd-out local initiatives which might be better suited for a local problem (Thiel et al., 2019; Weible et al., 2020). The potential for great efficiency exists since more groups at multiple societal levels active in producing public goods can result in the greater overall provision of those goods, as well as the level of production matching local demand. This involves looking at the full array of organizations involved in financing, producing, allocating, and maintaining systems that generate goods with shared production and consumption characteristics. It includes thinking past the standard assumption of hierarchical formal proscribed roles and refocusing attention to activities and functions. Polycentric systems are typically multi-scale, involving the community, local, regional, and national organizations, and are often organized in a mix of formal and informal networks that allow for cooperation across jurisdictional and other boundaries (Jordan et al., 2018; Kauneckis & Martin, 2020; Lubell & Morrison, 2017). As work has expanded to look at how governance systems interact to produce and manage public goods, polycentricity has increasingly been used to examine even highly centralized government systems (Dorsch & Flachsland, 2017). Climate change encompasses an expansive range of public goods, from the shared atmosphere utilized as an emissions dump to common water systems and the risk shared across communities along a shoreline facing rising sea levels. Communities, local, and regional governments facing climate impacts are likely attempting to manage climate-induced risks at different scales, across a variety of sectors, and working with different types of organizational partners regardless of actions at the national and international levels.

A polycentric approach to climate policy involves examining how authority and the ability to make decisions are distributed in a system, variation in the roles different types of organizations play in policy arrangements and processes, functional relationships between organizations, and full recognition of the diversity of organizations active in any particular policy arena, but especially in climate policy since the impacts involve multiple sectors and large spatial scales that include more than one political jurisdiction. Figure 14.1 illustrates the diversity of types and structures of climate policy arrangements across organizational actors active in a policy arena. In some cases, bottoms-up processes and partnerships between civil society and cities may be adequate to address climate risks that are highly localized. In others, federal government guidance can be the initiative that begins state-level climate planning efforts (SAPCCs, for example) that are then utilized for coordinating local actions. The associational networks that characterize

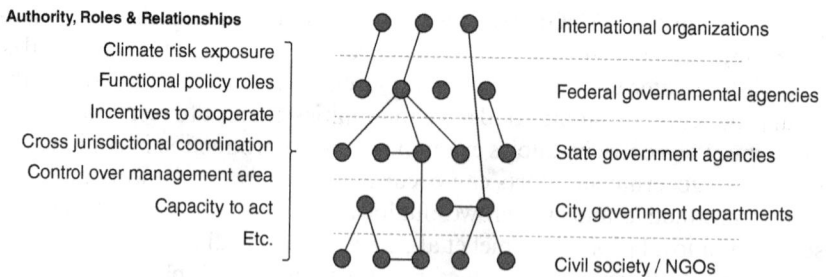

Figure 14.1 Configurations of Polycentric Policy Systems.

polycentric systems can even allow resources to flow in non-hierarchical arrangements, including international funding for city and NGO efforts, trace regional collaborations, and identify essential organizations that serve to bridge regions and sectors.

Equally as important, polycentricity allows unpacking the types of relationships between and across organizations. Researchers can examine how different jurisdictions exposed to similar levels of climate risk have responded, and how different organizations may take on similar or very diverse functional roles such as funding, implementation, or convener of local stakeholders. Any configuration can be examined to determine what incentives drove jurisdictions to cooperate or coordinate policy activities, how the ability to act distributed across different organizations can influence implementation, and how issues of power and control over part of the policy process determine outcomes. The polycentricity framework as a research tool allows analysis of the variety of emerging relationships and develops empirically testable hypotheses.

Research using polycentricity suggests normative elements such as greater efficiency, efficacy, and even legitimacy of decisions and outputs. These normative elements have made the approach particularly appealing for research on climate policy. The approach includes aspects of subsidiarity or the spatial matching of production systems for goods and services with recipients within specific jurisdictions. Polycentricity allows for decision-making analysis across highly differentiated local impacts within both mitigation and adaptation actions. In communities and local jurisdictions exposed to high risk, there may be independent activities that can be done to reduce exposure to climate impacts.

Similarly, in locations where residents are highly concerned about climate change for ethical or ideological reasons, they may be willing to bear the costs of emission reductions even without action at the national level. Smaller nested systems that operate independently of higher levels of authority can sometimes respond more quickly to changing conditions. The ability of polycentric systems to respond more closely and more quickly and be responsive to residents' preferences provides the potential better to fit the demand for public goods to production activities. The approach allows for a more detailed understanding of policy change and the responsiveness to environmental change (Greer & Scott, 2019; McGinnis, 1999; Ostrom, 2012).

There are numerous critiques of polycentricity as a normative approach (see discussions in Heikkila, et al., 2018; Jordan et al., 2015). The focus on localized nested systems means responding to larger-scale issues requires coordination across multiple jurisdictions, often heterogeneous, and each with its own activities, priorities, and resources to draw upon. While not insurmountable, these high coordination costs can hinder effective action at a scale appropriate to the nature of the problem. For example, for a community facing flood risk from increased precipitation building a retaining wall for a river may simply redistribute that flood risk to downstream communities without mitigation efforts. Effective risk mitigation for climate change requires coordination across both upstream and downstream communities, typically at huge scales. Another issue is that the design of solutions that match local circumstances may not be appropriate outside that jurisdiction or even at larger scales. For example, an adaptation strategy that relies on water users to supplement surface water by pumping groundwater aquifers may serve to reduce the risk of water shortages in that community, but not suit a community with a much deeper aquifer, while a strategy that reallocated water across the two communities would have reduced risk for both at a much lower cost. Misaligned decision-making authority can also lead polycentric systems to perform poorly. For example, having spending authority at a high level but revenue-raising authority at a lower level and implementation and design at a third can mean the implementation of action fails to match benefits to the costs. While having multiple local experiments in public goods production can expand policy options, it may make it difficult to scale up localized systems to larger-scale problems, address externalities from outside jurisdictions, and find ways to absorb the high coordination costs among multiple and diverse organizations.

More centralized systems are not necessarily an answer to the problems discussed above. In addition to coordination costs, scaling up effective solutions, and cross-jurisdictional externalities, centralized systems come with problems of their own. The benefits of a polycentric approach to analyzing existing arrangements are that it can capture hyper-localized initiatives, highlight structural connections across larger systems, and pinpoint new arrangements that bridge community groups, NGOs, and public agencies. As researchers have increasingly expanded the variety of roles that civil society, local, state, and federal agencies play in policy formulations and implementation, and the relationship across organizations, polycentricity has become a critical perspective through which to understand a wide range of policy areas (Thiel et al., 2019).

Emerging Multilevel Climate Policy

The power of subnational governments and networks that span regional and international actors has been found to be a critical element of climate policy (Bulkeley, 2010; Jörgensen et al., 2015a; Kauneckis & Martin, 2020). The importance of local jurisdictions, civil society actors, and production relationships across diverse organizations has long been a feature of the development, natural resource management, human rights, and other public policy domains. In climate policy, numerous networks across urban areas such as C40, International Urban Cooperation (IUC), 100 Resilient Cities, and others facilitate cooperation and information sharing across cities on issues of common concern, from heat mitigation strategies to flood risk. Data collected by Sastry and Kavya (2020) at both the Indian national and local levels within Kerala suggest significant subnational activities occurring. Although they report that much of it is centrally mandated by the national government, there are increasing city-initiated activities (Beerman et al., 2016). Climate adaptation actions include the first Heatwave Action Plan in India created independently by the city of Ahmedabad, Gujarat, following an extreme heat event in 2010 that left 1,344 dead (Azhar et al., 2014). In terms of carbon reductions, the Mayor of the City of Rajkot distinguished himself by making a unilateral commitment to reduce GHG emissions by 25%. Rajkot has even begun working on a climate resilience plan for the city (CapaCITIES, 2021). Among the states, Gujarat has set up the first state-level departmental climate office, while Kerala has created a US$ 220 million fund to address a variety of policy issues related to climate. While some states have focused on mitigation actions, such as Gujarat's policies to attract increased renewable energy investment, others have looked toward adaptation, including Assam's sustainable livelihood program around forests and Manipur's focus on water security. Other states, such as Odisha, subject to cyclones and the associated property and crop damage, have developed plans to address both mixes of mitigation and adaptation actions (Atteridge et al., 2012).

The potential for funding from international organizations and expanding energy production has so far motivated much of state and local activity. Sastry and Kavya (2020), in their survey of urban climate actions, identify six driving factors, including compliance with mandated actions through national regulatory policy from central governments. Much of these activities are attempts to align local with national policy prescriptions, often around hazard risk reduction or energy policy (Jörgensen et al., 2015b). Hazard reduction is typically reactive and responds to acute disasters such as heatwaves, floods, or droughts; rather than proactively looking forward to addressing future risk, however, it provides windows of opportunity to address emerging risks. Partnership with transnational and international funding agencies that provide implementation support is another common type. In this case, funding can often be the driver and may shift priorities toward the preference of the funding agency over those of a local community. Local leaders, civil society, and elected officials initiated some of the most interesting actions. Mayoral proclamations and commitments to emissions reductions fit in this category and represent manifestations of bottoms-up support in one form or another. Civil society and NGOs can also drive action, often with the goal of putting political pressure on elected officials to respond to local issues. Collective action to put political pressure on elected officials has a long history around forest and watershed conversation issues and increasing climate change (Sastry & Kavya, 2020).

Both cities and states align climate policies with other priorities, with development often the driving concern in India. However, this takes different forms across the more heavily industrialized states such as Gujarat and agricultural producers like Odisha. In highly centralized systems, motivations need to be aligned with national political actors to garner the resource to affect local change (Besley & Coate, 2003). There are likely to be a larger number of localized activities in decentralized systems covering a broader range of issues. Where hierarchies exist across levels of government and among multiple centers of authority, there can be a variety of interactions across scales.

In some cases, local priorities are misaligned with state and national agencies, leading to conflicting goals. Local actions may substitute for inactivity at other levels of government. In others, simultaneous policy action at multiple levels can provide additive benefits to a common goal (see discussion in Hooghe & Marks, 2003). These diverse activities and motivations, occurring at multiple levels from local to international, and interacting across scales can be examined through an analytic framework such as polycentricity. In nested overlapping centers of decision-making, different political and policy incentives manifest to drive action in response to climate risks and mitigation options (Cole, 2015; Ostrom, 2014; Rabe, 2007).

Implications for India's Climate Policy

Research on climate policy has shifted dramatically from a focus on the international and national stage to subnational, regional, and networked governance systems (Jordan et al., 2018; Kauneckis & Martin, 2020; Rabe, 2007). In the context of the United States, Australia, and many European nations, the most active sources of climate activity have occurred at the state and local levels (Bulkeley, 2010; Jordan et al., 2018; Massey et al., 2014; Rabe, 2007). Researchers have shifted the focus of investigations more to subnational governments for looking at the emergence of climate actions in the Indian context (Jogesh & Dubash, 2015; Jörgensen et al., 2015a). While less research has been done on developing nations' responses to increasing climate risk, it represents an area of research ripe for inquiry.

Polycentricity is a framework that offers a tool to capture the complexity of India's emerging climate policy. Many of the best illustrations of the utility of polycentricity can be found in applications to shared natural resources. Many natural resources have management regimes that are interconnected with both public agencies and NGOs having explicit management roles and with the communities that live and co-manage local forests, water, and grazing lands (see detailed descriptions of co-management arrangements in both Ciecierska-Holmes et al., 2020 and Meinzen-Dick et al., 2021). These are the same local systems experiencing the impacts of climate change and attempting to adapt. Even in sectors where the central government maintains management authority, such as water, public lands, energy, and irrigation system development, decisions are often made in negotiations with states and representatives of organized interests ranging from businesses to farmer groups (Jain, 1987; Jogesh & Dubash, 2015; Mitra, 2012). Local community water management groups have a long history and maintain a role in infrastructure development, maintenance, and water allocation (Chindarkar et al., 2018; Meinzen-Dick, 2007; Phadnis & Kulshrestha, 2012). Forest management mixes village-level panchayats and state-owned forests with various uses and management shared across different organizations (Bhattacharya et al., 2010; Sivaramakrishnan, 2002). As larger ecosystem-scale responses are needed to maintain livelihoods and habitat and improve ecological buffer capacity against the increasing frequency of extreme events, it will require even greater cooperation across many different organizations and management systems. Organizational and scalar diversity will likely only increase as it becomes necessary to address cross-sectoral issues.

From a research perspective, the polycentricity framework allows a deeper analysis of the flow of information, resources, authority, and other relationships across diverse organizational actors involved in policy formulation and implementation (Meinzen-Dick et al., 2021). This is critical for understanding effective action on the ground in the rich organizational network of NGOs and community-level organizations. The approach can also expand the range of prescriptive elements since it allows for looking at multiple dimensions of a management regime and a combination of responsibilities. The normative elements of polycentricity are also a critical component. There is clear evidence that small-scaled self-governing systems are more flexible and better respond to changing exogenous circumstances (Cole, 2015). However, effective action on global and encompassing issues such as climate impact is an open area of inquiry. While larger-scaled systems such as state and regional arenas must contend with national political agendas and political lobbying by sectors that may be less important in local settings, there are significant challenges in scaling up local activities to address the scale of climate impacts.

The framework also provides a diagnostic tool for understanding how complex management systems may fail to respond, not coordinate across scales or governing units, and create barriers to effectively responding to climate change. It allows researchers and practitioners to unpack the nature of specific policy problems rather than generalize why policy inaction may be occurring. Understanding how to focus on the nature of relationships between different types of organizations involved in producing the specific types of activities that reduce emissions, or buffer communities against the impacts of climate change, and moving away from a focus on national actors and often symbolic statements that may not be implemented allows focusing on the actual engagement around climate policy. The single most important contribution of the polycentricity approach is moving away from the common assumption that central governments are the only actor of relevance for climate policy (Cole, 2015; Ostrom, 2014). It puts the agency at multiple collective action levels, both formal and informal. While international negotiations between nations matter, incorporating the activities of subnational entities actively engaged and working on management more fully captures the range and scope of activities. The impacts of climate change are pervasive and will fundamentally transform the ecology, economy, and hazard risks across India. Having analytic tools that can examine activities at the micro-scale across the diversity of communities and governance arrangements provides insight into the local experiments that might be scaled up to national and regional policy, utilizes the power of networks that connect levels of governance, and provides a means of both understanding the impact of activities on community well-being and better responding to climate change.

Note

1 Public goods are those goods and services shared among a group of people as opposed to private goods.

References

Arcanjo, M. (2019) *The Future of Water in India.* Climate Institute. Retrieved November 11, 2021, from http://climate.org/the-future-of-water-in-India/

Atteridge, A., Shrivastava, M. K., Pahuja, N., & Upadhyay, H. (2012). Climate policy in India: What shapes international, national and state policy? *AMBIO*, 41(1), 68–77. https://doi.org/10.1007/s13280-011-0242-5

Azhar, G. S., Mavalankar, D., Nori-Sarma, A., Rajiva, A., Dutta, P., Jaiswal, A., Sheffield, P., Knowlton, K., & Hess, J. J. (2014). Heat-related mortality in India: Excess all-cause mortality associated with the 2010 Ahmedabad heatwave. *PLoS ONE*, 9(3), e91831. https://doi.org/10.1371/journal.pone.0091831

Banks, N., & Hulme, D., (2012). *The Role of NGOs and Civil Society in Development and Poverty Reduction.* Brooks World Poverty Institute, Working Paper 71, University of Manchester. http://dx.doi.org/10.2139/ssrn.2072157

Batley, R., & Rose, P. (2011). Analysing collaboration between non-governmental service providers and governments. *Public Administration and Development*, 31(4), 230–239. https://doi.org/10.1002/pad.613

Beer, C., Bartley, T., & Roberts, W. (2012) NGOs: Between advocacy, service provision, and regulation. In *Oxford Handbook of Governance*, ed. Levi-Faur, D. Oxford University Press.

Beermann, J., Damodaran, A., Jörgensen, K., & Schreurs, M. A. (2016). Climate action in Indian cities: An emerging new research area. *Journal of Integrative Environmental Sciences*, 8168(February). https://doi.org/10.1080/1943815X.2015.1130723

Besley, T., & Coate, S. (2003). Centralized versus decentralized provision of local public goods: A political economy approach. *Journal of Public Economics*, 87(12), 2611–2637. https://doi.org/10.1016/S0047-2727(02)00141-X

Bhattacharya, P., Pradhan, L., & Yadav, G. (2010). Joint forest management in India: Experiences of two decades. *Resources, Conservation and Recycling*, 54(8), 469–480. https://doi.org/10.1016/j.resconrec.2009.10.003

Bulkeley, H. (2010). Cities and the governing of climate change. *Annual Review of Environment and Resources*, 35(1), 229–253. https://doi.org/10.1146/annurev-environ-072809-101747

CapaCITIES. (2021). *Rajkot city discusses climate resilient city action plan.* Retrieved October 5, 2021, https://www.capacitiesindia.org/2018/06/05/rajkot-city-discusses-climate-resilient-city-action-plan/

Chandra, B., Mukherjee, M., Mukherjee, A., & Panikkar, K. N. (2016). *India's Struggle for Independence.* Gurgaon, India: Penguin.

Chaturvedi, A., Rattani, V., & Awasthi, K. (2019, September). State action plans on climate change need upscaling and capacity enhancement. *Down to Earth*. https://www.downtoearth.org.in/blog/climate-change/state-action-plans-on-climate-change-need-upscaling-and-capacity-enhancement-66796

Chindarkar, N., Chen, Y. J., & Wichelns, D. (2018). Government-civil society co-management contracts for rural water services: Lessons from India. *Policy Design and Practice*, 1(4), 269–280. https://doi.org/10.1016/S0047-2727(02)00141-X

Ciecierska-Holmes, N., Jörgensen, K., Ollier L. L., & Raghunandan, D. (Eds.). (2020). *Environmental Policy in India*. Abingdon, Oxon, New York: Routledge.

Cole, D. (2015). Advantages of a polycentric approach to climate change policy. *Nature Climate Change*, 5(2), 114–118. https://doi.org/10.1038/nclimate2490

Doman, M., & Shatoba, K. (2021, February 10). Tracing the path of destruction in India's Himalayas. *ABC Net*, www.abc.net.au/news/2021-02-11/satellites-capture-scale-of-indian-glacier-collapse/13137924?nw=0

Dorsch, M. J., & Flachsland, C. (2017). A polycentric approach to global climate governance. *Global Environmental Politics*, 17(2), 45–64. https://doi.org/10.1162/GLEP_a_00400

Dubash, N. K. (2021). *Varieties of climate governance: the emergence and functioning of climate institutions*. Taylor & Francis.

Dubash, N.K. & Jogesh, A. (2014). From Margins to Mainstream? State Climate Change Planning in India as a 'Door Opener' to a Sustainable Future, Centre for Policy Research (CPR), Climate Initiative, Research Report, New Delhi.

Eckstein, D., Künzel, V., & Schäfer, L. (2021). *Global climate risk index 2021. Who suffers most from extreme weather events, 2000–2019*. https://smartwatermagazine.com/news/germanwatch/who-suffers-most-extreme-weather-events-weather-related-loss-events-2019-and-2000#:~:text=The%20countries%20and%20territories%20affected,Myanmar%20and%20Haiti%20rank%20highest

Fadia, B. (1984). *State Politics in India. Volume I*. New Delhi: Radiant Publishers.

Fernandes, D., Jörgensen, K., & Narayanan, N.C. (2019). Factors shaping the climate policy process in India. In Environmental Policy in India, eds. Raghunandan, D., Jörgensen, K., Ollier, L., Ciecierska-Holmer, N. Taylor & Francis.

Freedman, P. (2000). Infrastructure, natural disasters, and poverty. In *Managing Disaster Risk in Emerging Economies*, eds. A. Kreimer and M. Arnold. Washington, DC: The World Bank.

Gogoi, E. (2015). *India's State Action Plans on Climate Change: Towards Meaningful Action*. Oxford Policy Management, New Delhi, available at https://www.opml.co.uk

Government of India (2018). India: Second Biennial Updated Report to the United Nations Framework Convention on Climate Change. Ministry of Environment, Forest and Climate Change, Government of India.

Government of India, Office of the Registrar General and Census Commissioner, Ministry of Home Affairs. (2001). *Census 2001*. Retrieved December 8, 2021, from https://censusindia.gov.in/2011-common/census_data_2001.html

Greer, R., & Scott, T. (2019). A network autonomy framework: Reconceptualizing special district autonomy in polycentric systems. *Perspectives on Public Management and Governance*, 2018, 59–76. https://doi.org/10.1093/ppmgov/gvz006

Harmel, R. (2016). Environment and Party Decentralization: A Cross-National Analysis. *Comparative Political Studies*, 14(1): 75–99.

Heikkila, T., Villamayor-Tomas, S., & Garrick, D. (2018). Bringing polycentric systems into focus for environmental governance. *Environmental Policy and Governance*, 28, 207–211. https://doi.org/10.1002/eet.1809

Hooghe, L., & Marks, G. (2003). Unravelling the central state, but how? Types of multi-level governance. *American Political Science Review*, 97(2), 233–243. https://doi.org/10.1017/S0003055403000649

IMF (International Monetary Fund). (2020). *World Economic and Financial Survey World Economic Outlook Database*. Retrieved December 1, 2021, from https://www.imf.org/external/pubs/ft/weo/2020/01/weodata/groups.htm

International Center for Not-For-Profit Law. (2021). *India*. Retrieved December 20, 2022 from https://www.icnl.org/resources/civic-freedom-monitor/india

IPCC (2021). Summary for policymakers. In *Climate Change 2021: The Physical Science Basis*. Contribution of Working Group I to the Sixth Assessment Report of the Intergovernmental Panel on Climate Change, eds. Masson-Delmotte, V., P. Zhai, A. Pirani, S. L. Connors, C. Péan, S. Berger, N. Card, Y. Chen, L. Goldfarb, M. I. Gomis, M. Huang, K. Leitzell, E. Lonnoy, J. B. R. Matthews, T. K. Maycock, T. Waterfield, O. Yelekçi, R. Yu, and B. Zhou. Cambridge University Press, pp. 3–31. Cambridge University Press, Cambridge, United Kingdom and New York, NY, USA.

Jain, R. B. (1987). The role of bureaucracy in policy development and implementation in India. *Asian Journal of Social Science*, 15(1), 20–39. https://www.jstor.org/stable/24491120

Jalan, B. (2000). *India's Economic Policy: Preparing for the Twenty-First Century*. Viking Press.

Jogesh, A., & Dubash, N. K. (2015). State-led experimentation or centrally-motivated replication? A study of state action plans on climate change in India. *Journal of Integrative Environmental Sciences*, 12(4), 1–20. https://doi.org/10.1080/19438 15X.2015.1077869

Jordan, A., Huitema, D., Hildén, M., van Asselt, H., Rayner, T., Schoenefeld, J., Tosun, J., Forester, J., & Boasson, E. (2015). Emergence of polycentric climate governance and its future prospects. *Nature Climate Change*, 5, 977–982. https://doi.org/10.1038/nclimate2725

Jordan, A., Huitema, D., van Asselt, H., & Forster, J. (2018). *Governing Climate Change: Polycentricity in Action?* Cambridge University Press.

Jörgensen, K. (2020). The role India's states play in environmental policymaking. In N. Ciecierska-Holmes, K. Jörgensen, O. Ll, & D. Raghunandan (Eds.), *Environmental Policy in India* (1st ed., pp. 39–59). Routledge.

Jörgensen, K., Jogesh, A., & Mishra, A. (2015a). Multi-level climate governance and the role of the subnational level. *Journal of Integrative Environmental Sciences*, 12(4), 235–245. https://doi.org/10.1080/1943815X.2015.1096797

Jörgensen, K., Mishra, A., & Sarangi, G. K. (2015b). Multi-level climate governance in India: the role of the states in climate action planning and renewable energies. *Journal of Integrative Environmental Sciences*, 12(4), 267–283.

Kauneckis, D., & Martin, R. (2020). Patterns of adaptation response by coastal communities to climate risks. *Coastal Management*, 48(4), 257–274. https://doi.org/10.1080/08920753.2020.1773209

Krishnan, R., Sanjay, J., Gnanaseelan, C., Mujumdar, M., Kulkarni, A., & Chakraborty, S. (2020). Assessment of climate change over the Indian region: A report of the ministry of earth sciences (MOES), Government of India. https://reliefweb.int/report/india/assessment-climate-change-over-indian-region-report-ministry-earth-sciences-moes

Lubell, M., & Morrison, T. (2017). Institutional navigation for polycentric sustainability governance. *Nature Sustainability*, 4, 664–671. https://doi.org/10.1038/s41893-021-00707-5

Mall, R. K., Singh, R., Gupta, A., Srinivasan, G., & Rathore, L. S. (2006). Impact of climate change on Indian agriculture: A review. *Climatic Change*, 78(2), 445–478. https://doi.org/10.1007/s10584-005-9042-x

Massey, E., Biesbroek, R., Huitema, D., & Jordan, A. (2014). Climate policy innovation: The adoption and diffusion of adaptation policies across Europe. *Global*

Environmental Change, 29, 434–443. https://doi.org/10.1016/j.gloenvcha.2014.09.002

Mazdiyasni, O., AghaKouchak, A., Davis, S. J., Madadgar, S., Mehran, A., Ragno, E., Sadegh, M., Sengupta, A., Ghosh, S., & Dhanya, C. T. (2017). Increasing probability of mortality during Indian heatwaves. *Science Advances*, 3(6). https://doi.org/10.1126/sciadv.1700066

McGinnis, M. D. (1999). *Polycentricity and Local Public Economies: Readings from the Workshop in Political Theory and Policy Analysis*. University of Michigan Press.

Meinzen-Dick, R. (2007). Beyond panaceas in water institutions. *Proceedings of the National Academy of Sciences of the United States of America*, 104(39), 15200–15205. https://doi.org/10.1073/pnas.0702296104

Meinzen-Dick, R., Chaturvedi, R., Kandikuppa, S., Rao, K., Rao, J. P., Bruns, B., & Elididi, H. (2021). Securing the commons in India: Mapping polycentric governance. *International Journal of the Commons*, 15(1), 218. https://doi.org/10.5334/ijc.1082

Mekonnen, M. M., & Hoekstra, A. Y. (2016). Four billion people facing severe water scarcity. *Science Advances*, 2(2). https://doi.org/10.1126/sciadv.1500323

Mitra, S. K. (2012). *Politics in India: Structure, Process and Policy*. Oxford University Press.

Murali, Kanta (2017). *Caste, Class, and Capital: The Social and Political Origins of Economic Policy in India*. Cambridge University Press.

Nandi, J. (2021). India seventh most affected by climate change in 2019 globally: Report. *Hindustan Times*. https://www.hindustantimes.com/india-news/india-seventh-most-affected-by-climate-change-in-2019-globally-report-101611552192319.html

OECD (2019). *OECD Economic Surveys: India*. Retrieved October 15, 2021, from https://www.oecd.org/economy/India-economic-snapshot/

Ostrom, E. (2012). Nested externalities and polycentric institutions: Must we wait for global solutions to climate change before taking actions at other scales? *Economic Theory*, 1–17. https://doi.org/10.1007/s00199-010-0558-6

Ostrom, E. (2014). A polycentric approach to climate change. *Annals of Economics and Finance*, 15(1), 97–134. https://doi.org/10.2139/ssrn.1934353

Pandey, R., & Jha, S. K. (2012). Climate vulnerability index-measure of climate change vulnerability to communities: A case of rural Lower Himalaya, India. *Mitigation and Adaptation Strategies for Global Change*, 17, 487–506. https://doi.org/10.1007/s11027-011-9338-2

Parikh, S., & Weingast, B. (1997). A comparative theory of federalism: India. *Virginia Law Review*, 83(1521). https://heinonline.org/HOL/LandingPage?handle=hein.journals/valr83&div=56&id=&page=

Phadnis, S. S., & Kulshrestha, M. (2012). Evaluation of irrigation efficiencies for water users' associations in a major irrigation project in India by DEA. *Benchmarking: An International Journal*, 19(2), 193–218. https://doi.org/10.1108/14635771211224536

Picciariello, A., Colenbrander, S., Bazaz, A., & Roy, R. (2021). *The costs of climate change in India: A review of the climate-related risks facing India, and their economic and social costs*. ODI Literature review. London: ODI. www.odi.org/en/publications/the-costs-of-climate-change-in-India-a-review-of-the-climate-related-risks-facing-India-and-their-economic-and-social-costs

Pillai, A. V., & Dubash, N. K. (2021). The limits of opportunism: the uneven emergence of climate institutions in India. *Environmental Politics*, 1–25.

Rabe, B. (2007). Beyond Kyoto: Climate change policy in multilevel governance systems. *Governance, 20*(3), 423–444. https://doi.org/10.1111/j.1468-0491.2007.00365.x

Rohini, P., Rajeevan, M., & Mukhopadhay, P. (2019). Future projections of heat waves over India from CMIP5 models. *Climate Dynamics, 53*(1), 975–988. https://doi.org/10.1007/s00382-019-04700-9

Roxy, M. K., Ghosh, S., Pathak, A., Athulya, R., Mujumdar, M., Murtugudde, R., Terray, P., & Rajeevan, M. (2017). A threefold rise in widespread extreme rain events over central India. *Nature Communications, 8*(1), 1–11.

Sarthi, P., Agrawal, A., & Rana, A. (2015). Possible future changes in cyclonic storms in the Bay of Bengal, India under warmer climate. *International Journal of Climatology, 35*(7), 1267–1277. https://doi.org/10.1002/joc.4053

Sastry, M., & Kavya, M. (2020). *Drivers of Climate Action in Indian Cities*. https://www.teriin.org/sites/default/files/files/Drivers-Climate-Action-Report.pdf

Satterthwaite, D., Archer, D., Colenbrander, S., Dodman, D., Hardoy, J., Mitlin, D., & Patel, S. (2020). Building Resilience to Climate Change in Informal Settlements. *One Earth, 2*(2), 143–156. https://doi.org/10.1016/j.oneear.2020.02.002

Sen, S. (1992). Non-profit organisations in India: Historical development and common patterns. *Voluntas: International Journal of Voluntary and Nonprofit Organizations, 3*(2), 175–193. https://doi.org/10.1007/BF01397772

Sen, S. (1999). Some aspects of state–NGO relationships in India in the post-independence era. *Development and Change, 30*(2), 327–355. https://doi.org/10.1111/1467-7660.00120

Shukla, P. R., Sharma, S. K., Ravindranath, N. H., Garg, A., & Bhattacharya, S. (Eds.). (2003). *Climate Change and India Vulnerability Assessment and Adaptation*. Universities Press.

Singh, C., Deshpande, T., & Basu, R. (2017). How do we assess vulnerability to climate change in India? A systematic review of literature. *Regional Environmental Change, 17*, 527–538. https://doi.org/10.1007/s10113-016-1043-y

Singh, D., Horton, D. E., Tsiang, M., Haugen, M., Ashfaq, M., Mei, R., Rastogi, D., Johnson, N. C., Charland, A., & Rajaratnam, B. (2014). Severe precipitation in Northern India in June 2013: Causes, historical context, and changes in probability. *Bulletin American Meteorological Society, 95*(9), S58–S61. https://deeptis47.github.io/papers/Singh2014a.pdf

Singh, S., Kumar, R., Bhardwaj, A., Sam, L., Shekhar, M., Singh, A., Kumar, R., & Gupta, A. (2016). Changing climate and glacier-hydrology in Indian Himalayan Region: A review. *Wiley Interdisciplinary Reviews: Climate Change, 7*(3), 393–410. https://doi.org/10.1002/wcc.393

Sivaramakrishnan, K. (2002). Forest co-management as science and democracy in West Bengal, India. *Environmental Values, 11*(3), 277–302. https://doi.org/10.3197/096327102129341091

Swapna, P., Ravichandran, M., Nidheesh, G., Jyoti, J., Sandeep, N., & Deepa, J. S. (2020). Sea-level rise. in *Assessment of Climate Change Over the Indian Region*, eds. R. Krishnan, J. Sanjay, C. Gnanaseelan, M. Mujumdar, A. Kulkarni, S. Chakraborty. Singapore: Springer, pp. 175–190.

Thiel, A., Garrick, D., & Blomquist, W. (2019). *Governing Complexity: Analyzing and Applying Polycentricity*. Cambridge University Press.

Wei, T., Yang, S., Moore, J. C., Shi, P., Cui, X., Duan, Q., Xu, B., Dai, Y., Yuan, W., Wei, X., Yang, Z., Wen, T., Teng, F., Gao, Y., Chou, J., Yan, X., Wei, Z., Guo, Y., Jiang, Y., Gao, X., Wang, K., Zheng, X., Ren, F., Lv, S., Yu, Y., Liu, B., Luo, Y., Li,

W., Ji, D., Feng, J., Wu, Q., Cheng, H., He, J., Fu, C., Ye, D., Xu, G., & Dong, W. (2012). Developed and developing world responsibilities for historical climate change and CO2 mitigation. *Proceedings of the National Academy of Sciences of the United States of America*, 109(32), 12911–12915. https://doi.org/10.1073/pnas.1203282109

Weible, C. M., Yordy, J., Heikkila, T., Yi, H., Berardo, R., Kagan, J., & Chen, C. (2020). Portraying the structure and evolution of polycentricity via policymaking venues. *International Journal of the Commons*, 14(1), 680–691. https://doi.org/10.5334/ijc.1021

Index

Pages in *italics* refer figures; pages in **bold** refer tables and pages followed by n refer notes.

For Product Safety Concerns and Information please contact our EU
representative GPSR@taylorandfrancis.com
Taylor & Francis Verlag GmbH, Kaufingerstraße 24, 80331 München, Germany

www.ingramcontent.com/pod-product-compliance
Lightning Source LLC
Chambersburg PA
CBHW061623220326
41598CB00026BA/3857